STAR AND HYPERBRANCHED POLYMERS

PLASTICS ENGINEERING

Founding Editor

Donald E. Hudgin

Professor
Clemson University
Clemson, South Carolina

1. Plastics Waste: Recovery of Economic Value, *Jacob Leidner*
2. Polyester Molding Compounds, *Robert Burns*
3. Carbon Black-Polymer Composites: The Physics of Electrically Conducting Composites, *edited by Enid Keil Sichel*
4. The Strength and Stiffness of Polymers, *edited by Anagnostis E. Zachariades and Roger S. Porter*
5. Selecting Thermoplastics for Engineering Applications, *Charles P. MacDermott*
6. Engineering with Rigid PVC: Processability and Applications, *edited by I. Luis Gomez*
7. Computer-Aided Design of Polymers and Composites, *D. H. Kaelble*
8. Engineering Thermoplastics: Properties and Applications, *edited by James M. Margolis*
9. Structural Foam: A Purchasing and Design Guide, *Bruce C. Wendle*
10. Plastics in Architecture: A Guide to Acrylic and Polycarbonate, *Ralph Montella*
11. Metal-Filled Polymers: Properties and Applications, *edited by Swapan K. Bhattacharya*
12. Plastics Technology Handbook, *Manas Chanda and Salil K. Roy*
13. Reaction Injection Molding Machinery and Processes, *F. Melvin Sweeney*
14. Practical Thermoforming: Principles and Applications, *John Florian*
15. Injection and Compression Molding Fundamentals, *edited by Avraam I. Isayev*
16. Polymer Mixing and Extrusion Technology, *Nicholas P. Cheremisinoff*
17. High Modulus Polymers: Approaches to Design and Development, *edited by Anagnostis E. Zachariades and Roger S. Porter*
18. Corrosion-Resistant Plastic Composites in Chemical Plant Design, *John H. Mallinson*
19. Handbook of Elastomers: New Developments and Technology, *edited by Anil K. Bhowmick and Howard L. Stephens*

20. Rubber Compounding: Principles, Materials, and Techniques, *Fred W. Barlow*
21. Thermoplastic Polymer Additives: Theory and Practice, *edited by John T. Lutz, Jr.*
22. Emulsion Polymer Technology, *Robert D. Athey, Jr.*
23. Mixing in Polymer Processing, *edited by Chris Rauwendaal*
24. Handbook of Polymer Synthesis, Parts A and B, *edited by Hans R. Kricheldorf*
25. Computational Modeling of Polymers, *edited by Jozef Bicerano*
26. Plastics Technology Handbook: Second Edition, Revised and Expanded, *Manas Chanda and Salil K. Roy*
27. Prediction of Polymer Properties, *Jozef Bicerano*
28. Ferroelectric Polymers: Chemistry, Physics, and Applications, *edited by Hari Singh Nalwa*
29. Degradable Polymers, Recycling, and Plastics Waste Management, *edited by Ann-Christine Albertsson and Samuel J. Huang*
30. Polymer Toughening, *edited by Charles B. Arends*
31. Handbook of Applied Polymer Processing Technology, *edited by Nicholas P. Cheremisinoff and Paul N. Cheremisinoff*
32. Diffusion in Polymers, *edited by P. Neogi*
33. Polymer Devolatilization, *edited by Ramon J. Albalak*
34. Anionic Polymerization: Principles and Practical Applications, *Henry L. Hsieh and Roderic P. Quirk*
35. Cationic Polymerizations: Mechanisms, Synthesis, and Applications, *edited by Krzysztof Matyjaszewski*
36. Polyimides: Fundamentals and Applications, *edited by Malay K. Ghosh and K. L. Mittal*
37. Thermoplastic Melt Rheology and Processing, *A. V. Shenoy and D. R. Saini*
38. Prediction of Polymer Properties: Second Edition, Revised and Expanded, *Jozef Bicerano*
39. Practical Thermoforming: Principles and Applications, Second Edition, Revised and Expanded, *John Florian*
40. Macromolecular Design of Polymeric Materials, *edited by Koichi Hatada, Tatsuki Kitayama, and Otto Vogl*
41. Handbook of Thermoplastics, *edited by Olagoke Olabisi*
42. Selecting Thermoplastics for Engineering Applications: Second Edition, Revised and Expanded, *Charles P. MacDermott and Aroon V. Shenoy*
43. Metallized Plastics: Fundamentals and Applications, *edited by K. L. Mittal*
44. Oligomer Technology and Applications, *Constantin V. Uglea*
45. Electrical and Optical Polymer Systems: Fundamentals, Methods, and Applications, *edited by Donald L. Wise, Gary E. Wnek, Debra J. Trantolo, Thomas M. Cooper, and Joseph D. Gresser*
46. Structure and Properties of Multiphase Polymeric Materials, *edited by Takeo Araki, Qui Tran-Cong, and Mitsuhiro Shibayama*
47. Plastics Technology Handbook: Third Edition, Revised and Expanded, *Manas Chanda and Salil K. Roy*
48. Handbook of Radical Vinyl Polymerization, *Munmaya K. Mishra and Yusuf Yagci*

49. Photonic Polymer Systems: Fundamentals, Methods, and Applications, *edited by Donald L. Wise, Gary E. Wnek, Debra J. Trantolo, Thomas M. Cooper, and Joseph D. Gresser*
50. Handbook of Polymer Testing: Physical Methods, *edited by Roger Brown*
51. Handbook of Polypropylene and Polypropylene Composites, *edited by Harutun G. Karian*
52. Polymer Blends and Alloys, *edited by Gabriel O. Shonaike and George P. Simon*
53. Star and Hyperbranched Polymers, *edited by Munmaya K. Mishra and Shiro Kobayashi*

Additional Volumes in Preparation

Practical Extrusion Blow Molding, *edited by Samuel L. Belcher*

Polymer Viscoelasticity: Stress and Strain in Practice, *Evaristo Riande, Ricardo Díaz-Calleja, Catalina Salom, Margarit Prolongo, and Rosa Masegosa*

Handbook of Polyethylene: Structures, Properties, and Applications, *Andrew J. Peacock*

Polycarbonates Handbook, *edited by Don LeGrand and John T. Bendler*

STAR AND HYPERBRANCHED POLYMERS

edited by

MUNMAYA K. MISHRA
Ethyl Corporation
Richmond, Virginia

SHIRO KOBAYASHI
Kyoto University
Kyoto, Japan

MARCEL DEKKER, INC. NEW YORK • BASEL

Library of Congress Cataloging-in-Publication Data

Star and hyperbranched polymers / edited by Munmaya K. Mishra and Shiro Kobayashi.
p. cm. — (Plastics engineering; 53)
Includes bibliographical references and index.
ISBN 0-8247-1986-7 (alk. paper)
1. Polymers. I. Mishra, Munmaya K. II. Kobayashi, Shiro. III. Series: Plastics engineering (Marcel Dekker, Inc.); 53.
QD381.8.S7 1999
547′.7—dc21 99-26160
CIP

This book is printed on acid-free paper.

Headquarters
Marcel Dekker, Inc.
270 Madison Avenue, New York, NY 10016
tel: 212-696-9000; fax: 212-685-4540

Eastern Hemisphere Distribution
Marcel Dekker AG
Hutgasse 4, Postfach 812, CH-4001 Basel, Switzerland
tel: 41-61-261-8482; fax: 41-61-261-8896

World Wide Web
http://www.dekker.com

The publisher offers discounts on this book when ordered in bulk quantities. For more information, write to Special Sales/Professional Marketing at the headquarters address above.

Current printing (last digit):
10 9 8 7 6 5 4 3 2 1

PRINTED IN THE UNITED STATES OF AMERICA

Preface

The field of macromolecular engineering has grown very large indeed, too large to be covered in detail in a single book. This volume focuses on advances in one of the most important areas of polymer research today. The book is essentially divided into two parts that are interrelated in many respects, first examining star-branched polymers and then hyperbranched polymers. Information on new star polymers, hyperbranched polymers, dendritic polymers, and so on, is systematically provided. The strength of this book is in the design strategies that it offers for working with these important polymers, their characterization, properties, and application. The architecture and properties of dendrimers, starburst polymers, multi-arm star polymers, and comb polymers are discussed.

The book also covers the solution properties of the regular star polymers. This class of materials has been known for many years to polymer chemists; however, rapid development has occurred only since the proposal of the idea of dendritic polymers in the mid-1980s. The characteristics of the polymers discussed here include the size and shape of the molecules, their biological activities, their low viscosity in solution, their substrate-holding properties inside the molecule, etc. The unique properties of these polymers attract many chemists, not only in polymer chemistry but also in organic chemistry, biochemistry, medicine, organometallic chemistry, catalyst chemistry, and so on, for these new materials are expected to find applications in many areas.

We trust that this book will represent a vital source of information for researchers in macromolecular engineering. The book is directed to industrial and academic scientists interested in designing new polymers, "polymers of geometrical beauty" and technological importance, as well as to students entering the wild world of contemporary polymer research and applications.

Certainly, future editions will include new developments as research continues. To our regret we could not include a chapter on star-branched polymers via radical polymerization since the research is still scanty and scattered, but we hope to include it in future editions.

It would not have been possible to complete a project like this without the help and participation of numerous individuals. We gratefully acknowledge all the contributors who made this book possible. Last, with love and appreciation MKM would like to acknowledge his wife and family for their support throughout the preparation of this book.

Munmaya K. Mishra
Shiro Kobayashi

Contents

Contributors

Yves Gnanou, Ph.D. Laboratoire de Chimie des Polymères Organiques, ENSCPB–CNRS–Université Bordeaux I, Talence, France

Yoshio Imai, Ph.D. Department of Organic and Polymeric Materials, Tokyo Institute of Technology, Meguro-ku, Tokyo, Japan

Stefan Ingrisch, Ph.D. Lehrstuhl für Makromolekulare Stoffe, Technische Universität München, Garching, Germany

Koji Ishizu, Ph.D. Department of Polymer Science, Tokyo Institute of Technology, Meguro-ku, Tokyo, Japan

Jun-ichi Kadokawa, Ph.D. Department of Materials Science and Engineering, Yamagata University, Yonezawa, Yamagata, Japan

Masa-aki Kakimoto, Ph.D. Department of Organic and Polymeric Materials, Tokyo Institute of Technology, Meguro-ku, Tokyo, Japan

Larry W. Kelts Corporate Research Laboratories, Eastman Kodak Company, Rochester, New York

Jungahn Kim, Ph.D. Polymer Division, Korea Institute of Science and Technology, Cheongryang, Seoul, Korea

Young H. Kim, Ph.D. DuPont Central Research and Development Experimental Station, Wilmington, Delaware

Shiro Kobayashi, Ph.D. Department of Materials Chemistry, Graduate School of Engineering, Kyoto University, Kyoto, Japan

Youngjoon Lee Maurice Morton Institute of Polymer Science, The University of Akron, Akron, Ohio

Timothy E. Long, Ph.D. Department of Chemistry, Virginia Polytechnic Institute and State University, Blacksburg, Virginia

P. J. Lutz, Ph.D. Institute Charles Sadron, CNRS, Strasbourg, France

S. Plentz Meneghetti, M.D. Institute Charles Sadron, CNRS, Strasbourg, France

Munmaya K. Mishra, Ph.D. Research and Development Center, Ethyl Corporation, Richmond, Virginia

Thomas H. Mourey, Ph.D. Imaging Research and Advanced Development, Eastman Kodak Company, Rochester, New York

Oskar Nuyken, Ph.D. Lehrstuhl für Makromolekulare Stoffe, Technische Universität München, Garching, Germany

Roderic P. Quirk, Ph.D. Maurice Morton Institute of Polymer Science, The University of Akron, Akron, Ohio

D. Rein Cytotherapeutics, Providence, Rhode Island

J. Roovers, Ph.D. Institute for Chemical Process and Environmental Technology, National Research Council Canada, Ottawa, Ontario, Canada

S. Sivaram, Ph. D. Division of Polymer Chemistry, National Chemical Laboratory, Pune, India

Jean-Luc Six Laboratoire de Chimie des Polymères Organiques, ENSCPB–CNRS–Université Bordeaux I, Talence, France

Owen Webster, Ph.D. DuPont Central Research and Development Experimental Station, Wilmington, Delaware

Jeffrey A. Wesson Corporate Research Laboratories, Eastman Kodak Company, Rochester, New York

1

Synthesis of Branched Polymers: An Introduction

Roderic P. Quirk and Youngjoon Lee
Maurice Morton Institute of Polymer Science, The University of Akron, Akron, Ohio

Jungahn Kim
Korea Institute of Science and Technology, Cheongryang, Seoul, Korea

I. INTRODUCTION

Major developments in the science and technology of polymeric materials have resulted from the preparation and characterization of polymers with well-defined structures [1,2]. Well-defined polymers with low degrees of compositional heterogeneity can provide the information and insight necessary to understand and predict polymer structure–property relationships. Branching in polymers is a useful structural variable that can be used advantageously to modify the processing characteristics and properties of polymers. A branched polymer is comprised of molecules with more than one backbone chain; that is, it is a nonlinear polymer [3]. A branched polymer is characterized by the presence of branch points (junction points): (atoms or a small group from which more than two long chains emanate) and by the presence of more than two chain end groups.

Branching affects the crystallinity, crystalline melting point, physical properties, viscoelastic properties, solution viscosities, and melt viscosities of polymers [3–6]. However, it is difficult to predict the relationships between branching and properties based on the behavior of most branched polymers because the branching reaction generally occurs in a random fashion. As a consequence, the number and types of branches per macromolecule are difficult to define except on an average basis.

Fundamental understanding of the effects of chain branching on polymer properties requires the availability of a variety of branched polymers with well-defined structures and low degrees of compositional heterogeneity [1,6]. Living chain reaction polymerizations are particularly suited for the preparation of these "model" polymers since it is possible to vary and control important structural parameters such as molecular weight, molecular weight distribution, copolymer composition and microstructure, tacticity, chain end functionality and the number of branches per molecule. Although a variety of mechanistic types of living chain reaction polymerization have been developed [6], living anionic polymerization, especially using alkyllithium initiators, is the paradigm [7] from which examples will be drawn for illustration of the general methods. This review will first consider the general polymerization methods that have been developed to synthesize regular star-branched polymers, heteroarm star-branched polymers, and other types of branched polymers, including graft copolymers [8]. Regular star-branched polymers have a single branch point and all arms exhibit low degrees of compositional heterogeneity with respect to composition, molecular weight, and molecular weight distribution. Heteroarm star-branched polymers [6,9], also described as mikto-arm star polymers [6], also have a single branch point, but the arms differ in either molecular weight or composition. When the arms differ in composition, heteroarm star-branched polymers can be considered as a special type of graft copolymer [8]. Finally, a brief review of dendrimers and hyperbranched polymers will be presented. Dendrimers are highly branched, three-dimensional macromolecules with a branch point at each monomer unit [10–14].

II. SYNTHESIS OF BRANCHED POLYMERS

The methodology of living polymerization is ideally suited for the preparation of well-defined, star-branched polymers and copolymers with low degrees of compositional heterogeneity. Because termination and chain transfer reactions are absent and the chain ends may be stable for sufficient time periods (the laboratory time scale), these polymerizations have the following useful synthetic attributes for star polymer synthesis:

1. One polymer is formed for each initiator molecule, so that the number average molecular weight of polymers or block segments can be predicted from the reaction stoichiometry. Multifunctional initiators with functionality n can form stars with n arms.

2. If the rate of initiation is rapid or competitive with the rate of propagation, polymers (precursor arms) with narrow molecular weight distributions ($M_w/M_n \leq 1.1$) [15] are formed.
3. When all of the monomer has been consumed, the product is a polymer with reactive chain ends that can participate in a variety of post-polymerization reactions:
   a. block copolymerization by addition of a second monomer, and/or
   b. end-linking with multifunctional linking agents to form the corresponding star-branched polymers with uniform arm lengths.

In the following sections, the general methods for synthesis of regular star-branched polymers and heteroarm star-branched polymers will be described. Specific examples will be shown based on alkyllithium-initiated anionic polymerization.

A. Postpolymerization, End Linking with Multifunctional Linking Agents

1. *General Aspects*

The products of living polymerizations are polymers that retain their active, propagating chain ends when all of the monomer has been consumed. Under appropriate conditions, these polymers exhibit well-defined, predictable number average molecular weights and narrow molecular weight distributions, i.e., low degrees of compositional heterogeneity. These living polymers can be reacted with multifunctional linking agents to form star-branched polymers in which the number of arms corresponds to the functionality of the linking agent, as shown in Eq. (1) where P* is a living polymer chain, $L(X)_n$ is a multifunctional linking agent of functionality n, and $L(P)_n$ is a star-branched polymer containing n arms.

$$nP^* + L(X)_n \rightarrow L(P)_n \tag{1}$$

The main advantage of this methodology is that the arms of the resulting branched polymer are well defined because the precursor arms can be characterized independently from the star. Because of the well-defined arms, the number of arms can be readily determined by measuring the molecular weight of the star. In principle, a variety of well-defined, star-branched polymers with different numbers of arms can be prepared using this method by varying the functionality of the linking agents [$L(X)_n$].

2. Anionic Polymerization

A wide variety of multifunctional linking agents have been investigated for preparation of star-branched polymers via anionic polymerization [7,16,17]. Arm functionalities range from 3 to very high values. However, many of the reported linking reactions, such as those involving polyfunctional alkyl halides, are complicated by side reactions such as elimination and metal–halogen exchange that lead to compositional heterogeneity. In contrast, linking reactions with polyfunctional silyl halides are very efficient and free of these complicating side reactions.

a. Polyhalosilanes and Stannic Chloride. One of the most general and useful methods for preparation of star-branched polymers is the reaction of polymeric organolithium compounds with multifunctional electrophilic species such as silicon tetrachloride, as shown in Eq. (2). A slight excess of living arm, PLi, is generally employed to drive the reaction to completion and to minimize the formation of stars with less than the stoichiometric number of arms. This in turn requires that the product be fractionated to obtain pure, star-branched polymer.

$$4PLi + SiCl_4 \rightarrow SiP_4 + 4LiCl \tag{2}$$

These linking reactions are not complicated by side reactions; however, they are sensitive to the steric requirements of the linking agents and the organolithium chain ends. For example, early work by Morton and co-workers [18] showed that the reaction of poly(styryl)lithium with a less than stoichiometric amount of silicon tetrachloride produced a polymer product composed of 26% of the four-armed star and 74% of the three-armed star polymers. More efficient linking can be effected using poly(butadienyl)lithium chain ends. This was illustrated by Zelinski and Wofford [19], who reacted poly(butadienyl)-lithium with methyltrichlorosilane and silicon tetrachloride to efficiently form the corresponding three-armed and four-armed, star-branched polymers, respectively. Linking efficiency can also be improved by separating the Si–Cl groups with spacers such as methylene groups to reduce the steric repulsions in the linked product [20].

Linking with multifunctional silyl chlorides has been extended to stars with arm functionalities over one hundred by utilizing the linking reactions of poly(butadienyl)lithium with carbosilane dendrimers containing up to 128 Si–Cl bonds [21]. Star-branched polybutadienes with more than 250 arms were reported for the linking reactions of poly(butadienyl)lithium with the product from hydrosilation of 1,2-polybutadiene [22].

Fetters and co-workers [23,24] developed a general method for synthesis of heteroarm star-branched polymers by utilizing the decreased reactivity of poly(styrtyl)lithium compared to poly(butadieny)lithium in linking reactions with polyhalosilanes. Hetero, three-armed star polystyrenes and polybutadienes were prepared using the reaction sequence shown in Scheme 1. After the first step, the excess methyltrichlorosilane was removed by evacuation. The excess of P′Li in the second step must be removed by fractionation. This method has been extended to a variety of other heteroarm star polymers as detailed in a recent review [6].

The efficiency of postpolymerization linking reactions of polymeric organolithium compounds with polyhalosilanes is utilized in a number of commercial processes [7]. The melt viscosity (high shear) and cold-flow (low-shear) properties of linear elastomers with narrow molecular weight distributions are poor. These limitations are eliminated by postpolymerization branching reactions such as linking with polyhalosilanes [7,25,26].

Stannic chloride undergoes linking reactions with polymeric organolithiums analogous to the reactions of silicon tetrachloride, as illustrated in Eq. (3). In addition to the improvement in cold flow and

$$PLi + SnCl_4 \rightarrow P_4Sn + 4LiCl \quad (3)$$

processing properties expected for a branched polymer, it has been reported that the incorporation of tin-functionalized, star-branched styrene-butadiene rubbers and polybutadienes into formulations with carbon black improved tire performance [27,28].

The development of processible, clear, impact-resistant grades of polystyrene for rapid molding operations is also based on end-linking reactions [7]. These polymers (e.g., Phillips K-resins) are mixtures of block copolymers of styrene and butadiene formed by incremental additions of alkyllithium initiator and monomers followed by some form of linking reaction. Linking agents such as aliphatic dicarboxylic acids and polyhalosilanes have been described [7]. An analogous product (BASF Styrolux), which includes butadiene/styrene blocks with a tapered block structure, utilizes an oligofunctional coupling agent to form an unsymmetrical star polymer with about four arms [29].

b. Divinylbenzenes. The copolymerization of styrenes and dienes with difunctional monomers such as divinylbenzene would be expected to form

$$PLi + \text{excess } CH_3SiCl_3 \rightarrow PSi(Cl)_2CH_3 + LiCl + CH_3SiCl_3$$
$$PSi(Cl)_2CH_3 + \text{excess } P'Li \rightarrow P(P')_2SiCH_3 + 2LiCl$$

Scheme 1

a gel and the gel point would depend on the monomer reactivity ratios and the relative concentrations of the two monomers [30,31]. The gel point is preceded by the formation of branched molecules that then react with other branched molecules to ultimately form a gel. Thus, if the amount of difunctional monomer were chosen carefully, it would be expected that copolymerization with a difunctional monomer could lead to a branched polymer rather than a gel. Copolymerization of divinylbenzene with dienes or styrene/diene mixtures is used commercially to introduce random long-chain branching during the polymerization process [7,32]. Subsequent to DVB incorporation into a growing chain, other active chains can add to the resulting pendant vinylstyrene units, followed by further monomer addition to generate long-chain branching, as shown in Eq. (4).

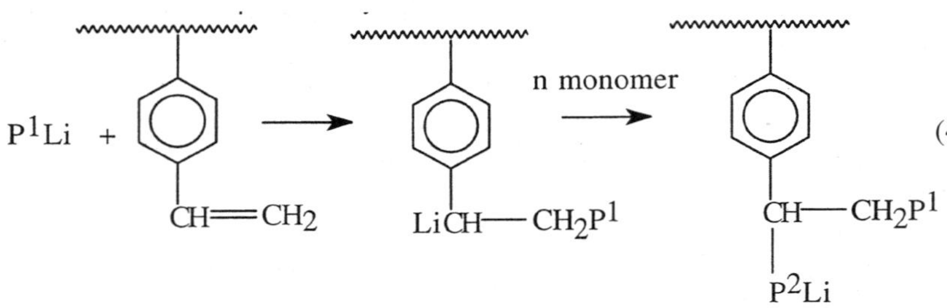

(4)

As described previously, introduction of branching improves both the cold flow (low shear) and melt visocity (high shear, processing) properties analgous to the effect of postpolymerization end-linking with polyhalosilanes.

Divinylbenzenes can also be used for postpolymerization end-linking reactions. The linking reactions of polymeric organolithium compounds with divinylbenzenes (DVB) provide a very versatile, technologically important, but less precise method of preparing star-branched polymers. As shown in Scheme 2, the linking reactions with DVB can be considered in terms of three consecutive and/or concurrent reactions: (a) crossover to DVB, (b) block polymerization of DVB, and (c) linking reactions of carbanionic chain ends with pendant vinyl groups in the DVB block [poly(4-vinylstyrene)] [3,7,33,34]. The uniformity of the lengths of the DVB blocks depends on the relative rate of the crossover reaction (a) compared to the block polymerization of DVB, (b) and the linking reactions (c). This block copolymerization–linking process has been described as formation of a DVB microgel nodule that serves as the branch point for the star-shaped polymer [33]. In principle, j molecules of divinylbenzene could link together $(j + 1)$ chains [33]. Although the number of arms in the star depends on the molar ratio of DVB to polymeric organolithium compound, the degree of linking obtained for this reaction is a complex func-

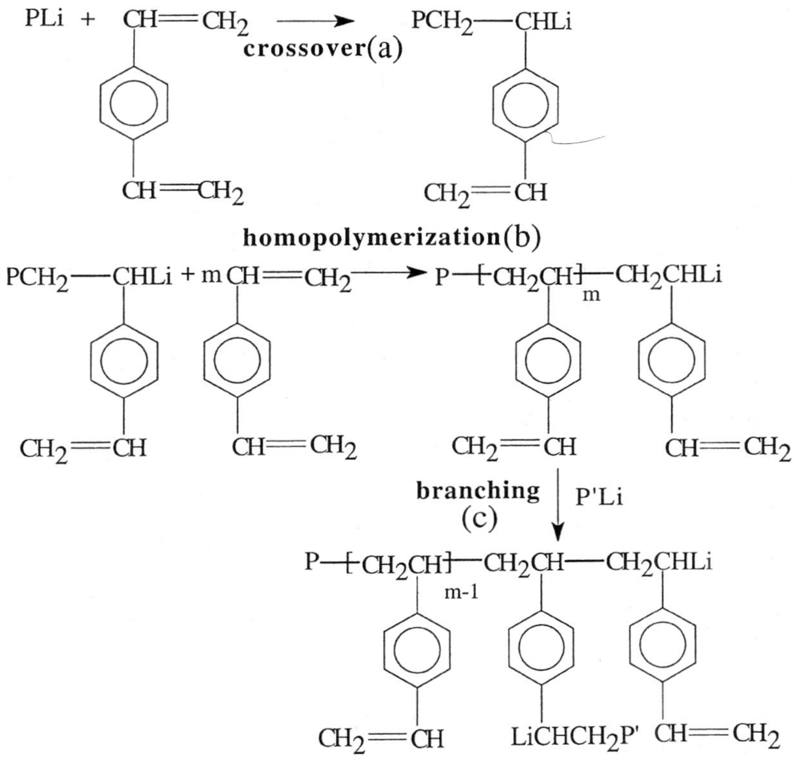

Scheme 2

tion of reaction variables [33–36]. It should be noted that these linking reactions are effected with technical grades of DVB that contain varying amounts of *p*-divinylbenzene, *m*-divinylbenzene, and *o*-, *m*-, and *p*-ethylvinylbenzene (EVB) [7].

For poly(styryl)lithium chains, the rate of crossover to DVB is comparable to the rate of DVB homopolymerization, and both of these rates are faster than the rate of the linking reaction of poly(styryl)lithium with the pendant double bonds in the poly(vinylstyrene) block formed from DVB. Therefore, it would be expected that the DVB block formed by crossover from poly(styryl)-lithium would be relatively uniform and that the linking reaction would generally occur after the formation of the DVB block. In general, the linking efficiency of poly(styryl)lithiums by DVB is quite high except for very low ratios of DVB/PSLi [7].

For poly(dienyl)lithium chain ends, the rate of crossover to DVB is much slower than the rate of DVB homopolymerization. Therefore, it would be expected that the DVB block length would be longer for poly(dienyl)lithiums compared with poly(styryl)lithium. As a consequence, although one obtains a higher degree of branching, the linking efficiency is lower for diene versus styrene stars. However, it is possible to obtain good linking efficiencies for dienyllithium chains using ratios of DVB/PLi ≥ 3 [7].

Although narrow molecular weight distribution stars can be obtained using the postpolymerization DVB linking chemistry, relatively broad distributions are typically obtained ($M_w/M_n = 1.2–1.4$) [35,36]. Polystyrene and polydiene stars can be prepared with arm functionalities as high as 40 and 56, respectively [7]. In general, the molecular weight distributions are broader for polydiene stars.

Postpolymerization end-linking reactions of poly(isoprenyl)lithium with DVB have been used to prepare viscosity improvers for motor oils, as shown in Scheme 3 [37]. With a DVB/PILi molar ratio of 3, relatively efficient link-

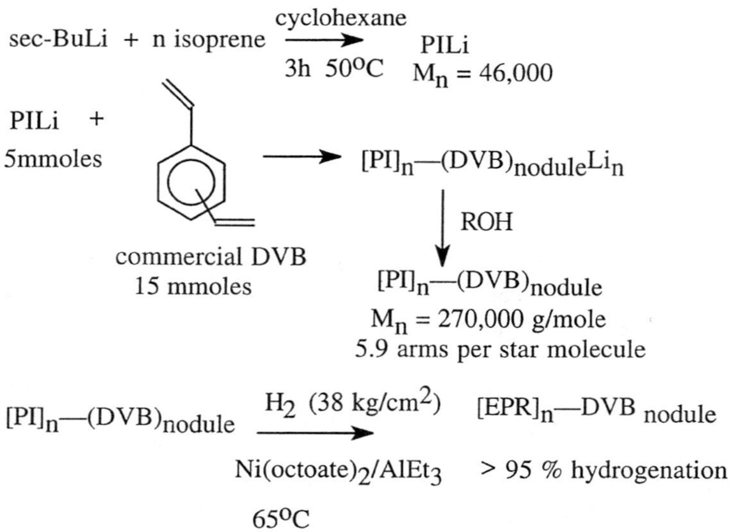

Scheme 3

ing is reported (~92%) to form stars with an average arm functionality of approximately 6.

B. Multifunctional Initiators

1. General Aspects

Living polymerization using a homogeneous, multifunctional initiator of functionality n can, in principle, form a star-branched polymer with n arms (see Eq. (5) for $n = 4$) and a low degree of compositional heterogeneity among the arms if the following conditions prevail:

1. All of the initiating sites must participate in initiating chain growth (quantitative, efficient initiation). This ensures uniform and predictable arm molecular weights.
2. The rate of initiation must be rapid or competitive with respect to the rate of propagation. This is required to obtain a narrow molecular weight distribution.
3. All propagating centers must be equally reactive with respect to chain growth (addition of monomer).

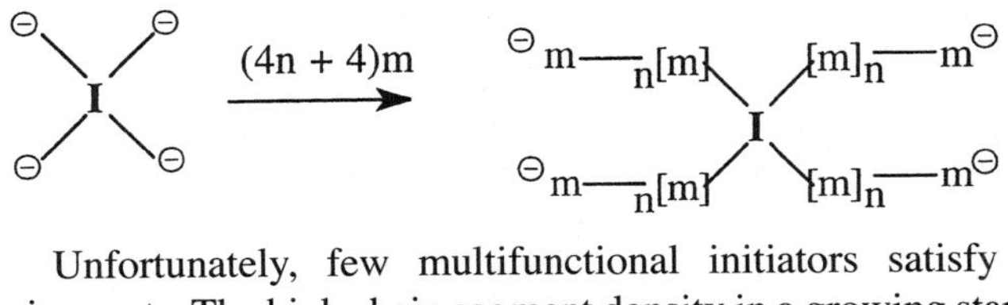

(5)

Unfortunately, few multifunctional initiators satisfy the preceding requirements. The high chain segment density in a growing star-branched molecule exacerbates complications arising from chain end/chain end interactions such as aggregation of ionic species, oxidation–reduction reactions of organometallic centers, and bimolecular termination reactions. As a consequence of these problems, few well-behaved multifunctional initiating systems are available.

2. Multifunctional Initiators from DVB

The ability to effect the concurrent polymerization and branching reactions of polymeric organolithium compounds with DVB [31] and form a soluble, nongelled product was extended by Burchard and co-workers [38] to form a multifunctional initiator. DVB was first polymerized using butyllithium in benzene to form soluble microgels of high molecular weight. These microgels with their attendant anionic groups were used as multifunctional initiators to polymerize monomers such as styrene. For example, when the ratio of [DVB]/[BuLi] was 2, the microgel exhibited $M_n = 1.9 \times 10^3$ g/mol ($M_w/M_n = 16.8$) and the resulting polystyrene star had a calculated star arm functionality of 6 [3].

This method has been extended by Rempp and co-workers [39,40] as a general "core-first" method to prepare star-branched polymers, as shown in Scheme 4. The "plurifunctional" metalorganic initiator (**A**) was prepared by potassium naphthalene-initiated polymerization of DVB in tetrahydrofuran (THF) at –40°C with [DVB]/[K^+] ratios of 0.5–3. Microgel formation was reported outside of this stoichiometric range or when *m*-DVB was used instead of either *p*-DVB or the commercial mixture. Within the prescribed stoichiometric ratios, star polymers (**B**) with arm functionalities varying from 8 to 42 were reported. As expected for this type of process, the polydispersities were described as being quite broad and were attributed primarily to a random distribution of core sizes and functionalities.

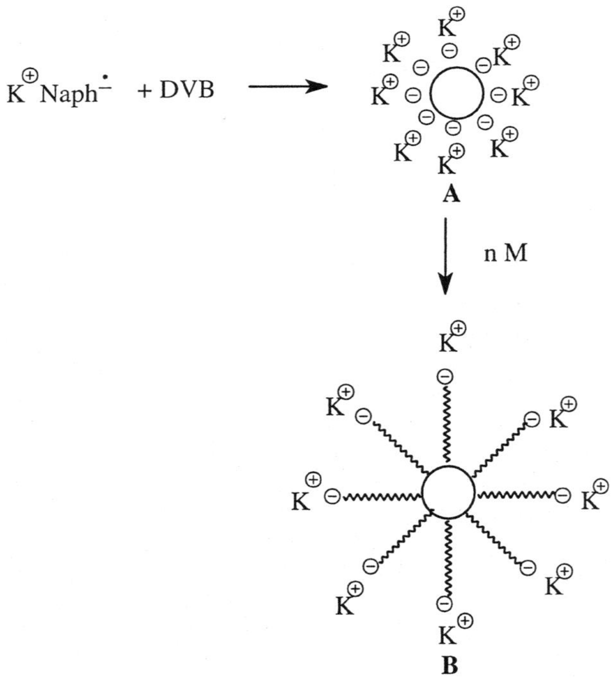

Scheme 4

3. *Trilithium Initiator Based on 1,1-Diphenylethylene*

One of the few hydrocarbon-soluble, trifunctional organolithium initiators has been prepared by the reaction of 3 moles of butyllithium with a trifunctional analog of 1,1-diphenylethylene, as shown in Scheme 5 [41].

The lack of oligomerization of 1,1-diphenylethylene units provides the basis for stoichiometric additions to this precursor. In the presence of small amounts of THF ([THF]/[BuLi] = 20), this initiator formed well-defined, three-armed star polymers that could be quantitatively carboxylated, as shown in Scheme 5.

4. *Heteroarm Star-Branched Polymers by Living Linking Reactions with DVB*

Eschwey and Burchard [42] recognized that the products of the reaction of DVB with polymeric organolithium compounds are living polymers (see Schemes 2 and 3) that could function as *polyfunctional, macromolecular initiators,* as indicated in Scheme 6 for linking of five polymeric organolithium chains (P^a). This type of process has been described as a living linking reaction [43]. Thus, after DVB linking of short poly(styryl)lithium chains to form the living, linked star polymer, **E,** additional styrene or isoprene monomer (M^b) was added to double the number of arms, in principle, and form a heteroarm,

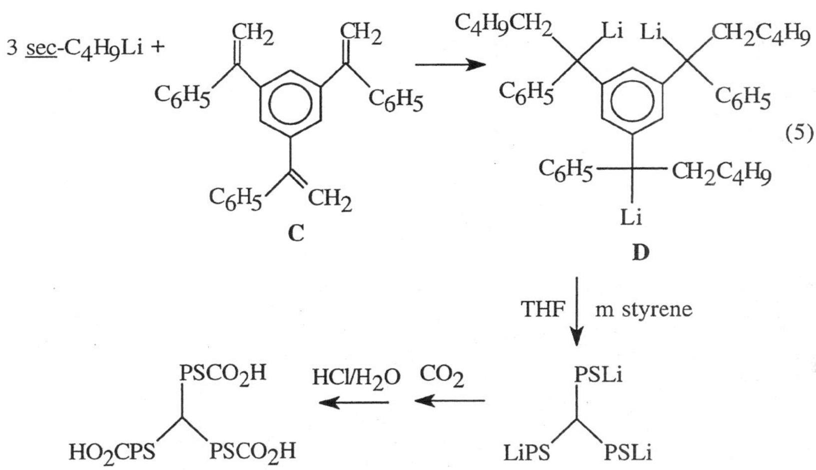

Scheme 5

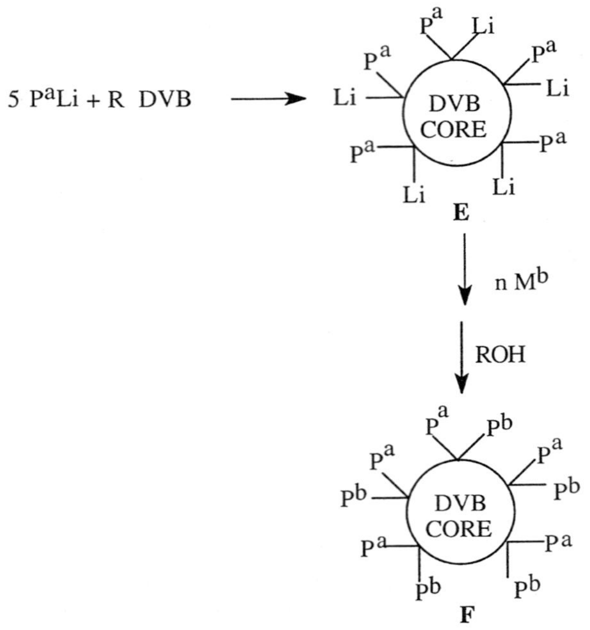

Scheme 6

star-branched polymer, **F.** Preliminary viscosity and light-scattering data were consistent with heteroarm, star polymer formation. However, the difficult determination of the uniformity of the arms and the efficiency of the reinitiation reactions was not established. This "core-first" heteroarm star synthesis method based on DVB linking of poly(styryl)lithium has been applied to the synthesis of *star*-polystyrene-*star*-poly(*n*-butyl methacrylate) [44] and *star*-polystyrene-*star*-poly(*t*-butyl methacrylate) [40,45].

Thus, the DVB linking methodology for the synthesis of both regular star-branched polymers (see Schemes 2 and 3) and heteroarm, star-branched polymers (Scheme 6) is very versatile, efficient, and rapid; these attributes are of considerable technological importance. Unfortunately, these methods lack the precision and control necessary for the synthesis of polymers with well-defined structures and low degrees of compositional heterogeneity. However, the insight gained from these methods has led to other general methods for the synthesis of both regular star-branched polymers and heteroarm star-branched polymers. For example, ethylene glycol dimethacrylate has been used as a multifunctional linking agent for lithium poly(methyl methacrylate) to make the corresponding regular star-branched polymers [46]. By using multifunctional 1,1-diphenylethylene derivatives instead of multifunctional styrene units for addition reactions with simple and polymeric organolithiums, general methods for the synthesis of well-defined multifunctional initiators (see Scheme 5), macromonomers, as well as regular and heteroarm, star-branched polymers have been developed, as shown in Scheme 7 [9].

III. DENDRIMERS AND HYPERBRANCHED POLYMERS

In the previous sections, star-shaped polymers were discussed. In this section, highly branched oligomers or polymers, as distinct from simple star-shaped polymers, will be considered. Several excellent, recent reviews are available on this subject [10–14,47,64,65].

Dendrimers are highly branched, three-dimensional macromolecules with a branch point at each monomer unit; they have structural features that are

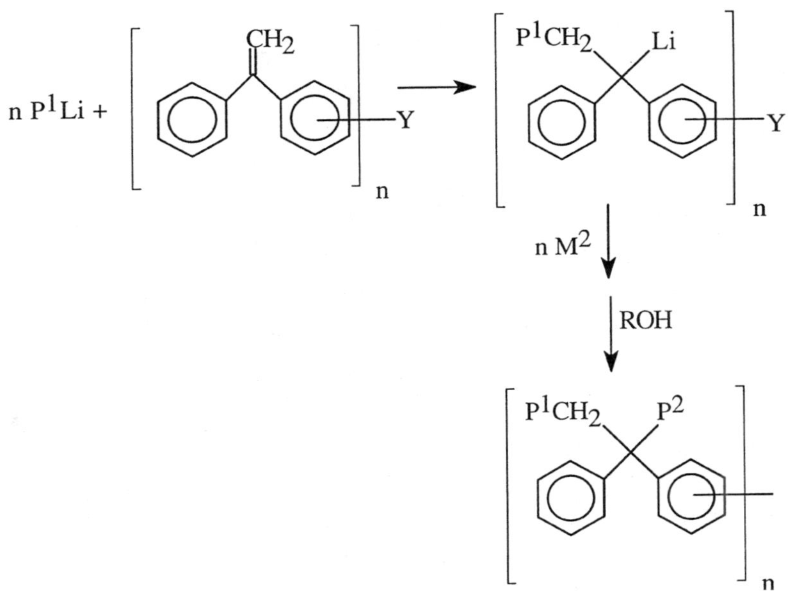

Scheme 7

analogous to the structure of trees [14]. Dendrimers have exact, monodisperse structures built layerwise in generations around a core moiety [10].

A. Synthesis of Dendrimers and Hyperbranched Polymers

1. *Dendrimers*

The synthesis of dendrimers generally involves numerous protection/deprotection and purification steps, but the general architectural growth patterns for the divergent and convergent routes are shown schematically in Schemes 8 and 9, respectively, where *f* represents the functionality of each structure, —— and ------- represent complementarily reactive functional groups, X is a latent functional group, C represents a complementary functional group, and · represents a chemical bond. It is important to note that because of the complementary functionality of the dendrimers and the growth units there are no dendrimer–dendrimer reactions and the AB_2-type units are prevented from reacting with

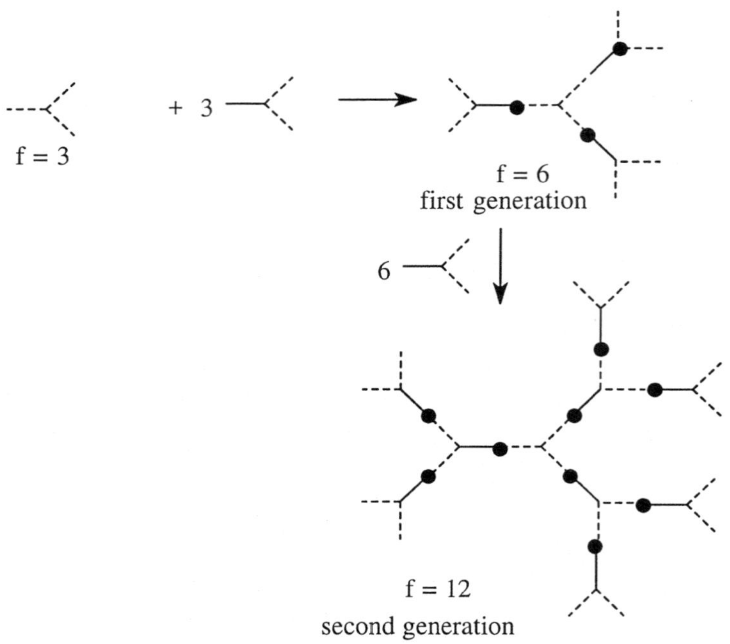

Scheme 8 Example of growth steps for formation of dendrimer (2 generations) by the divergent method.

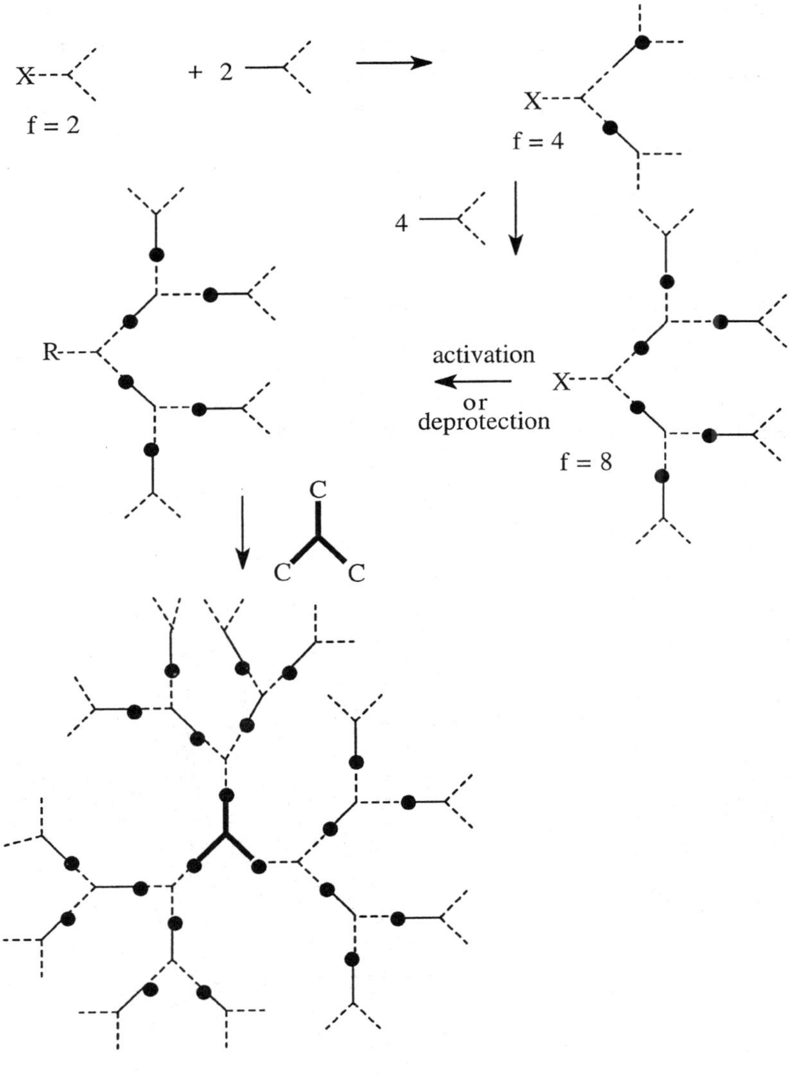

Scheme 9 Formation of dendrimer by the convergent method.

each other by suitable protecting groups on the B units. An important feature of dendrimers that is readily apparent even after the second generation is that they have high functionality and that these functional groups would be expected to be located primarily on the periphery of the dendrimers and to be accessible for chemical and physical interactions. It should be noted that the convergent synthetic route is very versatile since the final coupling step can be effected with different types of dendrons [14].

2. Hyperbranched Polymers

Hyperbranched polymers are prepared by the direct, one-step polycondensation of AB_x monomers with two different types of functional groups (A and B) that can react with each other to form a covalent bond, and the total number of reactive sites is $x + 1$ ($x \geq 2$) [48]. Hyperbranched polymers are highly branched like dendrimers, but their structure is not regular or highly symmetrical because linear segments can be formed, as shown in Scheme 10. It is important to note that in 1952 Flory [49] recognized the unique polymer structural type that

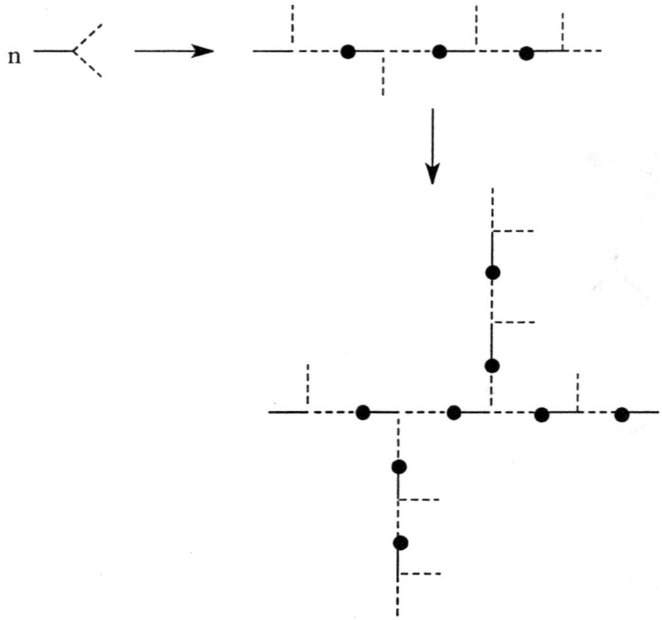

Scheme 10 Formation of hyperbranched polymers.

could be formed by self-condensation of AB_x monomers and even developed the mathematical description of molecular weight and molecular weight distribution as a function of the percentage of reaction of the functional groups. Hyperbranched polymers can be formed directly in a one-step reaction; therefore, to the extent that their properties approach those of dendrimers, they provide an expeditious and economical pathway to highly branched polymers.

3. Growth Reactions for Dendrimer Synthesis

Most dendritic polymers have been synthesized by condensation reactions. A wide variety of highly branched polymers, such as polyesters, polyamides, polyethers, polyphenylenes, polyarylacetylenes, and polycarbosilanes have been prepared by condensation mechanisms. Recently, synthetic methods to prepare polydendrons and hyperbranched polymers using vinyl monomers have been reported by several research groups [50–53]. The simplest application of vinyl polymerization involving dendrimers involves the polymerization of a vinyl monomer with an attached dendron (dendritic fragment) to form a polydendron. A schematic representation of a polydendron is shown in Figure 1b in comparison with a tridendron dendrimer (Figure 1a). The interest in the vinyl type of polydendron having dendrons (dendritic fragments) attached to the side chains comes from their potentially different topologies and properties.

Hawker and Frechet [50] prepared a dendritic macromonomer using the convergent-growth route to prepare a polyether dendron. The dendron core functional group (R in Scheme 9) was a hydroxymethyl group that was reacted with chloromethylstyrene to form the styrene-functionalized dendritic macromonomer. This macromonomer was copolymerized with styrene monomer using standard free-radical polymerization conditions.

Schluter and co-workers [51] have homopolymerized an analogous

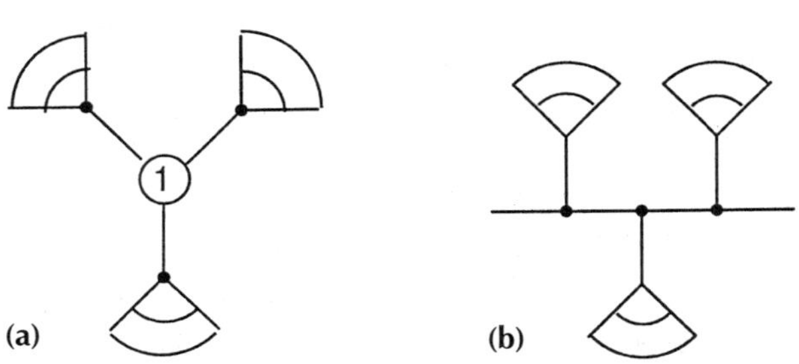

Figure 1 Comparison of the two-dimensional projections of a tridendron dendrimer (a) and a vinyl polymer (polydendron) carrying dendrons attached on the side chain (b).

series of styrene-functionalized dendritic macromonomers. For third-generation dendrons, the degrees of polymerization were 23–62. Light-scattering analysis indicated that these polymers were best described as Gaussian chains. It was concluded that the steric requirements of these side chains were insufficient to give cyclindical structures.

Xi and co-workers [52] prepared methacrylate-functionalized dendritic macromonomers by conversion of the hydroxymethyl group of the polyether dendron to the corresponding bromide and then functionalization with methacrylic acid. Free-radical polymerization of the macromonomer corresponding to a second-generation dendron resulted in formation of a relatively low molecular weight product ($X_n = 6–7$).

Schluter and co-workers [53] have also prepared a poly (*p*-phenylene) with dendritic branches by condensation polymerization. The structure of this polymer would be expected to be cylindrical because the backbone chain corresponds to a rigid rod.

A new block type of dendritic polymer has been prepared by Meijer and co-workers [54,55]. Polystyrene-*block*-dendron (PS-dendr-(COOH)*n*) was prepared by a divergent dendrimer synthetic methodology using an amine-functionalized polystyrene of well-defined structure as the core for the synthesis of a poly(propylene imine) dendrimer. A representation of the structure is shown in Figure 2. It was reported that the amphiphilic dendrimer exhibited a remarkable generation-dependent aggregation behavior in a variety of salt solutions regardless of pH.

4. Synthesis of Hyperbranched Polymers By Self-Condensing Vinyl Polymerization

Frechet and co-workers [56] have developed a new methodology for synthesis of hyperbranched polymers by combining both step growth polymerization and chain growth polymerization with dendrimer units formed by either the converging or diverging pathways, as illustrated in Scheme 11. The key element is this process is to use a monomer that has both a vinyl group and a pendant group that can be activated to form an initiating moiety.

B. Structure and Characterization

One of the primary differences between the star-branched polymers described previously and dendritic or hyperbranched polymers is their chain segment density distribution. For star-branched polymers, the chain segment density is highest at the core and decreases as the distance from the core increases. In contrast, for dendritic and hyperbranched polymers the chain segment density increases as the distance from the core increases (with increasing generations

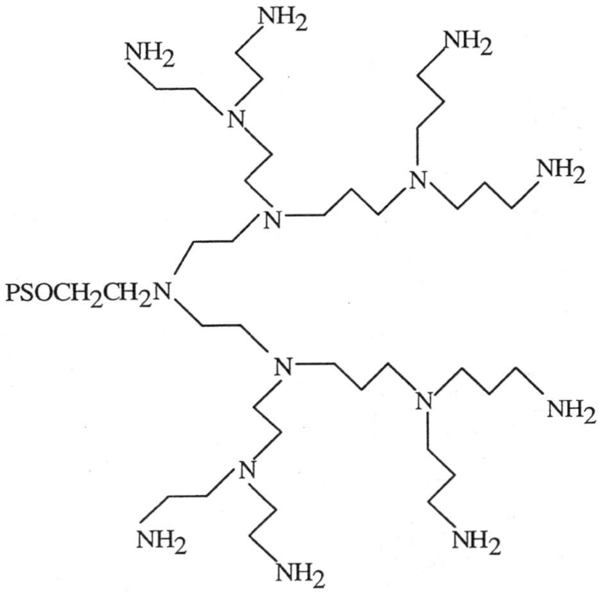

Figure 2 Structure of Ps-dendr-$(NH_2)_8$.

for dendritic polymers) [10]. In fact, it is calculated that a maximum number of dendrimer generations can be formed (starburst dense-packed generation), beyond which only defect structures can be generated [10,57].

We have discussed the synthetic methods for dendrimers and hyperbranched polymers. As shown in Schemes 8, 9 and 10, the molecular architecture of dendrimers and hyperbranched polymers will be remarkably affected by the core and its functionality, the branching-site multiplicity, the chain length depicted as a shell surrounding the central core, and so on. The size and the physical properties of the formed polydendrons will be greatly affected by these variables. Among these variables, the most important factor is the degree of branching (DB) represented in Eq. (6), as defined by Frechet and co-workers [58], which is directly related to polydendron size (hyperbranched polymer).

$$\mathrm{DB} = \frac{\sum \text{dendritic unit} + \sum \text{terminal unit}}{\sum \text{dendritic unit} + \sum \text{terminal unit} + \sum \text{linear unit}} \tag{6}$$

A variety of methods have been used to characterize the structures of dendrimers. In contrast to common polymers, which are polydisperse, dendrimers can be obtained as truly monodisperse materials as shown by mass

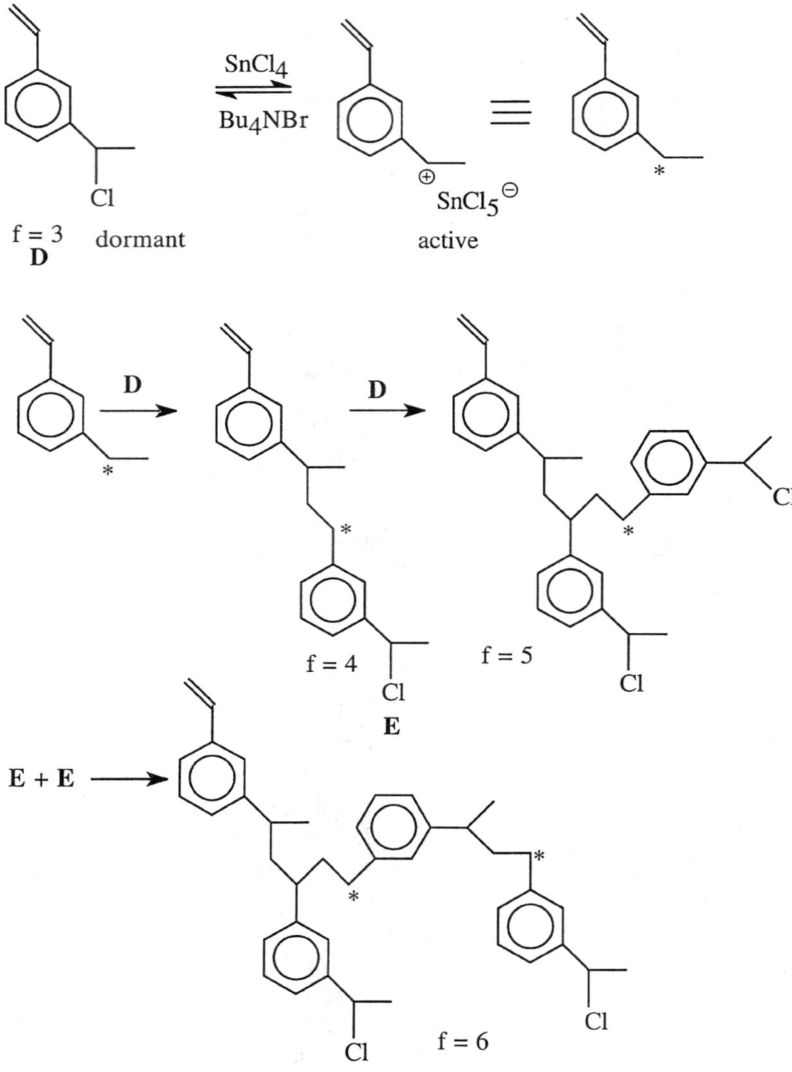

Scheme 11 Hyperbranched polymers by self-condensing vinyl polymerization.

spectrometry [14]. The degree of branching for a perfect dendrimer is 1.0 (100%). In contrast, hyperbranched polymers exhibit DB values in the range between 0 and 1 (0.4–0.7 for polyesters) and broad molecular weight distributions (e.g., $M_w/M_n = 2–10$ for polyesters) [13]. The molar mass can be investigated by low-angle laser light scattering, mass spectrometry, and size exclusion chromatography/universal calibration [13,14]. The structural perfection of dendrimers can be investigated by ^1H and ^{13}C NMR spectroscopy, electrospray ionization mass spectrometry, and matrix-assisted laser desorption/ionization mass spectrometry (MALDI). For example, the degree of branching can be investigated by NMR spectroscopy since the different units [see Eq. (6)] generally possess different chemical shifts because of the highly ordered and symmetrical branching sequence [14]. Shape-related properties can be characterized by light scattering, intrinsic viscosity measurements, and size exclusion chromatography.

The glass transition temperatures of aromatic polyesters were reported to be totally independent of the variation of the architecture or shape by comparison of dendrimer, hyperbranched, and linear analogs [59]. The thermogravimetric behavior was also found to be independent of architecture [59]. However, the glass transition temperatures of hyperbranched polyesters are very strongly dependent on the nature of the terminal groups, as is their solubility [59,60]. Thus, the solubility and thermal properties of dendrimers and hyperbranched polymers can be varied by varying the chain end groups.

C. Properties

Dendritic polymers exhibit very different properties compared with their linear analogs. For instance, they exhibit extremely high solubility in various organic solvents and low intrinsic viscosity in comparison to their linear analogs. These differences are probably a reflection of the large number of chain end functional groups as well as the influence of architectural differences [59].

Dendrimers do not exhibit linear Mark–Houwink plots [13]. The intrinsic viscosity increases with size and then decreases after about generation 4.

Many physical properties of hyperbranched aromatic polyesters, such as high solubility and low viscosity, resemble those of dendrimers. However, in contrast to dendrimers, hyperbranched polymers do exhibit a linear dependence of intrinsic viscosity on molecular weight with very low Mark–Houwink constants (0.3–0.4) [13]. In addition, their lack of reactivity toward catalytic hydrogenolysis greatly differs from that of the corresponding dendrimers, which are cleanly deprotected under mild conditions; rather their reactivity more resembles that of the corresponding linear polymers [59]. As shown in

Eq. (6), a hyperbranched polymer will resemble a dendrimer with increasing degree of branching. However, many questions still remain regarding the chemical and physical properties of hyperbranched polymers and dendrimers.

D. Applications

The synthesis, characterization, and applications of dendritic and hyperbranched polymers is an exciting active area of investigation. The "one-pot" synthesis of hyperbranched polymers suggests that their commercial exploitation will outpace that of dendrimers for many applications.

Hyperbranched polymers can be used as additives to conventional polymers, in blends, as thermosets, and as thermoplastics. In blends, it would be expected that miscibility could be varied by modification of end groups. It is considered that their addition to thermosets, blends, and additives would be expected to enhance mechanical properties and exhibit beneficial processing characteristics because of their nonentangled nature [13]. Thus, as an additive, a hyperbranched polyphenylene has been shown to reduce melt viscosity for polystyrene processing [48,61]. The processing characteristics and high functionality of hyperbranched polymers have been shown to impart desirable processing and curing characteristics to thermosets [62,63].

Investigation of the possible applications of dendrimers has been limited by their lack of ready availability [14]. It is anticipated that dendrimers may be used as a polymeric matrix for catalysts, as a membrane for gas separation, and as part of the diagnostic kit for MRI since many groups can be added to the core or to the periphery. The unique structure and solubility characteristics of dendrimers suggest that their use as dendritic micelles may be important. In addition, they may form interesting and useful monolayers [14].

REFERENCES

1. L. J. Fetters and E. L. Thomas, Model polymers for materials science, in *Material Science and Technology,* Vol. 12, VCH Verlagsgesellschaft, Weinheim, Germany, 1993, p. 1.
2. G. S. Grest, L. J. Fetters, and J. S. Huang, Star Polymers: Experiment, Theory and Simulation, *Adv. Chem. Phys.,* Vol. XCIV, I. Prigogine and S. A. Rice, Eds., Wiley, 1996, p. 67.
3. J. Roovers, in *Encyclopedia of Polymer Science and Engineering,* 2nd ed., J. I. Kroschwitz, Ed., Wiley, New York, 1985, Vol. 2, p. 478.
4. S. Bywater, *Adv. Polym. Sci.,* **30,** 89 (1979).
5. J. E. Mark, A. Eisenberg, W. W. Graessley, L. Mandelkern, E. T. Samulski, J. L. Koenig, and G. D. Wignall, *Physical Properties of Polymers,* 2nd ed., American Chemical Society, Washington, D.C., 1993.

6. M. Pitsikalis, S. Pispas, J. W. Mays, and N. Hadjichristidis, *Adv. Polym. Sci.,* **135,** 1 (1998).
7. H. L. Hsieh and R. P. Quirk, *Anionic Polymerization: Principles and Practical Applications,* Dekker, New York, 1996.
8. P. Dreyfuss and R. P. Quirk, Graft Copolymers, in *Encyclopedia of Polymer Science and Engineering,* 2nd ed., J. I. Kroschwitz, Ed., Wiley, New York, 1986, Vol. 6, p. 551.
9. R. P. Quirk, T. Yoo, and B. Lee, *J. Macromol. Sci.-Pure Appl. Chem.,* **A31,** 911 (1994).
10. D. A. Tomalia, A. M. Naylor, and W. A. Goddard III, *Angew. Chem. Int. Ed. Engl.,* **29,** 138 (1990).
11. G. R. Newkome, C. N. Moorefield, and G. R. Baker, *Adrichimica Acta,* **25**(2), 31 (1992).
12. J. M. J. Frechet, *Science,* **263,** 1710 (1994).
13. B. I. Voit, *Acta Polym.,* **46,** 87 (1995).
14. J. M. J. Frechet and C. J. Hawker, in *Comprehensive Polymer Science, Second Supplement,* S. I. Aggarwal and S. Russo, Eds., Pergamon Tarrytown, N.Y., New York, 1996.
15. L. J. Fetters, Monodisperse Polymers, in *Encyclopedia of Polymer Science and Engineering,* 2nd ed., J. I. Kroschwitz, Ed., Wiley, New York, Vol. 10, 1987, p. 19.
16. H. Hsieh, *Rubber Chem. Technol.,* **49,** 1305 (1976).
17. R. P. Quirk, D. J. Kinning, and L. J. Fetters, in *Comprehensive Polymer Science,* G. Allen and J. C. Bevington, Eds., Pergamon Press, Tarrytown, N.Y., Vol. 7, 1989, p. 1.
18. M. Morton, T. E. Helminiak, S. D. Gadkary, and F. Bueche, *J. Polym. Sci.,* **57,** 471 (1962).
19. R. P. Zelinski and C. F. Wofford, *J. Polym. Sci.: Part A,* **3,** 93 (1965).
20. J. E. L. Roovers and S. Bywater, *Macromolecules,* **5,** 384 (1972).
21. J. Roovers, L.-L. Zhou, P. M. Toporowski, M. van de Zwan, H. latrou, and N. Hadjichristidis, *Macromolecules,* **26,** 4324 (1993).
22. J. Roovers, P. Toporowski, and J. Martin, *Macromolecules,* **22,** 1897 (1989).
23. R. W. Pennisi and L. J. Fetters, *Macromolecules,* **21,** 1094 (1988).
24. N. Khasat, R. W. Pennisi, N. Hadjichristidis, and L. J. Fetters, *Macromolecules,* **21,** 1100 (1988).
25. G. Kraus and J. T. Gruver, *J. Polym. Sci.,* **A3,** 105 (1965).
26. F. C. Weissert and B. L. Johnson, *Rubber Chem. Technol.,* **40,** 590 (1967).
27. F. Tsutsumi, M. Sakakibara, and N. Oshima, *Rubber Chem. Technol.,* **63,** 8 (1990).
28. C. A. Sierra, C. Galan, M. J. Gomez Fatou, and V. Ruiz Santa Quiteria, *Rubber Chem. Technol.,* **68,** 259 (1995).
29. K. Knoll and N. Niessner, in *Applications of Anionic Polymerization Research,* R. P. Quirk, Ed., American Chemical Society, Washington, D.C., 1998, p. 112.
30. G. D. Karles, W. H. Christiansen, J. G. Ekerdt, I. Trachtenberg, and J. W. Barlow, *Ind. Eng. Chem. Res.,* **30,** 646 (1991).
31. H. Eschwey and W. Burchard, *J. Polym. Sci.: Symposium,* **5,** 1 (1975).

32. R. P. Zelinski and H. L. Hsieh (assigned to Phillips Petroleum Company), U.S. patent 3,280,084 (1966).
33. B. J. Bauer and L. J. Fetters, *Rubber Chem. Technol.,* **51,** 406 (1978).
34. R. N. Young and L. J. Fetters, *Macromolecules,* **11,** 899 (1978).
35. G. Quack, L. J. Fetters, N. Hadjichristidis, and R. N. Young, *Ind. Eng. Chem. Prod. Res. Dev.,* **19,** 587 (1980).
36. M. K. Martin, T. C. Ward, and J. E. McGrath, in *Anionic Polymerization. Kinetics, Mechanisms, and Synthesis,* J. E. McGrath, Ed., American Chemical Society, Washington, D.C., 1981, p. 557.
37. R. J. A. Eckert (assigned to Shell Oil Company), U.S. patent 4,116,917 (1978).
38. H. Eschwey, M. L. Hallensleben, and W. Burchard, *Makromol. Chem.,* **173,** 235 (1973).
39. P. Lutz and P. Rempp, *Makromol. Chem.,* **189,** 1051 (1988).
40. C. Tsitsilianis, P. Lutz, S. Graff, J.-P. Lamps, and P. Rempp, *Macromolecules,* **24,** 5897 (1991).
41. R. P. Quirk and H. Tsai, *Macromolecules,* **31,** 8016 (1998).
42. H. Eschwey and W. Burchard, *Polymer,* **16,** 180 (1975).
43. R. P. Quirk and F. Ignatz-Hoover, in *Recent Advances in Anionic Polymerization,* T. E. Hogen-Esch and J. Smid, Eds., Elsevier, New York, 1987, p. 393.
44. C. Tsitsilianis, P. Chaumont, and P. Rempp, *Makromol. Chem.,* **191,** 2319 (1990).
45. C. Tsitsilianis, S. Graff, and P. Rempp, *Eur. Polym. J.,* **27,** 243 (1991).
46. V. Efstratiadis, G. Tselikas, N. Hadjichristidis, J. Li, W. Yunan, and J. W. Mays, *Polym. Int.,* **33,** 171 (1994).
47. M. Johansson, E. Malmstrom, and A. Hult, *Trends Polym. Sci.,* **4**(12), 398 (1996).
48. Y. H. Kim and O. W. Webster, *Polym. Prepr., Am. Chem. Soc. Div. Polym. Chem.,* **29**(2), 310 (1988).
49. P. J. Flory, *J. Am. Chem. Soc.,* **74,** 2718 (1952).
50. C. J. Hawker and M. J. Frechet, *Polymer,* **33,** 1507 (1992).
51. I Neubert, E. Amoulong-Kirstein, A.-D. Schluter, and H. Dautzenberg, *Macromol. Rapid Commun.,* **17,** 517 (1996).
52. Y.-M. Chen, C.-F. Chen, W.-H. Liu, Y.-F. Li, and F. Xi, *Macromol. Rapid Commun.,* **17,** 401 (1996).
53. W. Claussen, N. Schulte, and A.-D. Schluter, *Macromol. Rapid Commun.,* **16,** 89 (1995).
54. J. C. M. van Hest, D. A. P. Delnoye, M. W. P. L. Baars, M. H. P. van Genderen, and E. W. Meijer, *Science,* **268,** 1592 (1995).
55. J. C. M. van Hest, M. W. P. L. Baars, C. Elissen-Roman, M. H. P. van Genderen, and E. W. Meijer, *Macromolecules,* **28,** 6689 (1995).
56. J. M. J. Frechet, M. Henmi, I. Gitsov, S. Aoshima, M. R. Leduc, and R. B. Grubbs, *Science,* **269,** 1080 (1995).
57. P. G. de Gennes and H. J. Hervet, *Phys. Lett. Paris,* **44,** 351 (1983).
58. C. J. Hawker, R. Lee and J. M. J. Frechet, *J. Am. Chem. Soc.,* **113,** 4583 (1991).
59. K. L. Wooley, J. M. J. Frechet, and C. J. Hawker, *Polymer,* **35,** 4489 (1994).

60. A. R. Brenner, B. I. Voit, D. J. Massa, and S. R. Turner, *Macromol. Symp.,* **102,** 47 (1996).
61. Y. H. Kim and O. W. Webster, *J. Am. Chem. Soc.,* **112,** 4592 (1990).
62. M. Johansson and A. Hult, *J. Coating Technol.,* **67** (Oct.), 35 (1995).
63. A. Hult, M. Johansson, and E. Malmstrom, *Macromol. Symp.,* **98,** 1159 (1995).
64. O. A. Matthews, A. N. Shipway, and J. Fraser Stoddart, *Prog. Polym. Sci.,* **23,** 1 (1998).
65. Y. H. Kim, *J. Polym. Sci.: Part A: Polym. Chem.,* **36,** 1685 (1998).

2

Star-Shaped Polymers via Anionic Polymerization Methods

S. Plentz Meneghetti and P. J. Lutz
Institute Charles Sadron, CNRS, Strasbourg, France

D. Rein
Cytotherapeutics, Providence, Rhode Island

I. INTRODUCTION

Anionic polymerization methods have been used to synthesize a wide variety of macromolecules including linear [1] and cyclic [2] homopolymers, linear copolymers [1], and functional polymers such as macromonomers [3]. These macromolecules are well defined with predetermined molar masses, sharp molar mass distributions, and low compositional heterogeneity. They serve as ideal compounds to establish the relation between the structure, the properties, and theory.

Branched macromolecules (Figure 1) are more compact than linear homologous ones because of their higher segment densities. The increased segment density results in a decreased tendency for these macromolecules to interpenetrate in solution as well as in bulk. In order to investigate this influence of structure (number of branches, length of branches) on macromolecular properties, well-defined star-shaped macromolecules are required. Statistically, branched polymers are not easily accessible by anionic polymerization. Comb-shaped polymers (Figure 1), constituted of a polymeric backbone carrying a number of side chains usually of the same length and identical chemical nature as the backbone (although this may not always be the case) can be synthesized.

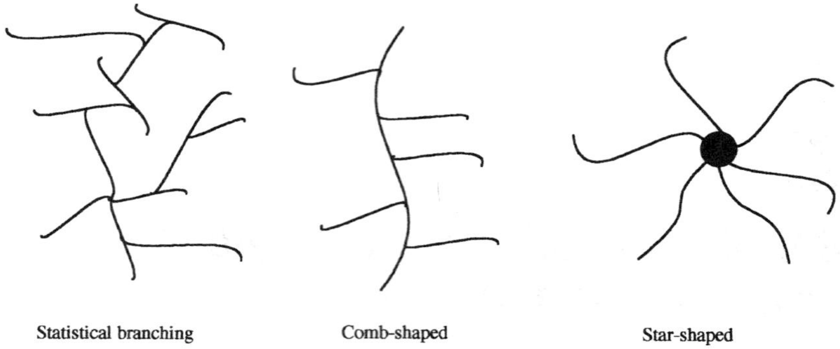

Figure 1 Various branched structures.

Among the various means to access these structures, the homopolymerization of macromonomers has proved to be one of the most efficient methods [4,5].

Star polymers (Figure 1) are characterized as the simplest case of branched species where all chains of a given macromolecule are connected to a single nodulus referred to as the core. The preparation methods and properties of star-branched polymers were examined in detail several years ago by Bywater [6] and more recently by others [7–11].

Star-shaped polymers have gained increasing interest because of their compact structure and high segment density, and because very efficient synthetic methods have made possible the functionalization of the outer branch ends. Until recently, anionic polymerization was one of the best methods to obtain well-defined star-shaped polymers of predetermined branch molar mass. This technique provided the long lifetime for the active sites necessary to allow the formation of star-shaped macromolecules. Anionic polymerization also limited the polymolecularity of the samples. Given the appropriate reaction conditions, the functionality of the core can be controlled in advance.

The various approaches to access star-shaped polymers via anionic polymerization will be presented and discussed in the following sections. Once these star-shaped materials have been characterized, they serve as ideal model compounds, and have been studied extensively with respect to theoretical predictions concerning their solution properties (dilute or semidilute regime) or their solid state properties [12–20].

Several anionic methods will be addressed that produce well-defined star-shaped polymers.

In the first method, referred to as the "arm-first" method, monofunctional living chains are used as initiator for the polymerization of bifunctional monomers to generate the star core. The "arm-first" method produces homopolymeric or copolymeric star-shaped polymers. The use of a plurifunctional low-molar-mass deactivating agent represents an interesting alternative method for preparing well-defined star-shaped polymers. That method will be discussed first, and its advantages and drawbacks will be presented. Recently, the "arm-first" method has been extended to the preparation of star-shaped polymers where polymer chains of different chemical nature are connected to the same nodulus.

In the second method, the "core-first" method, a polyfunctional core is used to initiate the polymerization of the branches of the star. This method allows for easy access to chain end functionalization by simple deactivation of the active sites. The extension of the "core-first" method to the preparation of functional star-shaped polymers in nonpolar solvents will be discussed.

Star polymers generated using the "core-first" method have a large distribution in functionalities. To minimize the distribution in functionalities of the stars, the "in–out" technique was developed. The "in-out" method is a combination of the two techniques mentioned above and first generates a small "arm-first" star with living active sites, then uses this core to initiate the polymerization of the star branches. The resulting stars can be functionalized, and the control over the distribution in functionalities is greatly improved.

These different star-shaped macromolecules were extensively characterized to confirm the expected structure. (The appropriate characterization methods are presented in another chapter of the book and are discussed here.) Briefly, star-shaped macromolecules exhibit a smaller hydrodynamic volume than linear homologous ones, as is expected from the high segment densities, which lead to higher elution volumes. Size exclusion chromatography (SEC) techniques based on calibration with linear samples were not applicable, but light-scattering methods and SEC with light-scattering detection on-line can be employed. Standard SEC is necessary to quantify unreacted linear branches present in the star molecule. In order to establish the average molar mass of the star-shaped macromolecule, the exact functionality must be known. In the case of star-shaped polymers based on copolymeric branches, the chemical composition of the branches must also be determined using the available classic methods. Star-shaped polymers are also characterized by a decreased radius of gyration, decreased viscosity, and higher translational diffusion coefficients relative to linear homologous macromolecules.

II. STAR-SHAPED POLYMERS VIA "ARM-FIRST" METHODS

In all "arm-first" methods a living monofunctional polymer of known length and low polymolecularity serves as a precursor. Subsequently, the active sites located at chain end can be used in one of two different ways:

Either they are reacted with a compound carrying a number of appropriate reactive functions, whereupon chemical links are formed.

Or they are used to initiate the polymerization of a small amount of an appropriate bisunsaturated monomer, whereupon small cross-linked cores are formed.

A. "Arm-First" Methods By Deactivation

1. Homopolymeric Branches

The living polymer precursor can be reacted stoichiometrically with a plurifunctional deactivator (an electrophilic compound if the sites are anionic), and chemical links are formed between the precursor chains and the deactivator.

The precursor chains become the star branches, and the deactivator becomes the core (Scheme 1). The difficulty is identifying compounds carrying a number of equally reactive and equally accessible electrophilic functions needed to control the average number of branches of the stars. The induced deactivation must be fast, quantitative, and free of any side reactions. The limitations encountered include the low functionality of the substances used as deactivators such as chloromethylated benzenes [21,22], trisallyloxytriazines [23], and silicon tetrachloride [24] (Figure 2).

The rate of electrophilic substitution also depends upon the carbanionic chain end: under identical conditions, living polyisoprene (PI) is much more reactive than living polystyrene (PS). As mentioned later in this discussion, Hadjicristidis has taken advantage of this to prepare a new type of star-shaped polymer where the same nodulus carries arms of different chemical nature.

In order to improve the functionalization yields and/or to increase the

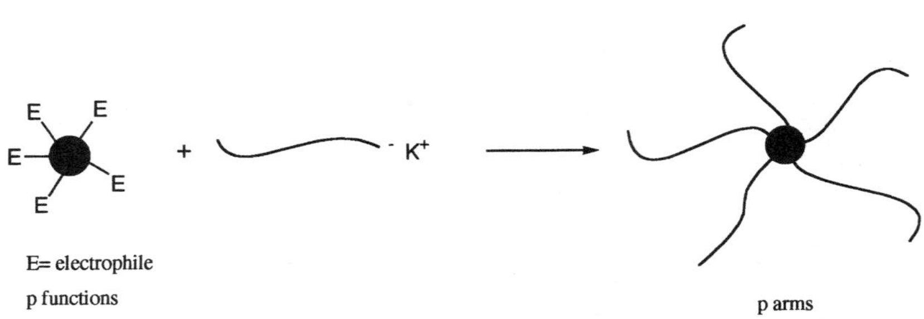

Scheme 1 Synthesis of star-shaped polymers via deactivation reaction.

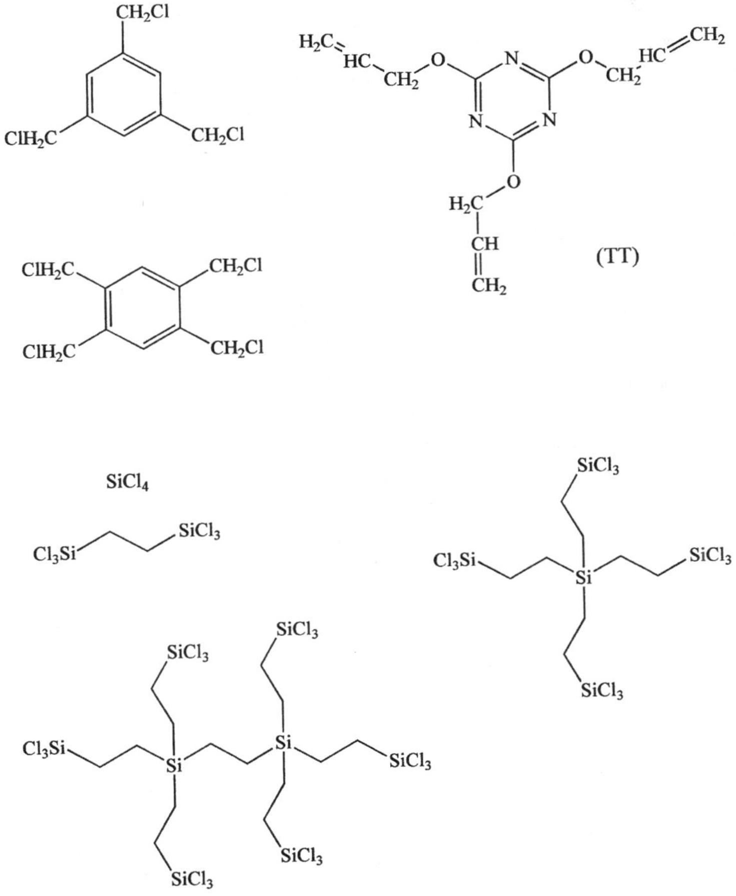

Figure 2 Structures of deactivating agents.

average functionality of these star-shaped polymers prepared with chlorosilane derivates, Fetters et al. [25] have synthesized and used a new generation of chlorosilane compounds of high functionality where the number of chlorine functions is limited to 2 or 3 per silicium and the silicium units are separated by ethylene spacers (Figure 2). In that case, silicon halides react quantitatively, by addition, with carbanions derived from styrene, dienes, and other living polymers. Well-defined star-shaped polymers with up to 128 arms have been obtained [26]. The solution properties of these star-shaped polymers were studied to confirm the expected structure. They have also been used as a model for semidilute and solid state behavior. The properties of these star-shaped polymers would not be affected even if several Si–Cl functions remained unreacted. The number of arms per star molecule is determined by the functionality of the electrophilic compound used provided the yield of the coupling reaction is close to quantitative. Since the fluctuations in length of the branches are averaged, the molar mass distribution in such star polymer samples is expected to be rather narrow. The accurate characterization of star-shaped polymer samples demonstrates the efficiency of the method: the molar mass of the star molecules is very close to the expected value. Therefore, the linking reaction attains high yields despite the bulkiness of these star molecules toward the end of the coupling process.

Similarly, tetrakis[4-(1-phenylvinyl)phenylplumbane] (Figure 3) was also used efficiently as a linking agent: well-defined tetrafunctional PS star-shaped polymers could be obtained in good yields. The procedure was extended to the synthesis of star-shaped polymers containing PS and poly(methylmethacrylate) (PMMA) branches and to the preparation of model networks exhibiting tetrafunctional crosslinking points [27].

Anionic "deactivation" methods have proved their efficiency in the synthesis of star-branched oligomers where the central core is obtained upon reaction of oligobutadienyllithium with diesters and epoxidized soybean oil [28].

2. *Copolymeric Branches*

The method described above was applied to the synthesis of star block copolymers (Figure 4), in which each branch was a polystyrene-*b*-polydiene copolymer. Under selected conditions of composition, branch length, and concentration, they exhibit bicontinuous mesomorphic phases referred to as double-diamond structures [29], which had never been observed before.

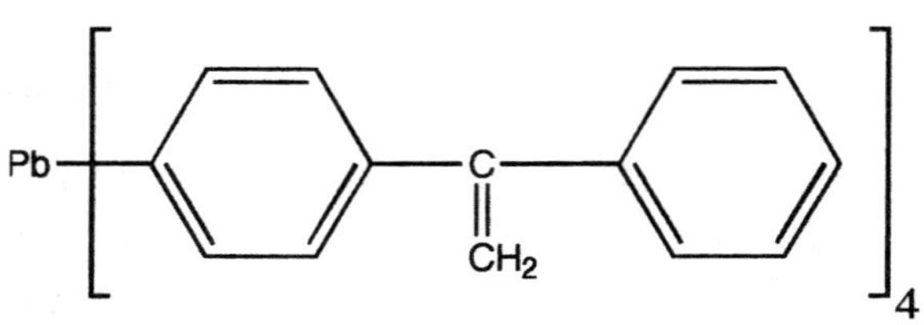

Figure 3 Tetrakis (4-(1-phenylvinyl) phenylplumbane).

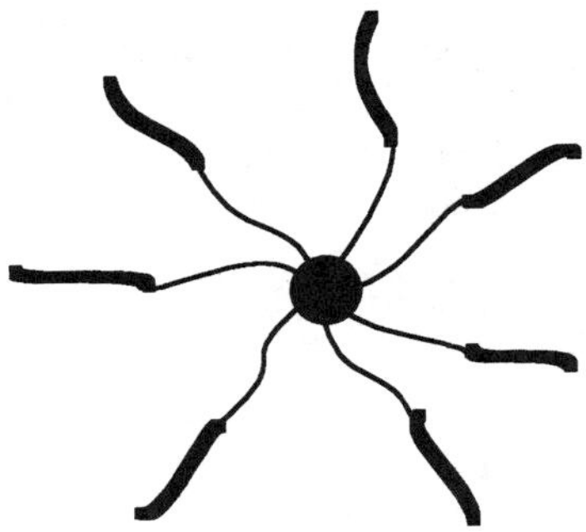

Figure 4 Star-shaped polymers with copolymeric branches.

3. *Functional Star-Shaped Polymers*

Burchard (Scheme 2) recently developed an approach to synthesize star-shaped polymers with tertiary amine groups at the outer end of the branches [30]. The polymerization of styrene was initiated with ((3-dimethylamino)propyl)lithium and the living carbanions were reacted stoichiometrically with tris(allyloxy)-1,3,5-triazine (TT).

Well-defined three functional star-shaped polymers were obtained using this technique. In some cases, fractionation was necessary to remove the unreacted branches. The reaction of the tertiary amino end group of these functional star-shaped polymers with an appropriate bifunctional low-molar-mass compound resulted in reversible gelation of the reaction medium. These compounds were also used in several theoretical studies focused on the comparison of the solution properties of linear, stars, and ring polymers. Similar approaches were used to attach star-shaped polymers with hydroxyl end groups. The use of functional cleavable protected initiators made this method difficult to perform. Three-arm poly(butadienes) (PBd) with one, two, and three functional end groups were prepared using ((3-dimethylamino)propyl)lithium and *sec*-butyllithium as initiators and methyltrichlorosilane as the linking agent [31]. The association phenomena of star-shaped poly(butadienes) were studied in detail after transformation of the dimethylamino end groups into sulfozwitterionics.

4. *"Miktoarm" Star-Shaped Polymers*

The "arm-first" method based on a deactivation reaction was used to build a novel type of star-shaped polymers: "miktoarm" star-shaped polymers and "quaterarm" star-shaped polymers. It has been shown that the reaction of the

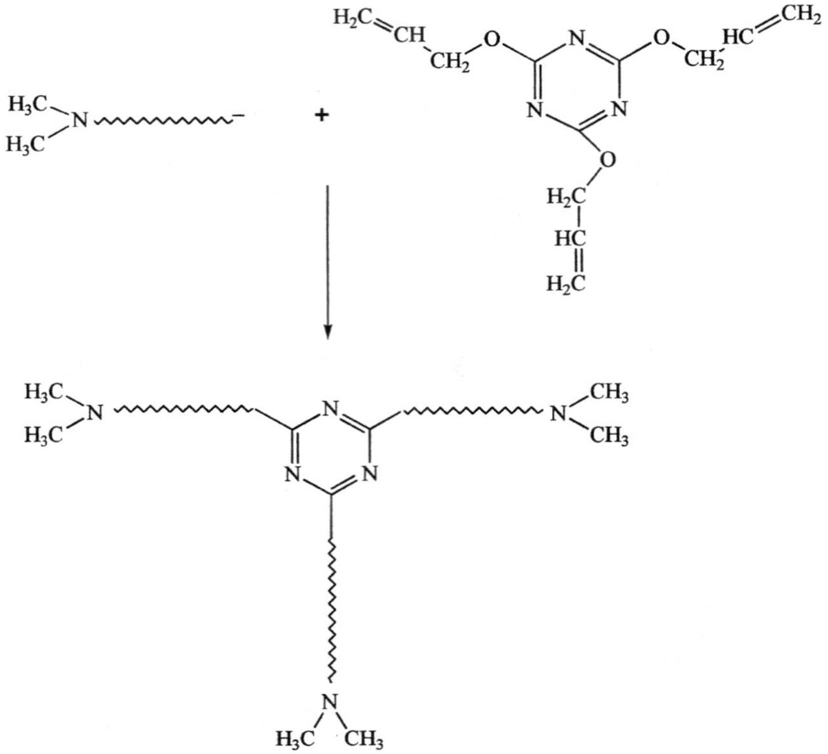

Scheme 2 Synthesis of functional star-shaped polymers via "arm-first" methods.

living chain end of PS could not be achieved quantitatively with the four functions of tetrachlorosilanes. Hadjichristidis [32] has taken advantage of that to prepare three and tetrafunctional star-shaped heteroarm polymers. The approach uses following plurifunctional compounds: tetrachlorosilanes ($f = 4$) or methyltrichlorosilanes ($f = 3$) (Figure 5).

3-"Miktoarm" star terpolymer resulted from the reaction of the methylchlorosilanes with a living PI first, then by adding methyltrichlorosilane in a great excess. The living PS is reacted with another function and finally with a small excess of poly(butadiene). In spite of the complexity of the procedure, a well-defined 3-"miktoarm" polymer was obtained [32]. One example is given in Figure 5.

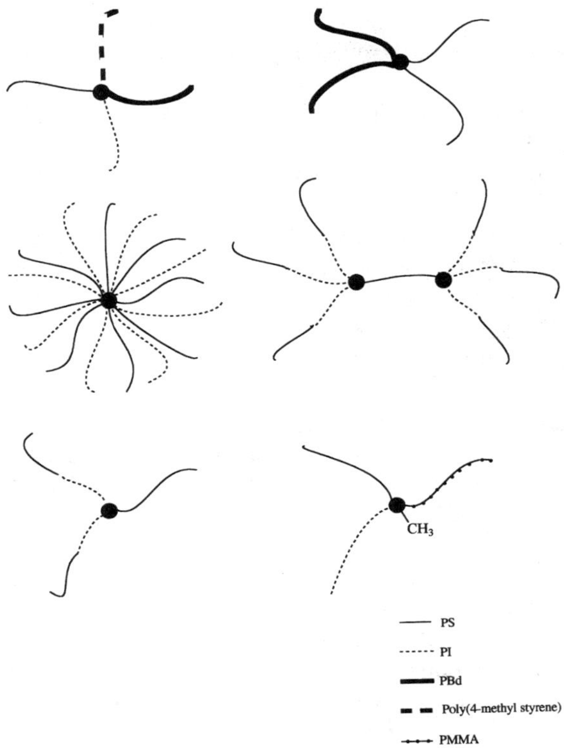

Figure 5 Various miktoarm and quaterarm star-shaped polymers.

5. "Quaterarm" Star-Shaped Polymers

Chlorosilanes were used applying an identical reaction procedure to prepare either "quaterarm" polymers of type ABCD or 4-"miktoarm" polymers of type A_2B_2. A, B, C, D correspond to PS, PBd, PI, and poly(4-methylstyrene), respectively [33]. That method was extended recently to well-defined poly(isoprene)/poly(butadiene) A_2B_2 copolymers [34]. The presence on the same nodulus of chains exhibiting different chemical structures leads to original solution properties. Roovers et al. [35] have examined in detail the solution properties and compared the specific behavior of these "miktoarm" star-shaped polymers to linear diblock copolymers. That strategy was extended to the preparation of

H-shaped structures (Figure 5) by simply replacing the ω-functional polymer of the first generation by an α,ω-bifunctional living polymer [36].

In order to study the influence of an increase of functionality on the properties of these "miktoarm" star-shaped polymers, Avgeropoulos et al. have synthesized and characterized 16-"miktoarm" star copolymers (Figure 5) [37]. Further work is now in progress to investigate the different properties of these "miktoarm" star-shaped copolymers [38].

The reaction of a living chain with 1,3-bis(1-phenylethenyl) benzene or 1,3-bis(1-phenylvinylbenzene) (MDDPE), a nonpolymerizable divinyl compound, can be considered as another approach to obtain A_2B_2 structures [39]. That point will be discussed later in the chapter.

Mason and Hogen-Esch [40] have also reported the preparation of star-shaped polymers bearing on the same nodulus chains of different chemical nature: a PMMA chain bearing a hydroxyl group at one chain end was prepared first. After transformation, that chain end was used as a polymeric initiator for polymerization of the second monomer. The active chain ends of the block copolymer were then coupled with chlorosilane linking agents. That preparation method represents a combination of "arm-first" and "core-first" procedures. That point will be presented and discussed later.

A final "arm-first" method based on the preparation via anionic polymerization of star-shaped polymers will be mentioned. Here the anionic polymerization is only used to prepare a linear ω-functional polymer of controlled molar mass and functionality and not for the coupling reaction. In many cases, anionic polymerization is the only way to access these well-defined monofunctional precursor chains. In a second step, these functional polymers were reacted with plurifunctional compounds exhibiting antagonist functions. Some examples are given below.

A monofunctional poly(dimethyl)siloxane (PDMS) chain fitted with a SiH end group is reacted under appropriate conditions with a low-molar-mass compound exhibiting antagonist Si-vinyl functions or the reverse [41] (Scheme 3). That method has also been used successfully in combination with anionic polymerization to prepare polystyrene-arm-styrene-arm-2-vinylpyridine star copolymers [42].

Triisocyanates have also been used efficiently in the preparation of well-defined trifunctional PS star-shaped macromolecules, with PS exhibiting hydroxyl or amino groups at one end, obtained by anionic polymerization, being used as precursor [43]. These triisocyantes have also been applied for the synthesis of well-defined PEO star-shaped molecules [44].

A further recent development along that line is the use of functional den-

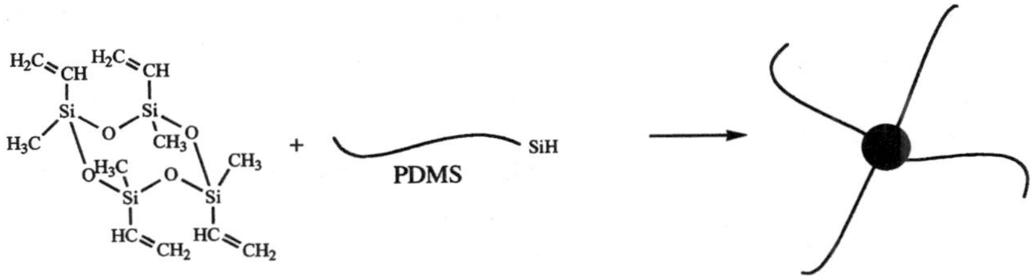

Scheme 3 Synthesis of star-shaped poly(dimethyl)siloxane via "arm-first" methods.

drimers as electrophilic deactivators, to synthesize star polymers with high numbers of arms.

As an example Yen and Merrill [45] have synthesized star-shaped polyethylene oxide (PEO) molecules using that method. Polyamidoamine dendrimers were used in that case as a central core.

B. Star-Shaped Polymers by Anionic Block Copolymerization

1. Homopolymeric Branches

The monocarbanionic precursor chains can also serve as a efficient initiator for the polymerization of a small amount of a bis-unsaturated monomer, such as divinylbenzene (DVB) or ethylene glycol dimethacrylate (EGDMA) (Scheme 4).

That procedure was mentioned first by Milkovich [46], developed extensively by Rempp and coworkers [47–49], and later extended by Fetters [50], McGrath [51], Teyssié [52], and Hadjichristidis [53]. Upon polymerization of

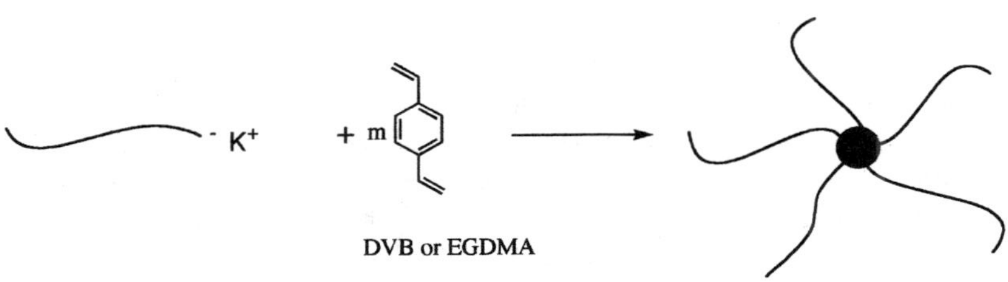

Scheme 4 Star-shaped polymers by anionic block copolymerization.

DVB (or EGDMA), small, tightly crosslinked nodules are formed, each of them being connected with the precursor chains that have contributed to its initiation. The branches shield the crosslinked cores to prevent crosslinking between individual star molecules.

In the case of DVB, no gelation has ever been observed in the reaction medium during the formation of the star molecules unless the cores grow very large and constitute more than 40 wt % of the star molecules [54]. The average length of the branches is given by the molar mass of the precursor chains. Although the average functionality of the crosslinked cores is not directly accessible, it can be determined from the ratio of the molar mass of the stars to the precursor polymer (taking into account the weight fraction of the cores). The formation of star polymers by the reaction of monocarbanionic chains with bifunctional polymerizable compounds is a kinetically and mechanistically complex reaction implying the participation of several simultaneous and competitive steps. Some aspects of that point have been discussed by Worsfold [55] and will also be discussed briefly in the fourth section of this chapter.

The average number of branches per star molecule [49] is influenced by several factors, the most important of which are the overall concentration of the reaction medium and the proportion of bis-unsaturated monomer. The proportion of bis-unsaturated monomer is expressed as the mole ratio of monomer to the active sites, [DVB]/[LE]. Depending on the average length of the precursor chains, a given [DVB]/[LE] mole ratio results in different weight-percents in the cores of the star molecules.

If the average number of arms per core is higher than 4, the polydispersity of a sample can be assigned exclusively to fluctuations in the number of branches. Fractional experiments have shown that, provided the experimental conditions have been optimized, the distribution of molar masses within such star polymer samples is reasonably narrow. Furthermore, as shown by SEC, the size of the distribution of the hydrodynamic volumes in star polymer samples is low [56].

These methods were also applied successfully to the preparation of star polymers exhibiting elastomeric branches (polybutadienyl, polyisoprenyl chains). The crossover copolymerization reactions between the poly(isoprenyllithium) chain ends and DVB and correlatively the homopolymerization of different isomers of DVB were studied in great detail [50]. It was confirmed that the extent of branching increases with reaction time, and that the maximum extent of branching is not reached until the later stage of the reaction. That result was confirmed recently by other methods on the same type of samples [57].

As mentioned by Teyssié and Hadjichristidis, star-branched polymers have been prepared from the reaction of a living poly(methyl methacrylate) chain prepared by anionic polymerization with ethylene glycol dimethacrylate (EGDMA). The active PMMA chain end initiates the polymerization of the two double bonds within EGDMA. Since no macroscopic gelation of the reaction medium was observed, the structure can be compared to a star-shaped macromolecule exhibiting a small crosslinked microgel in its center. A systematic investigation of that reaction is now under progress in different groups.

Quirk has recently applied that "arm-first" procedure to prepare functionalized star-branched PMMA using a protected hydroxy-functionalized alkyllithium initiator [58].

2. *Copolymeric Branches*

The anionic "arm-first" methods can also be applied to the synthesis of star block copolymers [59]. The procedure is identical except that living diblock copolymers (arising from sequential copolymerization of two appropriate monomers, added in the order of increasing nucleophilicity) are used as living precursor chains. The active sites subsequently initiate the polymerization of a small amount of a bis-unsaturated monomer (DVB in most cases) to generate the cores. If polystyrene and polyisoprene (or polybutadiene) are selected, the resulting star block copolymers behave as thermoplastic elastomers because of their different glass transition temperatures.

Self-organized honeycomb morphologies [60] have recently been identified with well-defined hompolymeric star-shaped PS (or on polydisperse star-shaped polymers or copolymers). That morphology was observed first on polystyrene–polyparaphenylene block copolymers and may be attributed for linear or branched polystyrenes to the specific behavior of PS in carbon disulfide [61].

III. STAR-SHAPED POLYMERS VIA "CORE-FIRST" METHODS

The "arm-first" methods are efficient at synthesizing well-defined star-shaped macromolecules. Difficulty arises, however, in the functionalization of the outer chain ends, which is only possible through the use of functional initiators to generate the precursor chains [58]. "Core-first" methods were developed extensively in polar solvents to access star-shaped macromolecules exhibiting functional groups at the outer chain ends. Once such species are obtained, they can serve as valuable intermediates in the elaboration of a large scope of macromolecular architectures. Attempts to prepare polyfunctional initiators have been described by Nagasawa and co-workers [62], who synthesized "core-first" star polymers starting from 1,3,5-tris(α-methoxystyrene), and more recently by Tung and Lo [63] in nonpolar solvents. Popov and Gehrke have worked on the same subject [64]. The most commonly used "core-first" method is actually derived from Burchard's method [65] and requires an efficient soluble polyfunctional initiator that can be used in polar solvents (Scheme 5).

A. Basic Principles of "Core-First" Methods

Several factors influence the functionality and the solubility of the plurifunctional initiator, including the solvent and counter-ion. In polar solvents, the solubility of the polyfunctional initiator is greater, the ionic associations are minimized, and the rate of initiation is greater compared to nonpolar solvents. Divinylbenzene (DVB) was chosen as the bifunctional compound for the preparation of a plurifunctional initiator in nonpolar solvents [65]. Motivation for the development of the "core-first" method for the polymerization of oxiranes in polar solvents was driven by the need to rend available functional poly(ethylene oxide) star-shaped macromolecules. The initiator systems developed by Burchard are based on lithium as a counterion, which is why it could not be applied to the preparation of multifunctional PEOs under mild conditions.

The "core-first" method is based on the preparation of soluble initiator

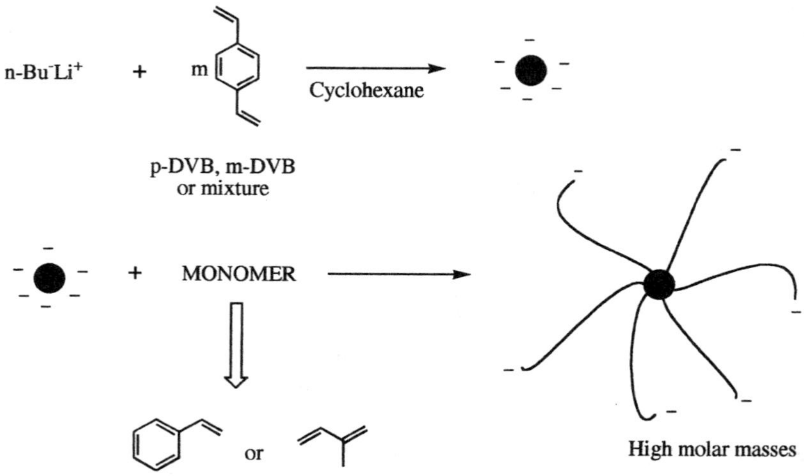

Scheme 5 "Core-first" method based on *sec*-butyl lithium and divinylbenzene.

molecules containing metal organic sites that are subsequently used to polymerize the branches [66]. These plurifunctional initiators have a strong tendency to aggregate and form gel particles. This tendency to aggregate, however, diminishes as the polymerizing branches grow in size and protect the plurifunctional cores. The extent of aggregation can have dramatic consequences for the nature and polydispersity of the resulting polymer samples. After the polymerization of the first generation of branches is complete, a second monomer can be initiated and polymerized or the ends can be functionalized.

B. Functional Star-Shaped Poly(ethylene Oxides)

Some examples of preparation via anionic polymerization of PEO star-shaped polymers have been reported in the literature, but the functionality of these star-shaped polymers is generally low [67]. The "core-first" method was first applied to the preparation of PEO star-shaped macromolecules (Scheme 6). In the first step, naphthalene potassium (systematic name: potassium dihydronaphthylide) was reacted with DVB in highly dilute solution keeping the mole ratio [DVB]/[K] below 3 to avoid the formation of a microgel. At these mole ratios, no aggregate of the multifunctional cores is observed. Oxirane is then added to the solution of initiating poly(DVB) cores. The system quickly becomes biphasic with the lower phase being highly viscous. As the polymerization progresses, the medium becomes homogeneous. Protonic deactivation of the sites yields alcohol functions at the outer end of the branches.

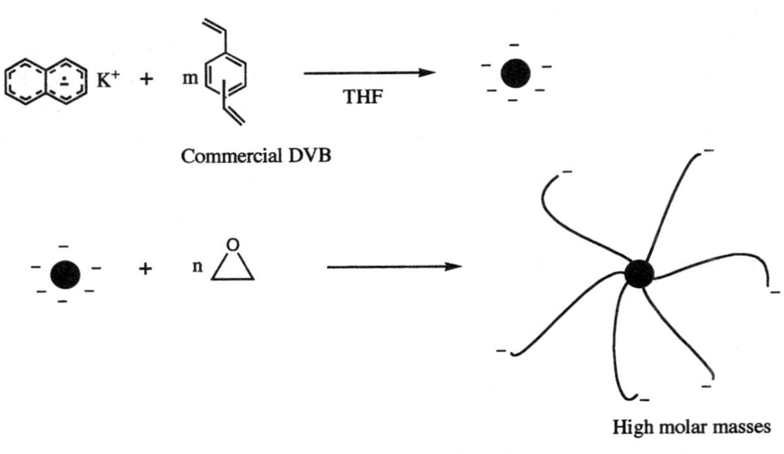

Scheme 6 Reactional scheme for "core-first" methods in polar solvents.

Samples have been synthesized under different conditions and for various ratios of [DVB] to [K] and characterized by SEC and light scattering. With the "core-first" methods, the individual branches cannot be characterized. The average molar mass, however, can be calculated from the ratio of monomer converted to potassium concentration, and the average number of branches per star molecule can be correlated to the [DVB]-to-[K] ratio of the core [67]. The functionality of these compounds is rather large. They can be used for a great number of applications. Similar results along that line were obtained by replacing naphthalene potassium by cumyl potassium, in the preparation of the polyfunctional initiator. Thus the poly(DVB) core is obtained by direct addition on DVB and not by electron transfer processes [68]. Here again, functional star-shaped PEOs can be obtained but they exhibit a large distribution in functionalities. In spite of the hydrophobic poly(DVB) core that constitutes up to 5% of the overall weight, these star-shaped PEOs are soluble in methanol or even in water. The molar masses of the star PEO macromolecules are systematically measured in methanol, a good solvent of PEO, in which the association phenomena are limited.

As mentioned earlier these star-shaped polymers obtained from polyfunctional initiators are characterized by the presence of active sites located at the outer end of the branches. Thus a large variety of structures can be prepared from these functional star-shaped polymers. Some examples are given in Scheme 7. One should be aware that the distribution in functionalities is large and that advance control of that functionality is difficult.

This is why, recently, another approach was developed to allow better control of that functionality: DVB was replaced by diisopropenylbenzene (DIB) in the preparation of the core: First a soluble poly(DIB) is obtained upon polymerization of DIB under well-defined conditions. This poly(DIB) carries one double bond per monomer unit in the polymer chain. These double bonds are reacted with a stoichiometric amount of cumyl potassium under selected conditions ($T = 40°C$) to avoid propagation. The active sites created on the poly(DIB) chain are thus used to initiate the polymerization of oxiranes [69].

C. Polystyrene Star-Shaped Polymers

1. *Synthesis*

Some preliminary experiments have shown that the poly(DVB) cores may contain unreacted pendant double bonds originating from incomplete DVB polymerization. The presence of these unreacted sites has no consequence for the oxirane polymerization. These growing sites cannot attack the pendant unsaturations. This is not the case for living polystyrenes; here the growing sites can

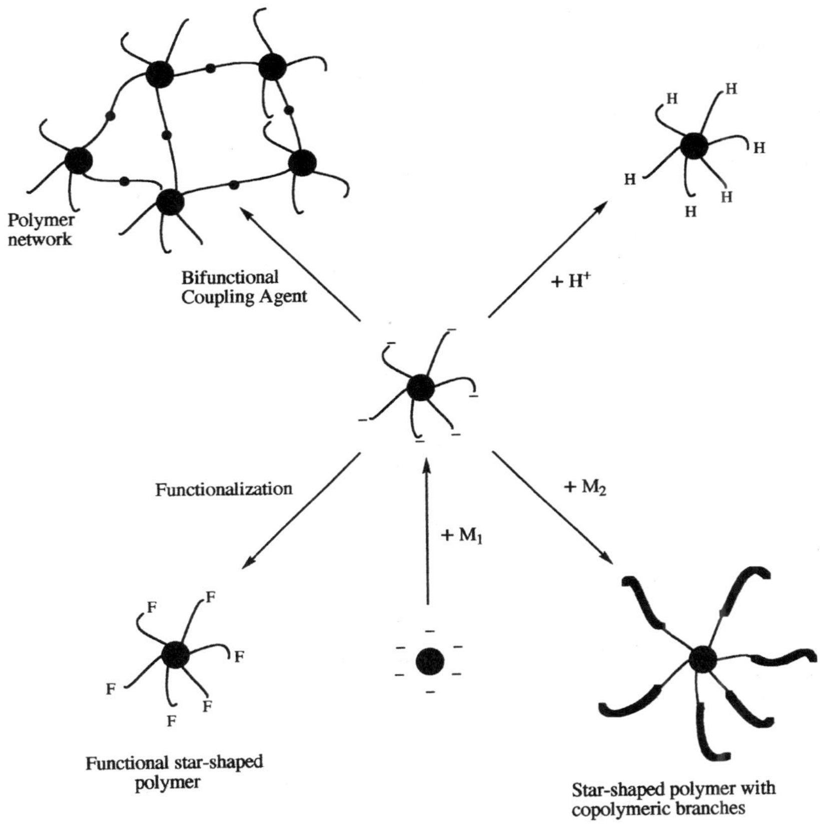

Scheme 7 Functional star-shaped polymers in macromolecular engineering.

react with the remaining double bonds leading to irreversible chemical crosslinking reactions.

The efficiency of the "core-first" method for the synthesis of functional PS star-shaped molecules was tested on a large domain of molar masses and functionalities. The resulting polymers were again characterized by SEC and light scattering [66]. The average molar mass of the branches is not directly accessible, but is determined by the ratio of the weight of the monomer introduced to the total number of sites available, assuming total conversion. Additional information can be obtained from SEC traces. Occassionally, there is a shoulder that was assigned to linear polymer adjacent to the main peak. The corresponding molar mass value measured by SEC via on-line light scattering is exactly the same as the theoretical value calculated from the ratio of monomer to initiator, and it originates from the unconsumed naphthalene potassium.

The number of branches is difficult to predetermine since it depends upon several parameters, including the ratio [DVB] to [K]. The average functionality of these star-shaped macromolecules calculated from the ratio of the value of the molar mass of the star molecule obtained by light scattering to the calculated value for the branch is rather high. The values obtained by SEC techniques based on the calibration with the linear samples are far below that obtained by light scattering. This result is not surprising based on the reduced hydrodynamic volume of a star molecule compared to that of the linear sample of the same molar mass. Size exclusion chromotagraphy apparatus fitted with three detectors in series (low-angle light scattering, photometer, continuous viscometer) is efficient for rapid and proper charaterization of star-shaped polymers [70]. It is now possible to measure simultaneously the diffusion at different angles, giving direct access to the molar mass and the radii of gyration (Wyatt technology) [71].

The core formation process may be modified by replacing naphthalene potassium with butyllithium (BuLi) or with cumyl potassium and keeping all other parameters unchanged. This core was then used to polymerize styrene. The resulting samples were characterized by SEC and light scattering. The average molar masses were higher than those for the star-shaped polymers prepared under standard conditions. In some cases visible microgel particles were present in the solution, which led to further examination of the initiator obtained by reaction of DVB with BuLi. High quantities of pendant double bonds were dectected by nuclear magnetic resonance (NMR) spectroscopy. It is not surprising that the growing polystyrene active chain ends attack these unsaturations and forms bridges between the individual stars. This is not the case when naphthalene potassium is used as an initiator since no remaining pendant double bonds are present. That implies that the core formation process is different when electron transfer processes are used instead of direct addition.

2. *Solution Properties*

It is well established that the properties of star-shaped molecules diverge from those of their linear counterparts as the number of arms increases. Thus it is important to examine the solution properties of the star-shaped polymers obtained by "core-first" methods. As expected, the molar mass distribution of these star-shaped polymers, for a given arm length, is rather large, confirming the large fluctuation in the number of carbanionic sites in the initiating cores. The formation of these cores by random coupling between radical sites may be responsible for the broad distribution of functionalities. The dilute viscosity behavior of these star-shaped polystyrene solution was examined. The intrinsic viscosity of these fractions of the PS star were nearly independent over a large range of functionality for three different series of "core-first" star-shaped polystyrenes. Dynamic light-scattering measurements were also performed and the dilute and semi dilute solution behavior were studied in detail [72]. The results were compared to those obtained for star-shaped polymers synthesized by classic "arm-first" methods and published by Zilliox [73] and Bauer et al. [74].

D. Extension of the "Core-First" Method to Other Polymers

This "core-first" method was thus extended to a number of other monomers such as styrene, methacrylates, vinyl-2-pyridine, and so on, with the counterion being lithium, sodium, or potassium depending on the monomer involved.

Star block copolymers in which each branch is an amphiphilic star block copolymers can be obtained in a similar way. The polymerization of the second monomer can be initiated by the carbanions of the first one, that is, in the order of increasing electroaffinity. Typical examples include styrene/butyl methacrylate [75] and styrene/vinylpyridine [76].

E. Synthesis of Networks

1. *Polystyrene Networks*

The main advantage of the "core-first" method is the easy introduction of functions at the outer end of the branches. Linking reactions can also be performed and should lead to network structures. Because of the high functionality of these star-shaped polymers not all of the functions are required to obtain networks exhibiting good mechanical properties. The so-called model networks are usually obtained by reacting the active chain ends of bifunctional polymers with an appropriate low-molar-mass compound [77]. That method was extended to the preparation of a polystyrene network in which the bifunctional polymer precursor was replaced with the functional star-shaped macromolecule (Scheme 7). Two approaches were developed: Either the living star-shaped polymer is reacted with a stoichiometric amount of DVB, DVB acting as a coupling agent, or the living star-shaped polymer is reacted first with ethylene oxide to provide the outer ends of the branches of the star molecules with hydroxyl groups. In the next step, these hydroxyl-terminated star-shaped molecules are reacted with appropriate diisocyanate compounds. In both cases these networks were characterized in the usual manner: amount of extractable material, the equilibrium degree of swelling, and the uniaxial elastic modulus.

The results were compared to those of networks obtained by classic end-linking procedures [78]. Gel synthesized by that crosslinking procedure were also studied by small-angle neutron scattering and compared to solutions of the same star-shaped macromolecules [79].

2. Functional Star-Shaped Polymers and Hydrogels

PEO is well known for its good water solubility but the use of PEO was limited either to the linear soluble polymers or to hydrogels. The "core-first" method enables easy preparation of functional star-shaped polyethylene oxide. This easy access to water-soluble multifunctional PEO molecules has widely extended the application of these polymers [80].

F. "Core-First" Methods in Nonpolar Solvents

The prerequisite to obtain dienes with high 1,4-cis content is the use of lithium as a counter-ion and the presence of a nonpolar solvent. Also important is the fact that in nonpolar solvents the associations between metal–organic sites are far more pronounced: If the growing polymer chains are plurifunctional, the reaction medium has a strong tendency to gel due to the physical association. These associations can be disrupted by protonic deactivation, and if there is no chemical crosslinking, the medium becomes soluble.

Burchard [65] was the pioneer for the preparation of functional star-shaped polymers in nonpolar solvents by an anionic "core-first" method. To build the cores pure *p*-DVB was reacted with *n*-butyllithium in dilute cyclohexane solution. Suspensions of small crosslinked poly(DVB) noduli were obtained that contained numerous lithium organic sites. In a second step, styrene (or isoprene) was added to the living cores and polymerized. The polymeric species obtained exhibit huge molar mass distribution and rather large polydispersity indices. Even if these star-shaped polymers could exhibit active sites at the outer end of the branches, the efficiency of initiation of a second generation of monomers or of functionalization was never given by the authors.

IV. STAR-SHAPED POLYMERS VIA "THREE-STEP" PROCEDURES

The so-called in–out procedure is a combination of the preceding two methods. It is meant to allow better control of the polydispersity of the samples and to provide the possibility of functionalization at the outer end of the branches. The underlying idea is to use the "arm-first" method to build the cores and to grow second-generation branches from these "living" cores.

A. "In–Out" Methods Based on Divinylbenzene

The basic idea of the "in-out" method is to take advantage of the reaction of the copolymerization of DVB with living chains to generate well-defined cores. The presence of these chains on these cores should improve the solubility and limit the aggregation of the particles.

This method, first disclosed by Funke [81], was extended and improved by Rempp et al. [82].

Funke [81] started from poly(*tert*-butylstyrene) of low molar mass made with *sec*-butyllithium in a cyclohexane solution. A small amount of DVB was then added to generate the living cores. Subsequently, second-generation branches of polydiene or polystyrene were grown from these living cores. Funke has studied the influence of the isomer of DVB and the diameter of the particles and has extended that reaction to other bifunctional monomers such as diisopropenylbenzenes. He has shown that the reaction is not limited to the surface of the microgel but also includes vinyl groups inside the particle. The accessibility of these double bonds to long ω-functional PS is in question.

Rempp has used a living PS precursor of low molar mass that is made either in a polar solvent, tetrahydrofuran (THF) [82], or in a nonpolar solvent [57]. Here again a living precursor initiates the polymerization of a small amount of DVB, whereupon the cores are formed, as in an "arm-first" process (Scheme 8). Each core contains as many active sites as there are branches surrounding it. In spite of their low molar mass, the protection the branches exerts on the cores is efficient and prevents the formation of aggregates. The living star polymer solution is molecularly dispersed. Subsequently, the active sites located in the cores are used to initiate the polymerization of another suitable monomer, whereupon a new set of branches (of different chemical nature or not) is generated from the cores.

In some cases, such as for the polymerization of acrylates, an intermediate addition of 1,1-diphenylethylene (DPE) was necessary to prevent side reactions during the polymerization of these monomers. A rapid change in color (from red to yellow) demonstrates the accessibility of the sites.

Several possibilities are to be considered here: If these second-generation branches are much longer than the first generation arms, the species obtained can be considered to be a regular star polymer. The "effective core," to which the long second-generation branches are attached, includes the short first-generation branches. In principle, provided no deactivation occurs, the number of

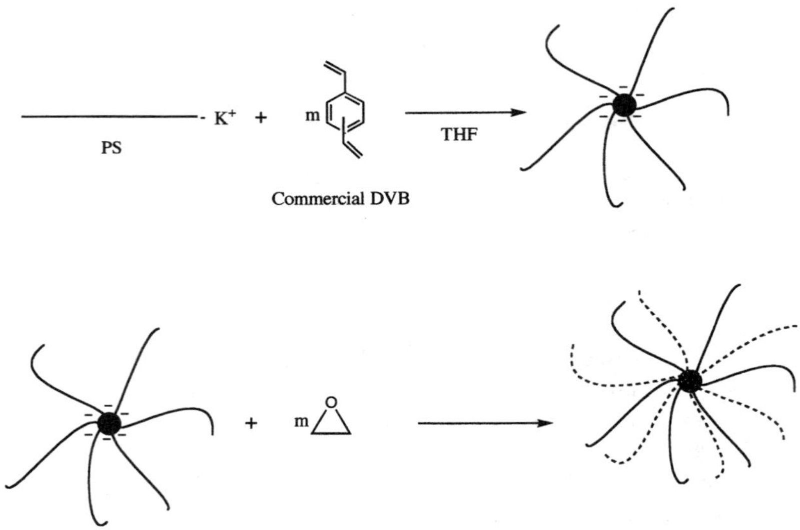

Scheme 8 "In–out" methods.

second-generation branches is equal to the functionality of the initiating cores. The lower polymolecularity of these samples reflects the narrower distribution of functionalities within the "arm-first" star precursors. This is the main advantage of the three-step process as compared with the "core-first" method. One should be aware that unreacted double bonds may be present in the noduli even in polar solvents, resulting in further crosslinking.

The second-generation branches can be functionalized, as in the "core-first" methods; they can also be used to initiate the polymerization of another appropriate monomer, to yield star block copolymers.

If the two kinds of branches are of different chemical nature, but of similar length, the resulting heterostar copolymer molecule is composed of a crosslinked core carrying equal numbers of branches of two different kinds. A variety of branches can be attached to a primary PS star molecule. Branches of poly(ethylene oxide) [68,83], poly(alkyl methacrylate) [84], poly(*tert*-butyl acrylate) (PtBA) [82], poly(2-vinylpyridine) [85], and others have been developed, with special emphasis on amphiphilic heterostar molecules. The presence of a population of linear precursors remains; however, they can be easily removed by fractional precipitation. Since polymers of different chemical nature are usually incompatible, there is a question of the conformation of such species in solution as well as in bulk. The branches may exhibit different thermomechanical properties. As an example, in "in-out" PS/PtBA stars, PS is amorphous and glassy whereas poly(tBA) is characterized by a rather low glass transition temperature.

B. "In–Out" Methods in Nonpolar Solvents

The three-step method has limitations. For example, cores resulting from "arm-first" processes based on DVB usually contain some unreacted unsaturations. If the active sites of the growing second-generation branches are able to attack these unsaturations, intermolecular linkages between individual star molecules are formed, resulting in irreversible gelation of the reaction medium.

If the cores of the initial star molecules are of pure poly(divinylbenzene), the use of styrene or dienes as the monomer for the second-generation branches may generate side reactions, but the formation of bridges between individual star molecules can be prevented to some extent. When DVB is added to a solution of living polystyrene, small cross-linked cores are formed, each of them being linked to the *f* precursor chain that contributed to its initiation. The formation of these cores is rather slow in nonpolar solvents and has been demonstrated by kinetic experiments [57]. Moreover, once the star molecules have reached their plateau molar mass, their core may still contain residual dangling unsaturations. If at that time a second generation of monomers is added to the living sites created on the DVB nodulus, gelation of the reaction medium occurs rapidly. This gel cannot be destroyed by a simple addition of methanol. The presence of these residual double bonds was confirmed and quantitated by ultraviolet spectrometric measurements. One method of eliminating the double bonds could be through deactivation by a stoichiometric amount of BuLi, resulting in an increase of the average functionality of the core. This reaction has been investigated in detail [57]. The addition of BuLi to a living star-shaped polymer efficiently reduces the number of dangling unsaturations present in the core. The study has revealed that the reaction is not very rapid and may not be quantitative. If BuLi is in excess, free active sites are present in the medium, resulting in linear chains beside the star-shaped polymer. This may be disadvantageous for the syntheses of well-defined double star polymers.

C. "In–Out" Methods Based on Diphenylethylene Derivatives

The so-called in–out method based on DVB provides good access to rather well-defined star-shaped polymers with branches of different chemical nature on the same nodules. The use of compounds such as DVB to generate the cores also has some drawbacks: Because of the uncontrolled polymerization of DVB, the functionality is difficult to control. Furthermore, the remaining double bonds may be problematic. Until recently, the chlorosilane chemistry was not typically used for the direct preparation of functional star-shaped polymers.

Quirk [86] has developed an alternative approach, based on nonpolymerizable compounds such as diphenylethylene (DPE) to prepare well-defined heteroarm star-shaped polymers (an example is given below in Scheme 9). This reaction involves first the synthesis of a linear precursor with the active chain end coupled with 1,3-bis(1-phenylvinylbenzene) or 1,3-bis(1-phenylethenyl) benzene (MDDPE) to form a living dianion. When butadiene is added, polymerization of the second generation of branches occurs, yielding the desired "heteroarm" star-shaped polymer. In the case of styrene, four-armed star polystyrenes have been synthesized. Their properties were compared to those of four-armed polystyrenes synthesized from classic deactivation of a living polystyrene chain with $SiCl_4$. In order to improve the yield of the different reactions, *sec*-butoxide was introduced.

Lambert et al. [87] have taken advantage of that coupling reaction with DPE derivatives to prepare a novel type of star polymer, exhibiting on the same nodules PS, PEO, and PMMA (or ε-caprolactone) chains (terpolymers). In addition, at least one chain end can be functionalized. They first prepared an ω-function linear PS chain and then reacted it with an appropriate DPE derivative containing a protected hydroxyl function. Thus a second generation of branches was grown, the polymerization initiated by active sites created on DPE. Poly(ethylene oxide) and PMMA chains were obtained. Finally the "protected" OH function was modified and served to initiate the anionic ring opener polymerization of ethylene oxide or ε-caprolactone.

Similarly, Hadjichristidis [88] has synthesized "miktoarm" star-shaped terpolymers exhibiting PMMA branches. As mentioned earlier, the chlorosilane chemistry does not apply efficiently to the synthesis of star polymers containing PMMA branches (living PMMA does not react with chlorosilanes). For this reason, Hadjichristidis used the DPE method to synthesize such species.

PI chains were prepared first and reacted with a large excess of trimethylchlorosilane (TMCS) under conditions to react with one chain per TCMS. After purification, PS was added to introduce one PS chain. The monofunctional chlorosilane is reacted with a stoichiometric amount of a bifunctional initiator, such as α,ω-1,1,4,4-tetraphenyldilithiobutane, to introduce one active site per polymer chain. In a subsequent step, the active sites are used to initiate the polymerization of MMA.

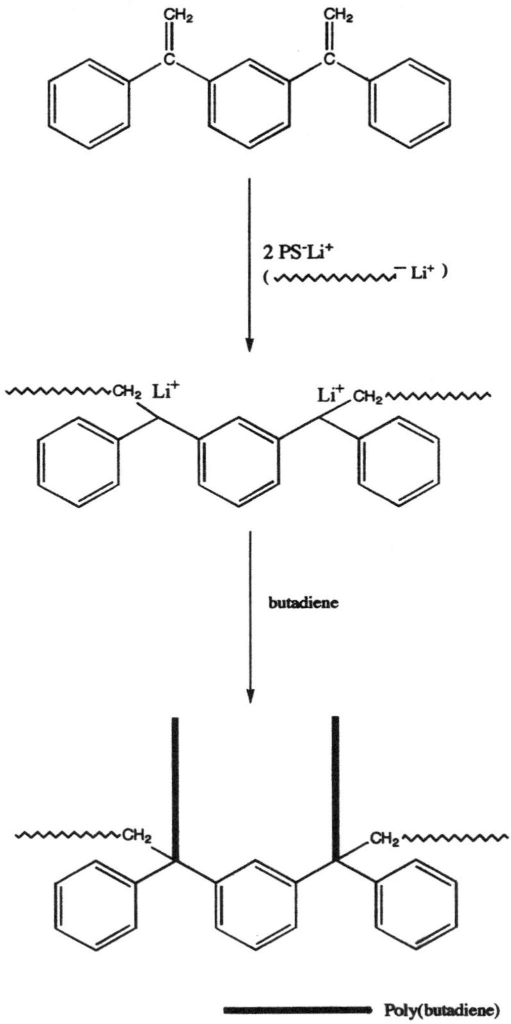

Scheme 9 Star-shaped polymers based on DPE.

The possibility of the preparation of terpolymers with PMMA branches opens new perspectives in the domain of copolymers containing hydrophobic and hydrophilic segments (or branches). The PMMA chain may be replaced by poly(*tert*-butyl acrylate) chains, which can be easily transformed into water-soluble polyacrylic acid.

Star polymers exhibiting three different branches (PS, PDMS, and poly(*tert*-butyl methacrylate) were prepared in a similar way, the first being the preparation of PDMS macromonomers containing a terminal nonpolymerizable DPE entity. Thus, a living PS is reacted with the PDMS macromonomers to create active sites on DPE. These sites served subsequently as initiators for the anionic polymerization of *tert*-butyl methacrylate [89]. Stadler and coworkers have used that reaction to generate polystyrene-arm-polybutadiene-arm-poly(methyl methacrylate) [90].

It is not possible to cite here all the other attempts to synthesize star polymers; however, the most important methods, yielding well-characterized species, have been quoted. The use of fullerenes as the central core provides an original and easy access to well-defined star-shaped polymers where the central fullerene entity is far better defined in size and functionality, as for example, in the DVB nodulus. As in DVB, "arm-first", "core-first" and "in–out" have been developed. One must remain aware that the intrinsic properties of the fullerene core may be affected by the presence of the arms [91].

As mentioned in the introduction, the homopolymerization of macromonomers can yield similar species, where each monomer unit of the backbone carries a graft [4,5,92].

V. CONCLUSION

The discovery of the living character of the anionic polymerization by M. Szwarc [93] has contributed tremendously to the domain of macromolecular engineering and especially to the synthesis of star-shaped polymers. Not only can the molar mass and the molar mass distribution of the branches be controlled in advance, but anionic polymerization has also demonstrated its efficiency in controlling the functionality of the star-shaped polymers. Earlier attempts to apply anionic polymerization to build star-shaped structures were focused essentially on "arm-first" methods. Well-defined star-shaped polymers exhibiting homopolymeric branches could thus be obtained. These samples have been studied to confirm expected structures or have served as models. Two decisive improvements have been made in the domain of application of anionic polymerization to the synthesis of well-defined star-shaped polymers:

the development of polyfunctional initiators ("core-first" method) giving access to star-shaped polymers functionalizable at the outer end of the branches and the availability of star-shaped polymers, functionalizable or not, where chains of different chemical nature are connected on the same nodulus ("in–out" methods).

The so-called core-first method has been extensively used for the synthesis of various kind of star-shaped polymers, water-soluble or not. The functionalizable outer end of the branches offers an original access to a large scope of macromolecular architecture: star-shaped polymers with copolymeric branches, functional star-shaped polymer networks, etc. The "in–out" method combines the advantage of the "arm-first" method and the "core-first" method allowing good control of the structure and the presence of functionalizable outer end of the branches. Different unsaturated compounds have been used to generate the core, such as DVB and DPE, the latter compound giving access to star-shaped terpolymers.

Further advances along that line include the development of anionic polymerization methods with the aim of attaining better control of the functionality in "core-first" methods and developing original strategies to access so-called heteroarm star-shaped polymers.

Anionic polymerization methods also have unique performance in synthezing well-defined star-shaped polymers.

Several others methods also exhibiting a living character are now being used for the construction of branched macromolecular architectures. One of them, group transfer polymerization, is briefly mentioned in the text. Special efforts are being made presently to apply living radical polymerization to the synthesis of star-shaped polymers. The performance of the different methods has been compared recently [94].

REFERENCES

1. P. Rempp and P. J. Lutz, *Makromol. Chem., Macromol. Symp.,* **62,** 213 (1992).
2. P. Rempp, C. Strazielle, and P. Lutz, *Encyclopedia of Polymer Science and Engineering,* Second Ed. Wiley, New York, 1987, Vol. 9, pp. 183–195.
3. P. Rempp and E. Franta, *Adv. Polym. Sci.,* **58,** 1 (1984).
4. S. Kobayashi and H. Uyama, *Macromolecular Design, Concept, and Practice,* M. K. Mishra, Ed., Polymers Frontiers International, Inc., New York, 1994, Vol. 1, pp. 1–38.
5. P. Rempp, P. Lutz, P. Masson, Ph. Chaumont, and E. Franta, *Makromol. Chem.,* Suppl. **13,** 47 (1985).
6. S. Bywater, *Adv. Polym. Sci.,* **30,** 89 (1979).

7. K. M. Martin, T. C. Ward, and J. E. McGrath, *Anionic Polymerization,* American Chemical Society, Washington, D.C., 1972, p. 121.
8. D. H. Rein, P. Rempp, and P. J. Lutz, *Macromol. Chem., Macromol. Symp.,* **67,** 237 (1993).
9. J. Roovers, *Trend Polym. Sci.,* **2,** 294 (1994).
10. G. S. Grest, L. J. Fetters, J. S. Huang, D. Richter, *Adv. Chem. Phys.* **XCIV,** 67 (1996).
11. P. Rempp and P. Lutz, *Polymeric Material Encyclopedia,* First Ed., CRC Press, Inc., Boca Raton, FL, 1996, vol. 10, pp. 7880–7885.
12. M. Daoud and J. P. Cotton, *J. Phys.* **43,** 531 (1982).
13. K. Huber, and W. Burchard, and L. J. Fetters, *Macromolecules,* **19,** 1404 (1986).
14. K. Huber, S. Bantle, W. Burchard, and L. J. Fetters, *Macromolecules,* **19,** 1404 (1986).
15. J. Roovers, *Polymer,* **26,** 1091 (1985).
16. J. Roovers, *Macromolecules,* **20,** 148 (1987).
17. J. F. Douglas, J. Roovers, and K. F. Freed, *Macromolecules,* **23,** 4168 (1990).
18. L. J. Fetters, A. D. Kiss, D. S. Pearson. G. F. Quack, and F. J. Vitus, *Macromolecules,* **26,** 647 (1993).
19. H. Benoit, J. F. Joanny, G. Hadziioannou, and B. Hammouda, *Macromolecules,* **26,** 5790 (1993).
20. G. Merkle, W. Burchard, P. Lutz, K. F. Freed, and J. Gao, *Macromolecules,* **26,** 2736 (1993).
21. F. Wenger and S. P. S. Yen, *Polym. Prep., Am. Chem. Soc. Div. Polym. Chem.,* **2,** 295 (1961).
22. T. G. Ngguyen and H. H. Kausch, *Makromol. Chem., Rapid Commun.* **6,** 391 (1985).
23. J. Herz, M. Hert, and C. Strazielle, *Makromol. Chem,* **160,** 213 (1972); C. Strazielfe, and J. Herz, *Eur. Polym. J.,* **13,** 223 (1977).
24. M. Morton, T. E. Helminiak, S. D. Gadkary, F. Bueche, *J. Polym. Sci.,* **57,** 471 (1962).
25. N. Hadjichristidis, A. Guyot, and L. J. Fetters, *Macromolecules,* **11,** 668 (1978); N. Hadjichristidis and L. J. Fetters, *Macromolecules,* **13,** 191 (1980).
26. J. Roovers, L. Zhou, P. M. Toporowski, M. Zwan, H. Iatrou, and N. Hadjichristidis, *Macromolecules,* **26,** 4324 (1993).
27. G. Beinert and J. Herz, *Makromol. Chem.,* **181,** 59 (1980).
28. F. Schué, J. Couve, R. Dobreva, J. Sledz, P. Nicol, *Macromol. Symp.,* **85,** 365 (1994).
29. D. B. Alward, D. J. Kinning, E. L. Thomas, and L. J. Fetters, *Macromolecules,* **19,** 215 (1986).
30. W. Burchard and G. Merkle, *J. Phys. Chem.,* **96**(10), 3915 (1992).
31. M. Pitsikalis and N. Hadjichristidis, *Macromolecules,* **28,** 3904 (1995).
32. H. Iatrou and N. Hadjichristidis, *Macromolecules,* **25,** 4649 (1992).
33. H. Iatrou and N. Hadjichristidis, *Macromolecules,* **26,** 2479 (1993).

34. J. Allgaier, R. N. Young, V. Efstratiadis, and N. Hadjichristidis, *Macromolecules,* **29,** 1794 (1996).
35. C. Vlahos, Y. Tselikas, N. Hadjichristidis, J. Roovers, A. Rey, and J. Freire, *Macromolecules,* **29,** 5599 (1996).
36. A. Avgeropoulus and N. Hadjichristidis, *J. Polym. Sci., Part A: Polym. Chem.,* **35,** 813 (1997).
37. A. Avgeropoulus, Y. Poulas, N. Hadjichristidis, and J. Roovers, *Macromolecules,* **29,** 6076 (1996).
38. N. Hadjichristidis, *Proceedings International Symposium on Ionic Polymerization, Paris,* 1997.
39. R. P. Quirk and F. Ignatz-Hoover, *Recent Advances in Anionic Polymerization,* T. Hogen-Esch and J. Smid, Eds., Elsevier, New York, 1987, p. 393.
40. J. P. Mason and T. E. Hogen-Esch, *Polym. Prepr., Am. Chem. Soc. Polym. Chem.,* **31**(1), 510 (1990).
41. S. Oulad Hammouch, G. J. Beinert, and J. E. Herz, *Polymer,* **37**(15), 3353 (1996).
42. I. M. Khan, Z. Gao, K. Khougaz, and A. Eisenberg, *Macromolecules,* **25,** 3002 (1992).
43. T. Kitano, T. Yamamoto, Y. Okemoto, S. Itsumo, and K. Ito, *Polym. J.,* **19,** 1013 (1987).
44. G. Zhou, X. Chen, and J. Smid, J. E. Gloss, Ed. American Chemical Society, Washington D.C., 1996, Vol. 2, p. 31.
45. D. R. Yen and E. Merrill, *Polym. Prepr., Am. Chem. Soc. Polym. Chem.,* **38**(1), 599 (1997).
46. R. Milkovich, Canadian Patent 716645 (1965).
47. D. Decker and P. Rempp, *C.R. Acad Sci.,* Paris, **261,** 1977, (1965); J. G. Zilliox, D. Decker, and P. Rempp, *C.R. Acad Sci.,* Paris, **262,** 726 (1966).
48. J. G. Zilliox, P. Rempp, J. Parrod, *J. Polym. Sci. C,* **22,** 145 (1968).
49. D. J. Worsfold, J. G. Zilliox, and P. Rempp, *Can. J. Chem.,* **47,** 3379 (1969).
50. R. N. Young and L. J. Fetters, *Macromolecules,* **11,** 899 (1978).
51. K. M. Martin, T. C. Ward, and J. E. McGrath, *Anionic Polymerization, ACS Symp. Ser.,* **166,** 557 (1981).
52. Huynh-Ba-Gia, R. Jerome, and Ph. Teyssié, *J. Polym. Sci., Polym. Chem.,* **18,** 3483 (1980); R. Jérome, P. Bayard, R. Fayt, C. Jacobs, S. K. Varshney, and Ph. Teyssié, *Thermoplastic Elastomers, Polyacrylate-based Elastomers,* G. Hoden, N. R. Legge, R. P. Quirk, and H. E. Schroeder, Eds., Hanser, Munich, 1996, p. 521.
53. V. Eftstratiadis, G. Tselikas, N. Hadjichristidis, J. Li, W. Yuanan, and J. Mays, *Polym. Int.,* **33,** 171 (1994).
54. F. Afshar Taromi and P. Rempp, *Makromol. Chem.,* **190,** 1791 (1989).
55. P. E. Black and D. J. Worsfold, *J. Appl. Polym. Sci.,* **14,** 167 (1970).
56. A. Kohler, J. G. Zilliox, P. Rempp, J. Polacek, I. Koessler, *Eur. Polym. J.,* **8,** 627 (1972).
57. D. H. Rein, P. Rempp, P. J. Lutz, *Macromol. Chem. Phys.,* **199,** 569 (1998).
58. R. P. Quirk, Q. Zhuo, Y. Tsai, T. Yoo, and Y. Wang, *Macromolecular Engineering:*

Recent Advances, M. K. Mishra, O. Nuyken, S. Kobayashi, Y. Yagci, and B. Sar Eds., Plenum, New York, 1995, p. 197.
59. L. K. Bi and L. J. Fetters, *Macromolecules,* **8,** 90 (1975).
60. G. Widawski, M. Rawiso, and B. François, *Nature,* **369,** 387 (1994).
61. B. François, O. Pitois, and J. François, *Adv. Mater.,* **7,** 1041 (1995).
62. T. Fujimoto, S. Tani, K. Takano, M. Ogawa, and M. Nagasawa, *Macromolecules,* **11,** 673 (1978).
63. L. H. Tung and G. Y. S. Lo, *Macromolecules,* **27,** 1680 (1994).
64. G. Popov and G. Gehrke, *Plaste Kautsch.,* **27,** 65 (1980).
65. H. Eschwey, M. L. Hallensleben, and W. Burchard, *Makromol. Chem,* **173,** 235 (1973); W. Burchard and H. Eschwey, *Polymer,* **16,** 180 (1975).
66. P. Lutz and P. Rempp, *Makromol. Chem.,* **189,** 1051 (1988).
67. Y. Gnanou, P. Lutz and P. Rempp, *Makromol. Chem.,* **189,** 2885 (1988).
68. D. Rein, J. P. Lamps, P. Rempp, and P. Lutz, D. Papanagopoulos, and C. Tsitsilianis, *Acta Polym.,* **44,** 225 (1993).
69. K. Naraghi, Y. Ederlé, D. Haristoy, and P. J. Lutz, *Polym. Prepr., Am. Chem. Soc. Polym. Chem.,* **38**(1), 599 (1997).
70. F. Afchar-Taromi, Z. Gallot, and P. Rempp, *Eur. Polym. J.,* **25,** 1183 (1989).
71. P. J. Wyatt, *Anal. Chem. Acta,* **272,** 1 (1993).
72. U. Buchholz, G. Merkle, and P. Lutz, in preparation.
73. J. G. Zilliox, *Makromol. Chem.,* **156,** 121 (1972).
74. B. J. Bauer, L. J. Fetters, W. W. Graessly, N. Hadjichristidis, and G. F. Quack, *Macromolecules,* **22,** 2337 (1989).
75. C. Tsitsilianis, P. Lutz, S. Graff, J. P. Lamps, and P. Rempp, *Macromolecules,* **24,** 5897 (1991).
76. U. Buchholz, P. Lutz, M. Kunz, and W. Burchard, *Makromol. Chem.,* **194,** 1371 (1993).
77. P. Lutz, C. Picot, G. Hild, and P. Rempp, *Brit. Polym. J.,* **9,** 151 (1977).
78. P. Lutz, private communication.
79. E. Mendes, P. Lutz, J. Bastide, and F. Boué, *Macromolecules,* **28,** 174 (1995).
80. E. W. Merrill, *J. Biomater. Sci. Polym. Edn.,* **5,** 1 (1993).
81. O. Okay and W. Funke, *Makromol. Chem. Rapid Commun.,* **11,** 583 (1990); W. Funke and O. Okay, Macromolecules, **24,** 2623 (1991).
82. C. Tsitsilianis, S. Graff, and P. Rempp, *Eur. Polym. J.,* **27,** 243 (1991).
83. C. Tsitsilianis, D. Papanagopoulos, and P. Lutz, *Polymer,* **36,** 3745 (1995).
84. C. Tsitsilianis, P. Chaumont, and P. Rempp, *Makromol. Chem.,* **191,** 2319 (1990).
85. C. Tsitsilianis and D. Voulgaris, *Macromol. Chem. Phys.,* **198,** 997 (1997).
86. R. P. Quirk, B. Lee, and L. E. Schock, *Makromol. Chem. Macromol. Symp.,* **53,** 201 (1992); R. P. Quirk and T. Yoo, *Polym. Bull.,* **31,** 29 (1993); R. P. Quirk and Y. J. Kim, *Polym. Prep., Am. Chem. Soc. Polym. Chem.,* **37**(2), 643 (1996).
87. O. Lambert, Ph. Dumas, G. Hurtrez, and G. Riess, *Macromol. Rapid Commun.* **18,** 343 (1997).
88. S. Sioula, Y. Tselikas, and N. Hadjichristidis, *Macromolecules,* **30,** 1518 (1997).

89. T. Fujimoto, H. Zhang, T. Kazana, Y. Isono, H. Hasegawa, and T. Hashimoto, *Polymer,* **33,** 2208 (1992).
90. H. Hückstadt, V. Abetz, and R. Stadler, *Macromol. Rapid Commun.* **17,** 599 (1996).
91. Y. Ederlé, C. Mathis, *Macromolecules,* **30,** 2546 t4262 (1997).
92. Y. Ederlé, F. Isel, S. Grutke, and P. J. Lutz, *Macromol, Symp.,* **132,** 197 (1998).
93. M. Szwarc, *Carbanions, Living Polymers and Electron Transfer Process,* Interscience, New York, 1968: M. Szwarc, *Adv. Polym. Sci.,* **49,** 1(1983).
94. M. C. Crossman, K. H. Hunt, D. M. Haddleton, *Proceedings International Symposium on Ionic Polymerization Paris* (1997).

3

Branched Polymers via Group Transfer Polymerization

S. Sivaram
National Chemical Laboratory, Pune, India

P. J. Lutz
Institute Charles Sadron, CNRS, Strasbourg, France

Munmaya K. Mishra
Ethyl Corporation, Richmond, Virginia

INTRODUCTION

Controlled synthesis of polymers of acrylic and methacrylic esters with predictable molecular weights, narrow molecular weight distributions, and well-defined molecular architecture is of fundamental and practical significance. For the controlled synthesis of acrylic polymers many excellent initiators have been reported, which include biphenyl sodium [1], *t*-BuMgBr coupled with $MgBr_2$ or R_3AI [2,3], metalloporphyrin of aluminum [4], (1,1-diphenylhexyl)lithium/LiCl [5], and organolanthanide [6] complexes. Among these, initiators based on ketene silyl acetals have proved to be among the most attractive methods for the synthesis of poly(alkyl methacrylate)s with controlled molecular structures.

Group transfer polymerization (GTP) is a technique for the polymerization of acrylic monomers discovered by the scientists at Dupont in 1983 [7–10]. The technique gives "living" polymers, remarkably free of termination or transfer reactions, at room temperatures or above. This is in distinct contrast to anionic polymerization of methacrylic monomers, which can be performed in a truly "living" manner only at low temperatures (much below 0°C). GTP works best for methacrylates and is considerably more complex as a process for acrylic monomers. In view of the "living" nature of GTP chain ends, a variety of well-defined methacrylate random, block, graft, and star-branched as well as functionally terminated polymers can be prepared.

GTP is an example of Michael addition polymerization involving the addition of a silyl ketene acetal to α,β-unsaturated carbonyl compounds. A typical polymerization scheme is illustrated in Scheme 1, using methyl methacrylate as the monomer and (1-methoxy-2-methyl-1-propenoxy) trimethyl silane (MTS) as the initiator in the presence of an anionic catalyst.

The original patent issued to Dupont describes several monomers for GTP [11]. However, methyl methacrylate (MMA) is the preferred monomer for most studies [12–16]. Typical examples of MMA polymerization using various anionic catalysts are shown in Table 1.

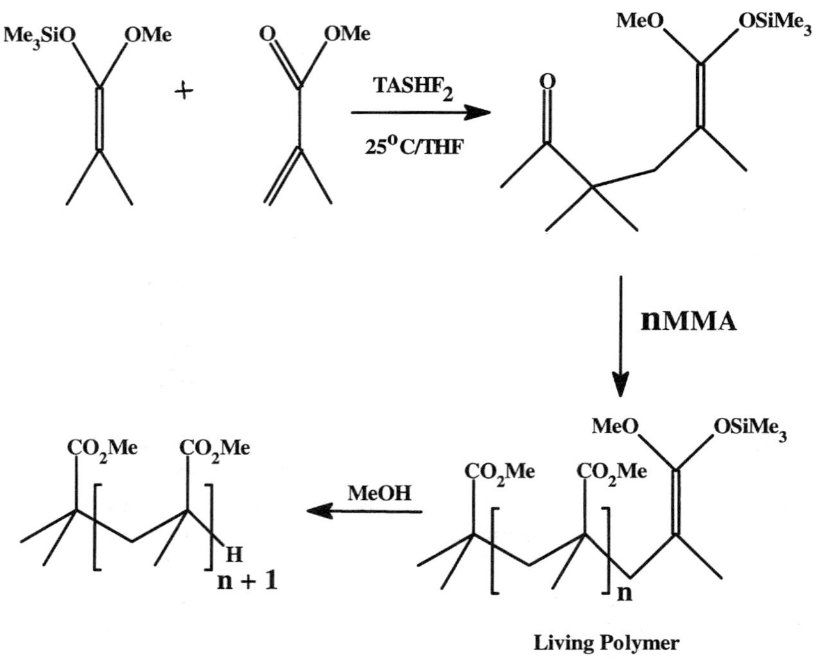

Scheme 1 GTP of MMA using MTS as initiator and $TASHF_2$ as catalyst.

Table 1 Effect of Catalyst and Solvent on the Group Transfer Polymerization of Methyl Methacrylate Initiated by MTS

Sr. No.	*Catalyst*	*Solvent*	$M_n \times 10^{-3}$ *(theory)*	$M_n \times 10^{-3}$ *(GPC)*	M_w/M_n
1	TBAF, H_2O	THF	60.0	62.3	1.15
2	$TASHF_2$	THF	10.1	10.2	1.17
3	$ZnBr_2$	$ClCH_2CH_2Cl$	3.4	6.02	1.20
4	TBABOAc	THF	4.78	4.86	1.32
5	TBAB	THF	4.78	4.50	1.38
6	$(Ph_4P)_2HF_2$	THF	28.4	43.0	1.32

MTS, 1-methoxy-2-methyl-1-propenoxyl trimethyl silane; TBAF, tetra-*n*-butylammonium [fluoride]; $TASHF_2$, tris(dimethylamino)sulfonium [bifluoride]; TBABOAc, tetra-*n*-butylammonium biacetate; TBAB, tetra-*n*-butylammonium benzoate; $(Ph_4P)_2HF_2$, tetraphenylphosphonium bifluoride.

POLYMER SYNTHESIS THROUGH GTP

In view of the "living" nature of GTP, the method is amenable for the synthesis of well-defined random, block, graft, and star-branched polymers as well as macromonomers, end-functionalized polymers, and telechelics. Some examples of typical random and block copolymers prepared by GTP are shown in Tables 2 and 3. Macromonomers have been prepared by terminating "living" GTP chain ends with electrophilic reagents bearing a polymerizable group, namely, methacryloyl fluoride [17] or *p*-vinylbenzyl tosylate [18]. Alternatively, a silyl-protected hydroxy-containing initiator was used to synthesize a

Table 2 Random Copolymers Prepared by GTP (Initiator: MTS Catalyst: $TASHF_2$)

	Monomers					
Sr. No.	*A*	*B*	*C*	$M_n \times 10^{-3}$ *(theory)*	$M_n \times 10^{-3}$ *(GPC)*	M_w/M_n
1	MMA (35)	*n*-BMA(65)		20.21	22.21	1.11
2	MMA(58)	*n*-BMA(17)	GMA (25)	4.09	4.29	1.10
3	MMA	AMMA		4.81	16.99	1.10
4	1(75)	DMA (25)		80.6	127.0	1.06
5	1(78)	EMA(22)	2	100.0	142.0	1.27

n-BMA, *n*-butyl methacrylate; GMA, glycidyl methacrylate; AMMA, allyl methacrylate; DMA, decyl methacrylate; EMA, ethyl methacrylate;

Table 3 Block Copolymers Prepared by GTP

	Monomers						
Sr. No.	*A*	*B*	*Initiator*	*Type*	$M_n \times 10^{-3}$ *(theory)*	$M_n \times 10^{-3}$	M_w/M_n
1	MMA(25)	2-EHMA (75)	MTS	A-B	—	41.50	1.30
2	MMA (90)	LMA (90)	MTS	A-B	7.14	6.65	1.06
3	LMA (10)	MMA (90)	MTS	A-B	7.12	6.54	1.14
4	*n*-BMA	MMA	SKA-OMC	A-B-A	—	7.2	1.21
5	*t*-BMA	MMA	SKA-OMC	A-B-A	4.5	4.3	1.34
6	MMA (10)	DMA (90)	MTS	A-B	135.0	185.0	1.18
7	*n*-BMA	TMS-HEMA	SKA-OMC	A-B-A	43.6	51.2	1.39
8	MMA	THPMA	MTS	A-B	8.6	7.73	1.08
9	MMA	DMAEM	MTS	A-B	9.44	8.69	1.07
10	MMA	2-EHA	MTS	A-B	12.2	18.2	2.09
11	MMA	tBA	MTS	A-B	10.9	15.9	1.91
12	MMA	BzMA	MTS	A-B	—	42.8	1.05
13	MMA	t-BMA	MTS	A-B	—	20.6	1.34
14	MMA	EA	MTS	A-B	9.5	11.1	2.33

2-EHMA, 2-ethylhexyl methacrylate; LMA, lauryl methacrylate; *n*-BMA, *n*-butyl methacrylate; *t*-BMA, *t*-butyl methacrylate; DMA, decyl methacrylate; TMS-HEMA, trimethylsilyloxyethyl methacrylate; THPMA, tetrahydropyranyl methacrylate; 2-EHA, 2-ethylhexyl acrylate; DMAEM, dimethylaminoethyl methacrylate; *t*-BA, *t*-butyl acrylate; BZMA, benzyl methacrylate; EA, ethyl acrylate; SKA-OMC, trimethyl silyl ketene acetal of octane—2-7-methylcarboxylate.

hydroxyl-containing poly(methyl methacrylate) [19,20]. When a GTP is initiated using a protected —OH or —CO_2H containing initiator and the "living" chain end is coupled using a bifunctional electrophile, telechelics can be synthesized. The best coupling agent for a GTP chain end appears to be terephthaloyl fluoride [7].

STAR-BRANCHED METHACRYLATE POLYMERS

The term "star" polymers denotes a polymer with branches emanating from a common source or core. The first star polymers were synthesized by coupling living polymer chain ends with polyfunctional molecules (e.g., polystyryl lithium with $SiCl_4$) [21,22]. Multiarm star polymers were synthesized by reacting a "living" chain end [e.g., poly(butadienyl) lithium] with divinylbenzene.

In general, star polymers can be synthesized by two different approaches, known as "arm-first" and "core-first" methods [23]. The "arm-first" method consists of either terminating the "living" chain end by plurifunctional electrophiles or reinitiation using bis unsaturated monomers. In the "core-first" method, a plurifunctional initiator core is first synthesized by reaction of an initiator with a bis-unsaturated monomer (e.g., reaction of *n*-BuLi with divinylbenzne). The plurifunctional initiator core is used to initiate further polymerization. The synthetic advantages and drawbacks of the two methods are summarized in Table 4.

Table 4 Synthesis of Branched Polymers

Technique	*Approach*	*Remarks*
Arm-first	Termination by plurifunctional electrophiles	• Functionality determined by electrophilic reagent • Knowledge of precise concentration of chain ends required
	Reinitiation using bisunsaturated monomers	• Functionality cannot be chosen at will • Suitable for $F_n = 5–15$ • Arm-first techniques cannot be used for branch functionalization at outer end
Core-first	Multifunctional initiation from a core of bisunsaturated polymer	• Functionality cannot be chosen at will • Suitable for $F_n = 20–1000$ • Branch ends can be functionalized

Both approaches have been used for the synthesis of star-branched (meth)acrylate polymers using classical "living" anionic polymerization. 1,3,5-bromomethyl benzene has been used for the termination of living anionic chain ends of methacrylate polymers [24]. The low reactivity of enolate chain ends, derived from MMA, especially, at low temperatures makes the nucleophilic substitution reactions somewhat less efficient, leading to less than quantitative coupling. On the contrary, greater success has been obtained with the sterically less hindered enolate derived from *t*-butylacrylate as monomer. Living cores from the reaction of lithium naphthalene with divinylbenzenes were prepared. The lithium chain ends were end-capped with diphenyl ethylene and were used to polymerize *t*-butyl acrylate in the presence of LiCl [25]. A star-branched poly(*t*-butyl acrylate) could be obtained with an average functionality (number of arms per molecule) 22–130. The polymer had broad/multimodal molecular weight distribution. Similarly, a 3-arm star poly(*t*-butylacrylate) was prepared by terminating the living chain end of poly (*t*-butylacrylate) with 1,3,5-triformyl benzene at –80°C in tetrahydrofuran (THF) [26].

GTP, on the other hand, offers good potential for the synthesis of star-and comb-branched polymers. The GTP chain ends are "living" even at temperatures as high as 70°C and, thus, are capable of participating efficiently in branching reactions. Both the "arm-first" and "core-first" approaches have been reported for the synthesis of multiarm star-branched acrylic polymers. However, very few details of experimental conditions and polymer characterization have been reported for star-branched polymers synthesized using the "arm-first" approach.

Treatment of a "living" GTP chain end with 1,3,5-tris(bromomethyl) benzene has been reported to yield a three-arm star poly(methyl methacrylate) (PMMA) [27] ((Scheme 2). However, no experimental methods were reported. Recently, a critical reexamination of this reaction revealed that the termination reaction is less than quantitative. Using MTS as initiator and tetra-*n*-butylammonium bibenzoate (TBABB) as catalyst, the PMMA obtained was found to be a mixture of linear and three-arm star-branched copolymers [28]. This is attributed to the fact that the reaction of the —CH_2Br group with the GTP chain end occurs in a stepwise manner and with every substitution, the reactivity of the remaining CH_2Br group is reduced, leading to less than quantitative conversion to three-arm star polymers. A three-arm heteroarm star-branched polymer via the combined anionic–GTP has been reported (Scheme 3) [29]. However, no details on experimental methods or polymer characterization are described.

Another variant of the "arm-first" method involves preparing the living polymer chain using GTP (polymer A), followed by reacting the living chain

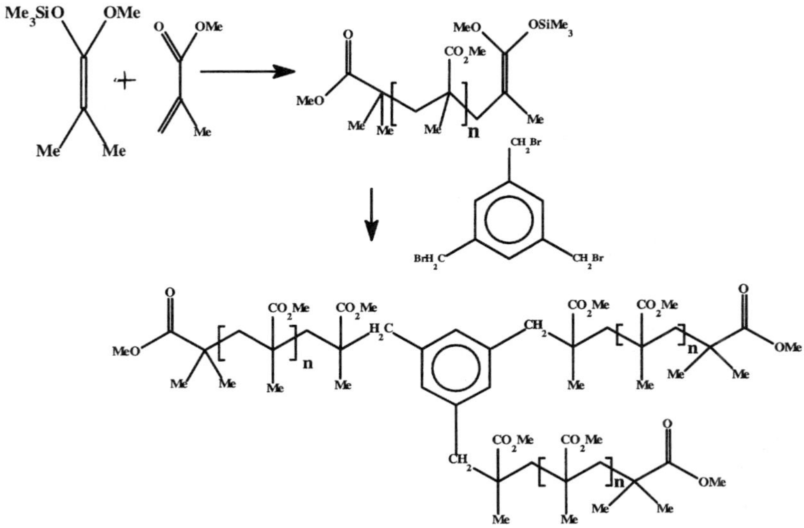

Scheme 2 Three-arm PMMA star.

end with a multifunctional linking agent, having at least two polymerizable groups (monomer B). This produces a star polymer having arms of polymerized monomer A attached to a crosslinked core of polymerized monomer B. The active group transfer sites in the core can be deactivated by reaction with a proton source. The multifunctional linking agent is typically ethylene glycol dimethacrylate. However, other multifunctional unsaturated monomers such as tetraethyleneglycol dimethacrylate, trimethylol propane trimethacrylate, and 1,4-butylene dimethacrylate can be used.

Similarly, one can prepare a "living" core by the reaction of GTP initiator with a multifunctional linking agent (monomer B) having at least two polymerizable groups. The resulting living core is then contacted with a monomer (A) to produce a star polymer having arms of polymerized monomer "A" attached to a crosslinked core of polymerized monomer "B."

These two methods for the synthesis of star-branched polymers are illustrated schematically in Scheme 4. In the formalism shown, each method produces a star of three arms, wherein each arm is made up of five monomer molecules. Thus, the number of arms can be expressed by the relationship:

$$\frac{1}{\dfrac{[IS]}{[M\text{–}M]} - 1} + 1$$

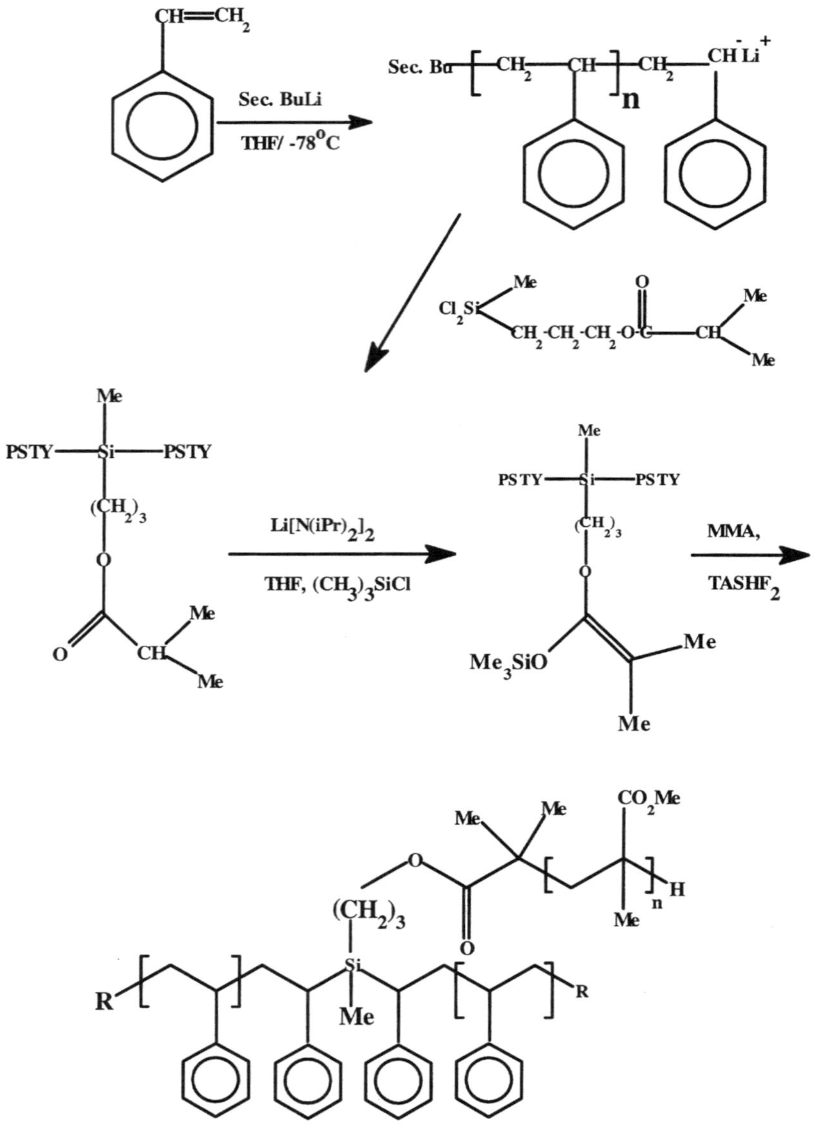

Scheme 3 Three-arm hetero star of poly(methyl methacrylate) and poly(styrene).

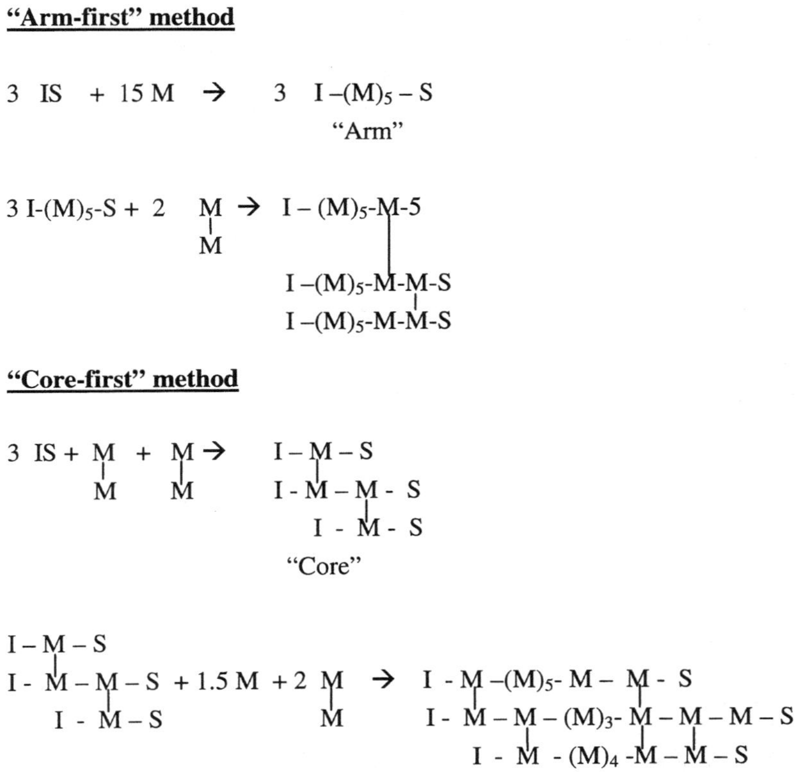

Scheme 4 Star formation using “arm-first” and “core-first” approach.

where [IS] is the moles of initiator and [M–M] = moles of difunctional monomer.

The size of the arms can be varied by changing the ratio of [M]/[IS]. Long arms are obtained when the [M]/[IS] ratio is large. The number of arms can be varied by changing the ratio of [IS]/[M–M]. If the ratio of [IS]/[M–M] is slightly higher than 1 (say 1.05), then the resulting star will have 21 arms:

$$\left[\frac{1}{(1.05/1.00)-1}+1\right]$$

If the ratio of [IS]/[M–M] is equal to or less than 1.00, the preceding equation cannot be used to calculate the number of arms. In such a case, a crosslinked core having a very large number of arms will result. These are called giant stars.

Table 5 Synthesis of Star-Shaped Poly (methyl methacrylate)s by "Arm-First" Method: Some Examples

Initiator	*1-Trimethylsiloxy-1-isobutoxy-2-methylpropene*			*(1-Methoxy-2-methyl 1-propenoxy) trimethyl silane (MTS)*	
Conditions	*1*	2	*3*	4	5
[I], mol	0.07	0.006	0.006	0.00123	0.00028
[MMA], mol	8.5	0.64	1.19	0.056	0.028
[EGDMA], mol	0.27	0.022	0.015	0.00319	0.00168
[I]/[EGDMA]	0.26	0.27	0.40	0.40	0.25
Catalyst	TBA HF_2	TBA HF_2	TBA HF_2	TBABB	TBAmCB
Solvent, g	1200 (glyme)	91 (glyme)	184 (glyme)	THF,25	THF,25
M_w (arm)	18,100	16,600	27,900	6,720	8,827
M_w/M_n (arm)	1.52	—	—	1.20	1.35
M_w (star)	$2.7.00 \times 10^5$	6.3×10^5	4.3×10^5	7.6×10^5	1.998×10^5
N_{arms}	149	38	15	113	23
M_w/M_n (star)	—	—	—	1.80	1.42
Reference	30	30	30	31	32

$TBAHF_2$, tetrabutyl ammonium hydrogen difluoride; TBABB, tetrabutyl ammonium bibenzoate; TBAmCB, tetrabutyl ammonium-*m*-chlorobenzoate.

Some typical examples of synthesis of giant stars are shown in Table 5 [30]. In Examples 4 of Table 5 [31], by varying the reaction conditions, star-branched PMMAs with varying arm lengths of average degree of polymerization between 15 and 150 could be prepared. The number of arms is also dependent on the molecular weight of precursor polymer at constant core-to-arm ratio. In our study, the lag time between completion of arm formation and addition of EGDMA was kept at 2 min to minimize free-arm formation due to premature termination. The polydispersity of the star is broader than the arm. The high polydispersities of the star-branched polymers are not caused by differences in the arm length of the stars but are a result of variation in the number of arms per microgel, that is, the polydispersity of the microgel.

In all the reactions, the addition sequence was as follows. Initially, the desired quantities of solvent, initiator, and catalyst were mixed at room temperature. The desired quantity of MMA was added over a period of 40 min followed by addition of ethylene glycol dimethacrylate over a period of 15 min. The reaction was quenched 30 minutes later. The reaction was accompanied by a significant exotherm.

The synthesis of star-branched polymers was performed with different catalyst concentrations (1.0 and 0.5 mol % based on initiator concentration). When 0.5 mol % of catalyst was used, the star polymer showed a trimodal distribution of molecular weights. However, at 1.0 mol % catalyst a star polymer free of linear arm polymer contamination was obtained.

The conditions used for the synthesis of star polymerization of methyl methacrylate has been extended to lauryl methacrylate (LMA). However, the linear living poly(LMA) chain end did not undergo star polymerization with EGDMA. However, when a second dose of catalyst was introduced after the complete homopolymerization of LMA, prior to addition of EGDMA, star formation was found to occur. Poly(LMA) stars with a polydispersity of 1.9 and possessing up to ten arms could be prepared [31].

Haddleton and Crossman have recently reported a more detailed study of methacrylic multiarm star copolymers by GTP [32]. Example 5 in Table 5 is taken from this work. It was shown that the molecular weight of randomly branched PMMA armed star polymers was controlled by (1) the concentration of living arm (P*) in solution prior to addition of EGDMA, (2) the ratio of P* to EGDMA, and (3) dilution of the star polymer core. The molecular weight of star increases linearly with the M_w arm. Increasing monomer-to-solvent ratio (increasing P*) leads to increase in the number of arms per star.

During the synthesis of star polymers, addition of certain silyl esters has been reported to improve conversion and reduce the wt % of unattached arms [33]. These silyl esters, whose pK_a value is equal to or lower than the pK_a value of the catalyst used in the polymerization are believed to enhance the livingness of the chain end. The beneficial use of such "livingness-enhancing agent" in the synthesis of well-defined block copolymers by GTP has been recently reported [34]. The beneficial action of the "livingness enhancer" is proposed to be the result of the formation of a complex between the silyl ester and the carboxylate anion that ensures maintenance of a low concentration of the "active" species in equilibrium with the "dormant" species (Scheme 5).

The addition of livingness enhancer is reported to improve the efficiency of the star-forming reaction. Addition of trimethyl silyl-3-chlorobenzoate (TMSCB) catalyst results in a 99.7% and 99.0% conversion of EGDMA and MMA. Similarly, when TMSCB was added to a polymerization using tetra-*n*-butylammonium acetate as catalyst, the conversions of MMA and EGDMA were 98.5% and 98% respectively. Furthermore, the wt % of unattached arms was only 21. In the absence of TMSCB, large quantities unattached arms (obtained via premature termination) of living ends were obtained.

A detailed study of the methacrylate star synthesis by GTP has been presented by Simms [35]. The purpose of the study was to define experimental conditions to minimize termination process of the living chain ends and to ensure that the maximum level of living arms could be carried into the core-

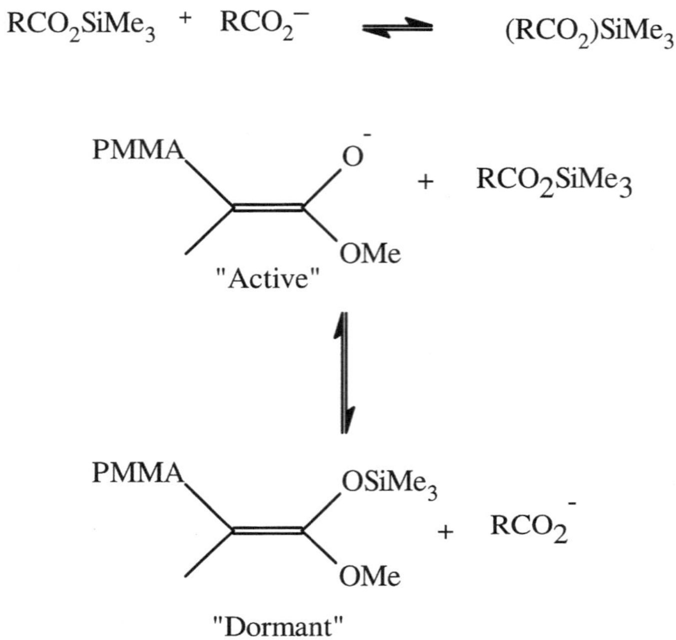

Scheme 5 Mechanism of silyl carboxylates livingness enhancer in GTP.

forming step. It was found that use of 0.2 mol % catalyst (based on initiator), minimizing the time between arm formation and core formation, adding core-forming monomer in about 5 min, and keeping monomer concentrations high, led to star polymer formation with a minimum of free arms.

It is well known that branched macromolecules have dimensions that differ from those of linear chains of the same chemical composition and molecular weight. For regularly branched materials, the effect of branching becomes noticeable through (1) a decrease in the mean square radius of gyration, (2) a decrease in the intrinsic viscosity, (3) an increase in the translational diffusion coefficient, and (4) a decrease in the second virial coefficient. However, in randomly branched materials the effects are more complicated. The complexity arises from the extraordinary width in the molecular weight distribution masking the branching effects [36]. The star-branched polymers prepared by the method of Spinelli [30] are expected to show a behavior intermediate between the regularly and randomly branched polymers. In this case, a soluble microgel is obtained with many dangling PMMA chains. The structure of such PMMA/EGDMA star-branched polymers was examined by Lang et al. [37], using static and dynamic light scattering, viscometry, and size exclusion chromatograph with a low-angle-laser light-scattering detector. Surprisingly, it was found that the ratio of hydrodynamic volumes of star-branched polymers to linear polymers at the same molecular weight is dependent on the number of arms but independent of the arm molecular weight. The capability of GTP to produce block copolymers was exploited by Spinelli and Hutchins to first generate a block copolymer of MMA with hydroxyethylmethyl methacrylate, which was crosslinked with a diisoscyanate to produce a hybrid star with a urethane core [38]. A block copolymer of MMA with 3(trimethoxy)silyl propyl methacrylate was synthesized by GTP with a M_n = 12,500. The block copolymer was hydrolyzed with water methanol in the presence of tetra-n-butylammonium fluoride. This resulted in a giant star with a polysiloxane core having M_w = 5,166,000 and M_n = 205,000 and an average number of arms of 300 [39].

Two examples of the core-first approach for the synthesis of star polymers by GTP have been reported. Trimethylolpropane triacrylate is converted to a silyl enol ether that is used to initiate the polymerization of ethyl acrylate (Scheme 6). A polymer with a M_n = 2190 and M_w/M_n = 1.39 was obtained [9]. A cyclic tetramer of methyl hydrogen siloxane was converted to a core containing four initiating groups using a Pt-catalyzed hydrosilylation reaction. The tetrafunctional initiator was used to initiate the polymerization of MMA to form a four-arm star PMMA (Scheme 7), with about 20 to 150 MMA repeat

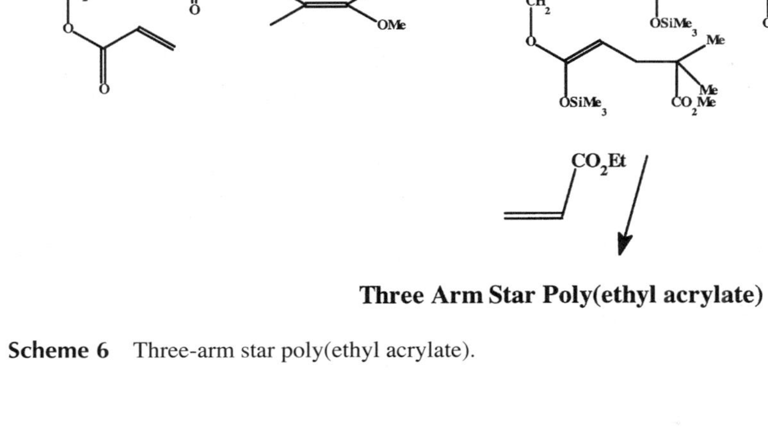

Scheme 6 Three-arm star poly(ethyl acrylate).

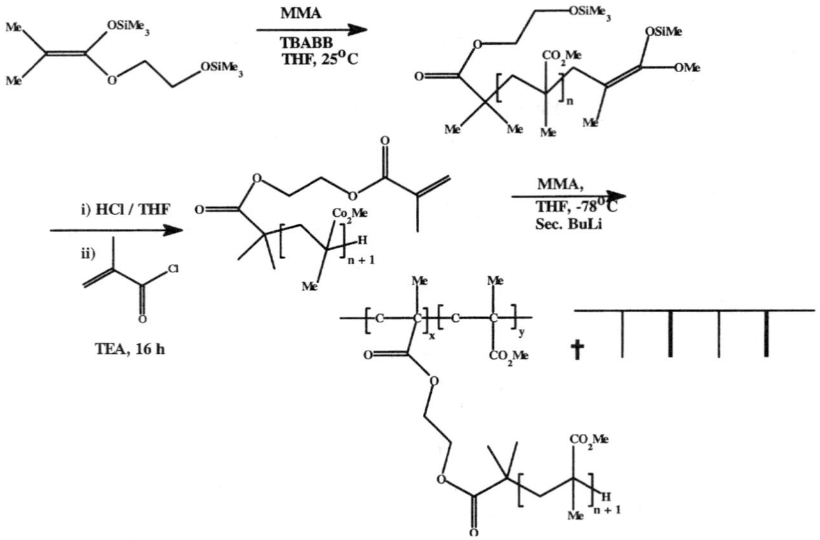

Scheme 7 Comb-shaped poly(methyl methacrylate).

units [40]. Star formation was established by chemically cleaving the arm from the core and determining the M_n of the cleared arm. It was found that the ratio of M_n of star (calculated) to arm (found) was about 4. However, since the gel permeation chromatography (GPC) data on the star polymer was based on an RI detector using linear PMMA as the standard, the determination of exact number of arms based on experimentally found M_n of star was not reliable.

COMB- AND LADDER-SHAPED POLYMERS

Comb-shaped polymers are derived from polymerizing or copolymerizing macromonomers. Macromonomers can be synthesized by a variety of synthetic techniques. Asami and co-workers prepared a methacrylate-terminated polystyrene by anionic polymerization. The macromonomer was then polymerized using GTP [41] to yield an oligomer with a polystyrene backbone and PMMA grafts. McGrath and co-workers prepared a poly(dimethyl siloxane) macromonomer end-capped with a methacrylate group. This macromonomer was polymerized by GTP to yield a comb-shaped polymer with PDMS branches [19].

Witkowski and Bandermann used GTP initiator bearing a styryl group to

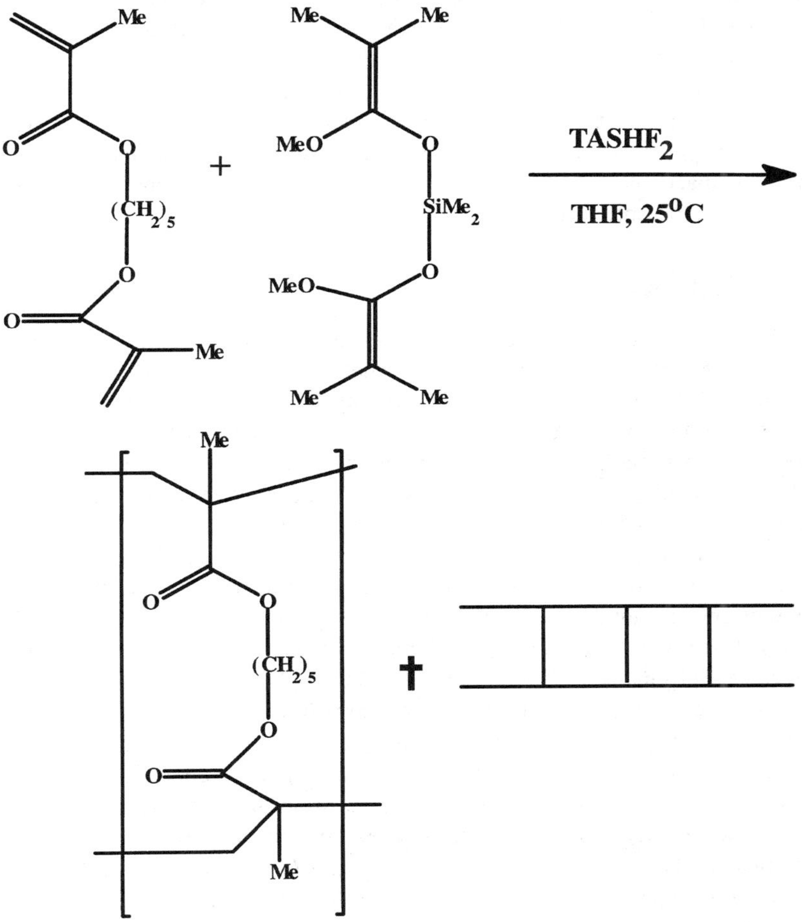

Scheme 8 Ladder-shaped poly(methyl methacrylate).

make PMMA macromonomer with a styryl head group. This was later copolymerized with styrene using free radical initiators [42].

McGrath and co-workers synthesized a novel all-PMMA comb polymer by a combination of GTP and anionic polymerization (Scheme 7) [19]. The resulting comb had a M_n = 2,48,600, with an average of 14 grafted PMMA chains per molecule. The grafted PMMA chain had a M_n = 6300 and M_w/M_n =

1.11. Additional examples of comb-grafted polymers via GTP have been reported by Heitz and Webster [43], Hertler et al. [44], and Jenkins et al. [45]. A ladder-type polymer can be prepared from GTP using a bis-methacrylate monomer and a difunctional initiator using dilute solutions [46]. If a monofunctional initiator is used, crosslinked polymer is obtained. The ladder polymer had a $M_n = 5100$ and a $M_w/M_n = 1.89$ and was free of gel (Scheme 8). The number of spacer atoms between methacrylate functions is critical for successful synthesis, the optimum being 4 to 7.

CONCLUSIONS

GTP, by virtue of its many desirable features, offers a versatile method for the synthesis of star-branched and comb-branched polymers of (meth)acrylic monomers. (The mechanism and kinetics have been examined by different authors [13] and the living character of the polymerization has been established.) The fact that GTP can tolerate many diverse type of functional groups implies that synthesis of functional stars will be possible. The largest success to date has been with the synthesis of multi-arm stars using an “arm-first” approach and using a difunctional acrylic monomer as the core-forming monomer. Very limited reports are available on the synthesis of well-defined stars with precise numbers of arms. This is due to two factors. First, access to plurifunctional GTP initiators is synthetically difficult. Second, enolates derived from GTP chain ends are unreactive in nucleophilic substitution reactions for the synthesis of star polymers using an “arm-first” approach and terminating the living end using multifunctional electrophilic reagents. Precisely those conditions that make GTP chain ends living, namely, a solvent of low polarity, a weak nucleophilic catalyst, and a high concentration of “dormant” species, renders such nucleophilic displacement difficult and less than quantitative. Additional studies are warranted in this area since GTP appears to be the synthetic method of promise for producing well-defined star-branched polymers of methyl methacrylate.

ACKNOWLEDGMENTS

Thanks are due to Dr. B. Sannigrahi, Dr. D. Baskaran, Dr. P. P. Wadgaonkar, and Dr. J. C. Sehra for their contribution to the ongoing program on the “controlled synthesis of polymers derived from (meth)acrylates” at National Chemical Laboratory, Pune. Financial assistance from the Indo-French Center for Promotion of Advanced Research (1208-3), New Delhi, is gratefully acknowledged.

REFERENCES

1. L. J. Fetters, *J. Polym. Sci.,* **C 26,** 1 (1969).
2. K. Hatada, K. Ute, K. Okamoto, and T. Kityama, *Polym. J.,* **20,** 505 (1986).
3. T. Kitayama, T. Shinozaki, E. Masuda, M. Yamamoto, and K. Hatada, *Polym. Bull.,* **23,** 2618 (1988).
4. M. Kuroki, T. Aida, and S. Inoue, *J. Am. Chem. Soc.,* **113,** 5903 (1991).
5. S. K. Varshney, J. P. Houteker, R. Fayt, R. Jerome, and Ph. Teyssie, *Macromolecules,* **23,** 2618 (1990).
6. H. Yasuda, H. Yamamoto, K. Yokota, S. Miyake, and A. Nakamura, *J. Am. Chem. Soc.,* **114,** 4908 (1992).
7. O. W. Webster, W. R. Hertler, D. Y. Sogah, W. B. Farnham, and T. V. Rajanbabu, *J. Am. Chem. Soc.,* **105,** 5706 (1983).
8. W. J. Brittain, *Rubber Chem. Technol.* **65,** 580 (1992).
9. O. W. Webster and B. C. Anderson, in *New Methods of Polymer Synthesis,* W. J. Mij, Ed., Plenum Press, New York, 1992, pp. 1–32.
10. W. R. Hertler, in *Macromolecular Design of Polymeric Materials,* K. Hatada, T. Kityama, and O. Vogl, Eds., Dekker New York, 1997, p. 109.
11. W. B. Farnham and D. Y. Sogah, U. S. Patent 4,414,372 (1983).
12. O. W. Webster, in *Encyclopedia of Polymer Science and Engineering,* N. Bikales, C. G. Overberger, and G. Menges, Eds., Wiley-Interscience, New York, 1987, Vol. 7, p. 580.
13. D. Y. Sogah, W. R. Hertler, O. W. Webster, and G. M. Cohen, Macromolecules, **20,** 1473 (1987).
14. R. P. Quirk and G. P. Bidinger, *Polym. Bull.,* **22,** 63 (1989).
15. I. B. Dicker, G. M. Cohen, W. B. Farnham, W. R. Hertler, E. D. Lignis, and D. Y. Sogah, *Polym. Prepr.* (Am. Chem. Soc., Div. Polym. Chem.), **28(2),** 106 (1987).
16. A. Pickering and A. J. Thorne, U. S. Patent 4,791,181 (1988).
17. G. M. Cohen, Polym. Prep. (Am. Chem. Soc., Div. Polym. Chem.), **29(2),** 46 (1988).
18. R. Asami, Y. Kondo, and M. Takaki, in *Recent Advances in Anionic Polymerization,* T. E. Hogen-Esch and J. Smid, Eds., Elsevier, New York, 1987, p. 381.
19. J. M. DeSimone, A. M. Hellstern, E. J. Siochi, S. D. Smith, T. C. Ward, V. J. Krukonis, and J. E. McGrath, *Macromol. Chem. Macromol. Symp.,* **32,** 21 (1990).
20. W. Radke and A. H. E. Muller, *Macromol. Chem. Macromol. Symp.,* **54/55,** 583 (1992).
21. J. A. Simms and H. J. Spinelli, in *Macromolecular Design of Polymeric Materials,* K. Hatada, T. Kitayama, and O. Vogl, Eds., Dekker, 1997, New York, p. 379.
22. M. Morton, T. E. Helminiak, S. D. Gadkary, and F. Beuche, *J. Polym. Sci.,* **57,** 471 (1962); B. J. Bauer and L. J. Fetters, *J. Polym. Soc.,* **32,** 21 (1990); Rubber Chem. Technol., **51,** 406 (1978).
23. P. Rempp, E. Franta, and J. E. Herz, *Adv. Polym. Sci.,* **86,** 145 (1988); P. Rempp, *Macromol. Chem. Macromol. Symp.* **60,** 209 (1992).
24. G. D. Andrews and L. R. Melby, in *New Monomers and Polymers,* B. M. Culbert-

son and C. Pittman, Eds., Plenum Press, London, 1984 p. 357; T. Q. Nguyen and H. H. Kausch, *Macromol. Chem. Rapid Commun.,* **6,** 391 (1985).
25. C. Tsitsilianis, P. Lutz, S. Graft, and J.-P. Lamps, *Macromolecules,* **24,** 5897 (1991).
26. European Patent 0 551 221 A1 (1993).
27. O. W. Webster, *Macromol. Chem. Macromol. Symp.,* **33,** 133 (1990).
28. S. Sivaram, unpublished results.
29. A. D. Jenkins, *Macromol. Chem. Macromol. Symp.,* **53,** 267 (1992).
30. H. J. Spinelli, U.S. Patent 4,627,913 (1984).
31. B. Sannigrahi, A. J. Raj, P. P. Wadgaonkar, J. C. Sehra, and S. Sivaram, unpublished results.
32. D. M. Haddleton and M. C. Crossman, *Macromol. Chem. Phys.,* **198,** 871 (1997).
33. L. V. Sneider and I. B. Dicker, U.S. Patent, 4,736,003 (1988).
34. B. Sannigrahi, P. P. Wadgaonkar, J. C. Sehra, and S. Sivaram, *J. Polym. Sci. A, Polym. Chem.,* **35,** 1999 (1997).
35. J. A. Simms, *Rubber Chem. Technol.,* **64(2),** 139 (1991); *Polym. Prepr.* (Am. Chem. Soc. Divn. of Polym. Chem.); **33** (1), 164 (1992).
36. P. J. Flory, *Principles of Polymer Chemistry,* Cornell University Press, Ithaca, N.Y. 1953.
37. P. Lang, M. S. Burchard, N. S. Wolfe, and H. J. Spinelli, *Macromolecules,* **24,** 1306 (1991).
38. C. S. Hutchins and H. J. Spinelli, U.S. Patent 4,851,477 (1989).
39. H. J. Spinelli, U.S. Patent 5,036,139 (1991).
40. Z. Zhu, J. Rider, C. Y. Yang, M. E. Gilmartin, and G. E. Wnek, *Macromolecules,* **25,** 7330 (1992).
41. R. Asami, M. Takaki, and Y. Moriyama, *Polym. Bull.,* **16,** 125 (1986).
42. R. Witkowski and F. Bandermann, *Macromol. Chem.,* **190,** 2173 (1989).
43. T. Heitz and O. W. Webster, *Macromol. Chem.* **192,** 2463 (1991); D. Y. Sogah, W. Hertler, and O. Webster, *Polym. Prepr.* (Am. Chem. Soc. Divn. Polym. Chem.), **25**(1), 3 (1984).
44. W. Hertler, D. Y. Sogah and F. P. Boettcher, *Macromolecules,* **23,** 1264 (1990).
45. A. D. Jenkins, E. Tsartolia, D. R. M. Walton, J. Hoska-Jenkins, P. Kratchovil, and J. Stejskal, *Macromol. Chem.,* **191,** 2511 (1990).
46. D. Y. Sogah, *Polym. Prepr.* (Am. Chem. Soc., Div. Polym. Chem.), **29(2)** 3 (1988); U.S. Patent 4,906,713 (1990).
47. D. T. Martin and S. Bywater, *Macromol. Chem.,* **193,** 1011 (1992).
48. A. H. E. Müller, *Macromolecules,* **27,** 1685 (1994).

4

Star-Shaped Polymers via Living Cationic Polymerization

Stefan Ingrisch and Oskar Nuyken
Lehrstuhl für Makromolekulare Stoffe, Technische Universität München, Garching, Germany

Munmaya K. Mishra
Ethyl Corporation, Richmond, Virginia

Star-shaped polymers steadily grow into a field of high potentiality by virtue of the unique properties they exhibit both in solution and in bulk. Nanoscale-ordered materials, low-viscosity paints, and thickeners are examples of technological applications that include star polymers. However, the synthesis of well-defined star polymers having a known number of arms with a precise molecular weight and narrow polydispersity is a challenge to the macromolecular engineer. Particularly, it was a major challenge to synthesize stars via cationic polymerizations per se until the discovery of living polymerization of isobutylene and vinyl ethers during the early 1980s. To date, three major synthesis techniques have been described and used for the synthesis of stars: (1) the use of multifunctional linking agents, (2) sequential copolymerization/linking with a divinyl monomer, and (3) use of multifunctional initiators. The use of multifunctional linking agents has been proven effective by anionic techniques for the preparation of a variety of star polymers with a varying number of arms. The detailed description of stars via anionic polymerization is given in other chapters. This chapter mainly focuses on the star polymers via cationic polymerization involving vinyl ethers and isobutylene monomers.

I. STAR-SHAPED POLYMERS CONTAINING POLYVINYL ETHERS

A. Amphiphilic Star-Shaped Block Copolymers By Reaction with Bifunctional Vinyl Ethers

Star-shaped polymers can be prepared by living polymerization in several ways; one typical method involves a reaction of a linear living polymer with a small amount of a divinyl compound [1]. This method is particularly suited to prepare star polymers with many arms. These multibranched polymers with interesting three-dimensional shapes are expected to possess properties that differ from those of linear polymers, particularly when they have functional end groups. Among the typical multiarmed polymers with functional groups thus far known are dendritic polymers. There are few examples of similar "functionalized" star-shaped polymers in anionic polymerization, because of the difficulty in preparing living polymers with polar pendant groups [2]. The hydrogen iodide/Lewis acid initiating system (HI/I_2, HI/ZnI_2, etc.) induces living cationic polymerization of not only alkyl vinyl ethers but also those having a pendant ester group that leads after hydrolysis to water-soluble polymers [3]. A recent study showed that a series of amphiphilic polymers could also be prepared by sequential living cationic polymerization of a functional vinyl ether and an alkyl derivative [4–6]. On the basis of these findings it was possible to synthesize amphiphilic star-shaped polymers **5** with pendant hydroxyl groups, new functionalized polymers with controlled spatial (three-dimensional) shapes [7].

As illustrated in Scheme 1, the synthesis starts from the sequential living cationic polymerization of 2-acetoxyethyl vinyl ether (AcOVE; $CH_2{=}CHOCH_2CH_2OCOCH_3$) and isobutyl vinyl ether [IBVE; $CH_2{=}CHOCH_2CH(CH_3)_2$] by HI/ZnI_2. The resulting living block polymer **3** (P*) is allowed to react with a small amount of bifunctional vinyl ether **1** to give star-shaped block copolymer **4.** Alkaline hydrolysis of the ester pendant in **4** then leads to an amphiphile **5,** where the arms consist of hydrophilic polyalcohol [poly(2-hydroxyethyl vinyl ether), poly(HOVE)], and hydrophobic poly(IBVE) segments.

An important feature of this methodology is that by applying opposite polymerization sequences for AcOVE and IBVE, two types of amphiphilic star-shaped polymers may be prepared (**5a** and **5b;** Scheme 1): namely, polymer **5a** carries hydrophilic polyalcohol segments in its outer layer, whereas **5b** does in its inner layer. Therefore, **5a** and **5b** are expected to be different in their properties (such as solubility), even though they are the same in segment composition and degree of polymerization. Amphiphilic star-shaped polymers **5**

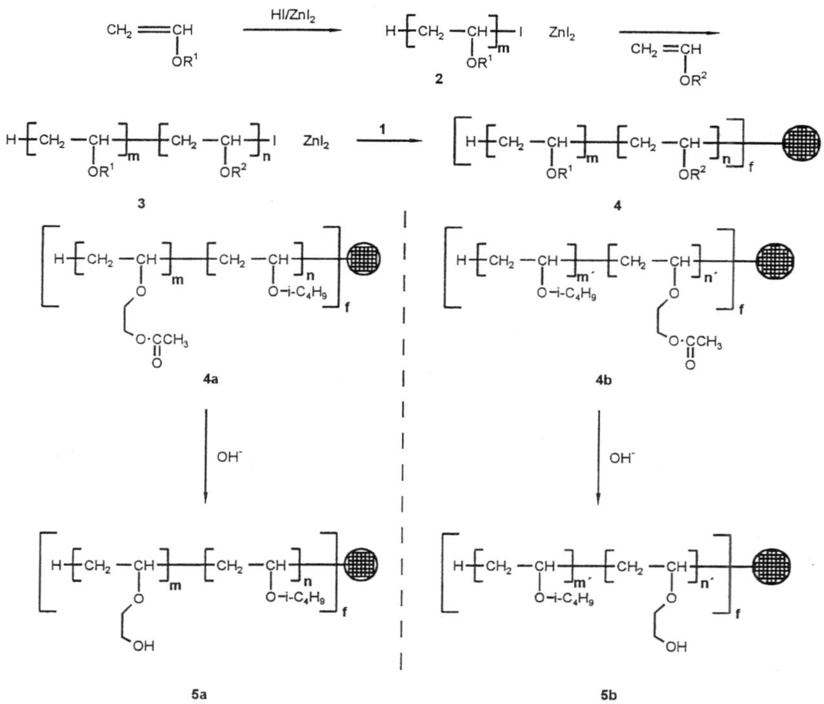

Code	R^1 [Outer Segment]	R^2 [Inner Segment]
2a, 3a, 4a	$CH_2CH_2OC(=O)CH_3$	$i\text{-}C_4H_9$
2b, 3b, 4b	$i\text{-}C_4H_9$	$CH_2CH_2OC(=O)CH_3$

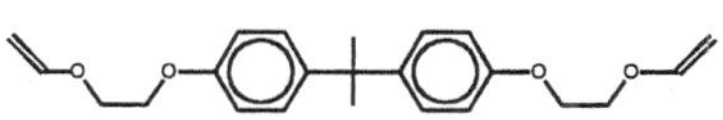

1

= Microgel Core of Poly(1)

Scheme 1

were synthesized and their solubility characteristics, specifically relative to those of the corresponding linear AB block polymers, were examined.

Linear amphiphilic polymers with carboxyl groups were also available via sequential cationic polymerization of a functional vinyl ether and an alkyl derivative [6]. The sequential living cationic polymerization of diethyl 2-(vinyloxy)ethylmalonate [VOEM; $CH_2{=}CHOCH_2CH_2CH(COOCH_2CH_3)_2$]

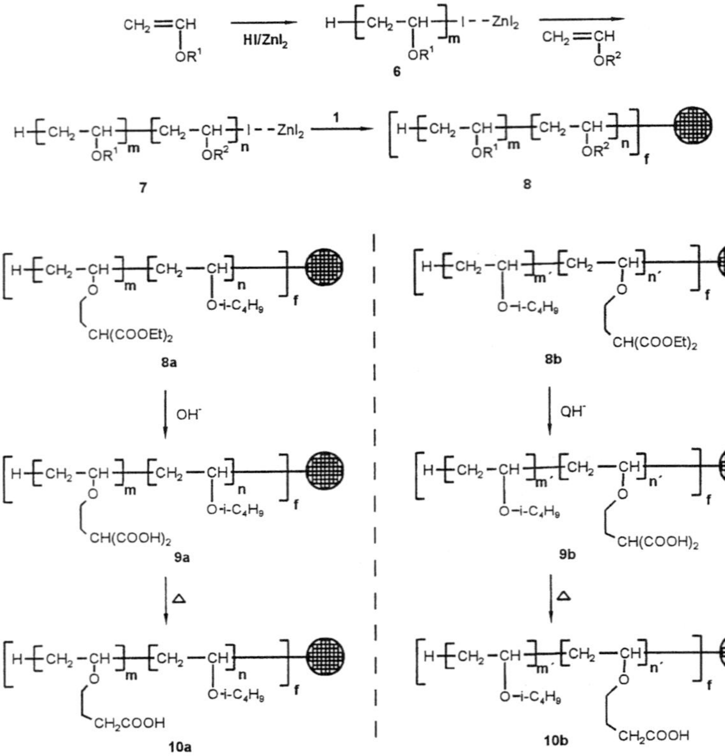

Code	R^1 [Outer Segment]	R^2 [Inner Segment]
6a, 7a, 8a	$CH_2CH_2CH(COOEt)_2$	i-C_4H_9
6b, 7b, 8b	i-C_4H_9	$CH_2CH_2CH(COOEt)_2$

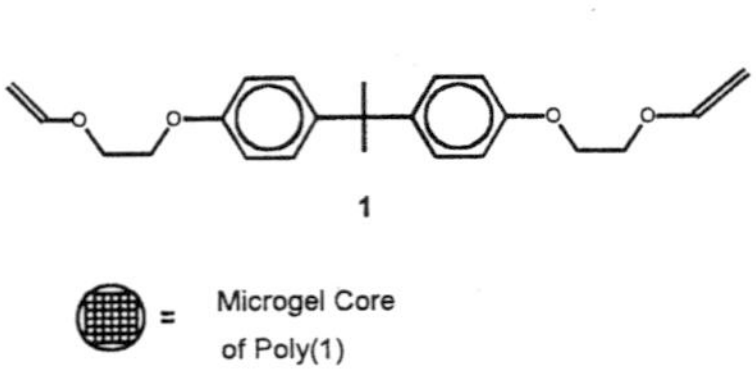

Scheme 2

and isobutyl vinyl ether [IBVE; $CH_2{=}CHOCH_2CH(CH_3)_2$] with HI/ZnI_2 affords a linear living block copolymer **7** as shown in Scheme 2. Then, **7** is allowed to react with a small amount of bifunctional vinyl ether **1** to give a star-shaped block copolymer **8.** Alkaline hydrolysis of the ester groups in **8,** followed by neutralization, leads to a diacid-type amphiphile **9,** and subsequent decarboxylation gives the monoacid form **10** (Scheme 2). The hydrophilic segments of **9** and **10** consist of poly(2-(vinyloxy)ethylmalonic acid) [poly(VOEMA)] and poly(4-vinyloxybutanoic acid) [poly(VOBA)]. As reported for the polyalcohol counterparts [7], two types of amphiphilic star-shaped copolymers with carboxyl groups may also be prepared (**9a** and **10a** versus **9b** and **10b**) by applying opposite polymerization sequences for VOEM and IBVE: the hydrophilic polyacid segments are located in the outer layer in **9a** and **10a,** whereas they are placed in the inner layer in **9b** and **10b.**

Star-shaped polymers with monodisperse arms may also be prepared by intramolecular crosslinking. For example, a living polymer of vinyl ether may be allowed to react with a small amount of divinyl ether **1** to form a block copolymer **11** in which a short segment of **1** is attached to the end of the living chain. Subsequent intermolecular reactions of the pendant vinyl groups of **11** with P* or with the living ends of **11** are expected to result in polymer linking (into **12,** for example), followed by intramolecular crosslinking that gives a star-shaped polymer **13** as illustrated in Scheme 3.

B. Heteroarmed Amphiphilic Star-Shaped Polymers of Vinyl Ethers

Amphiphilic star-shaped polymers with heteroarms of vinyl ethers can be prepared in a similar way on the basis of living cationic polymerization [8], where living polyvinyl ether chains, generated with the HI/ZnI_2 system, undergo linking reactions via a bifunctional vinyl ether into a star-shaped polymer. The initially formed star polymer (first star; **14** in Scheme 4) may still carry living growing sites within its microgel core. These "core" living sites may be utilized to initiate a second-phase living polymerization to grow new arms from the core to give a "second star" polymer **15** where the number of arms per molecule is doubled from the first star. When a second monomer differs from the first polymerized monomer, a "heteroarm" star polymer may be obtained where different arms are attached to a single core.

As illustrated in Scheme 5, treatment of a linear living polymer **2b** of isobutyl vinyl ether, prepared by the HI/ZnI_2 initiating system, with a small amount of divinyl ether **1** affords "living" star-shaped polymer **16.** From the living sites in its core, an ester-containing vinyl ether may be polymerized to give a heteroarm star-shaped polymer **17.** Alkaline hydrolysis of the ester pendant groups in **17** then leads to an amphiphilic heteroarm star polymer **18,** where separate sets of hydrophobic and hydrophilic arms are independently attached to a single core. The separation of hydrophilic and hydrophobic arms is an important feature of **18;** note that for **18,** the amphiphilicity stems from these two sets of independent homopolymer arm chains, whereas in the corresponding "star block" versions, each arm is an amphiphilic block polymer [9].

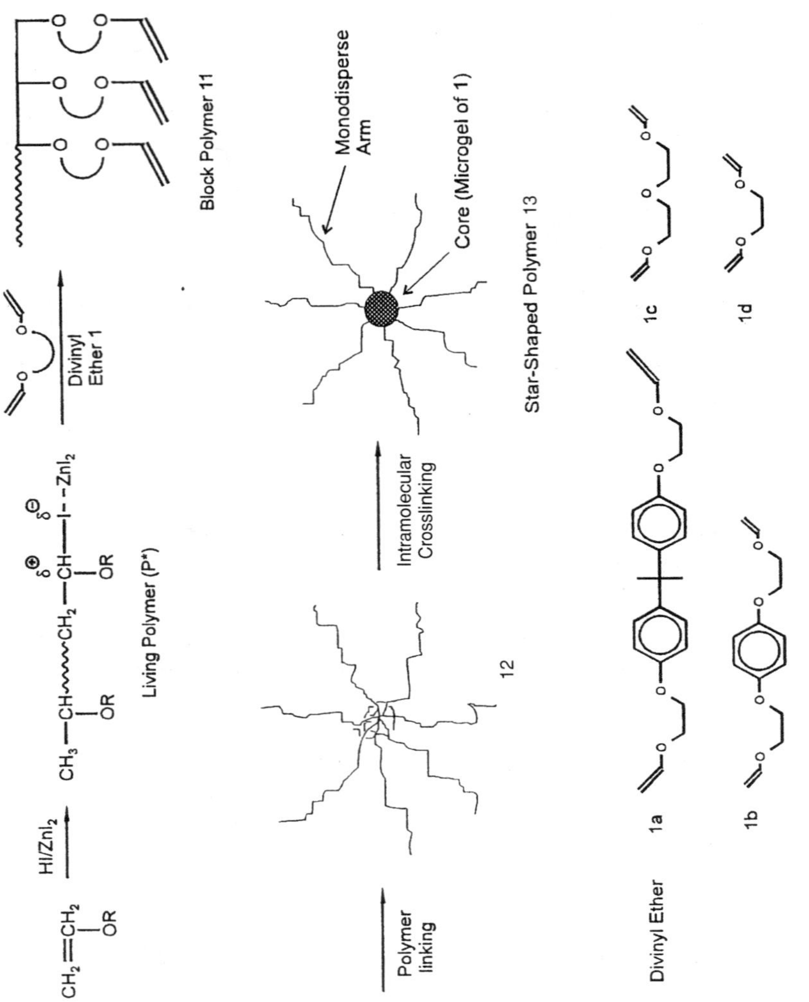

HI/ZnI2
Living Polymer (P*)
Divinyl Ether 1
Block Polymer 11
Polymer linking
12
Intramolecular Crosslinking
Monodisperse Arm
Core (Microgel of 1)
Star-Shaped Polymer 13
Divinyl Ether
1a
1b
1c
1d

Scheme 3

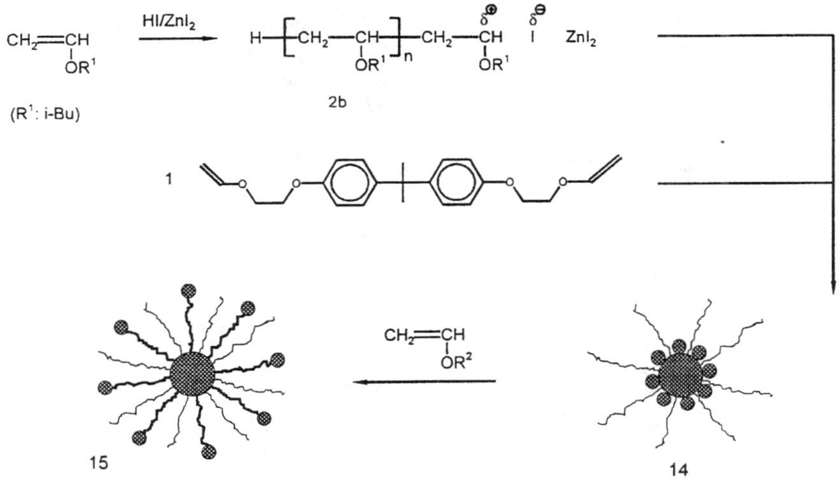

15a: $R^2 = OCOCH_3$

15b: $R^2 = CH(COOCH_2CH_3)_2$

15c, 15d: $R^2 = CH_2CH(CH_3)_2$

Scheme 4

Therefore, star polymers **18** are expected to possess properties that differ from those of the star block amphiphiles, where outer segments may render inner segments restricted in motion, and govern the properties of the star polymer.

C. Core-Functionalized Amphiphilic Star-Shaped Polymers of Vinyl Ethers with Hydroxyl Groups

Another type of functionalized star polymer would be "core-functionalized" star polymers, where not arms but the core carries functional groups. As illustrated in Scheme 6, a living polymer **2b** of isobutyl vinyl ether, prepared with the HI/ZnI_2 initiating system, is allowed to react with a mixture of a divinyl ether **1** and 2-acetoxyethyl vinyl ether. The simultaneous feeding of **1** and AcOVE permits both monomers (as linking agents) to react in parallel with **2b** to undergo a near random copolymerization from the living site of **2b,** and thereby form a block polymer **19.** Subsequent linking reactions among several

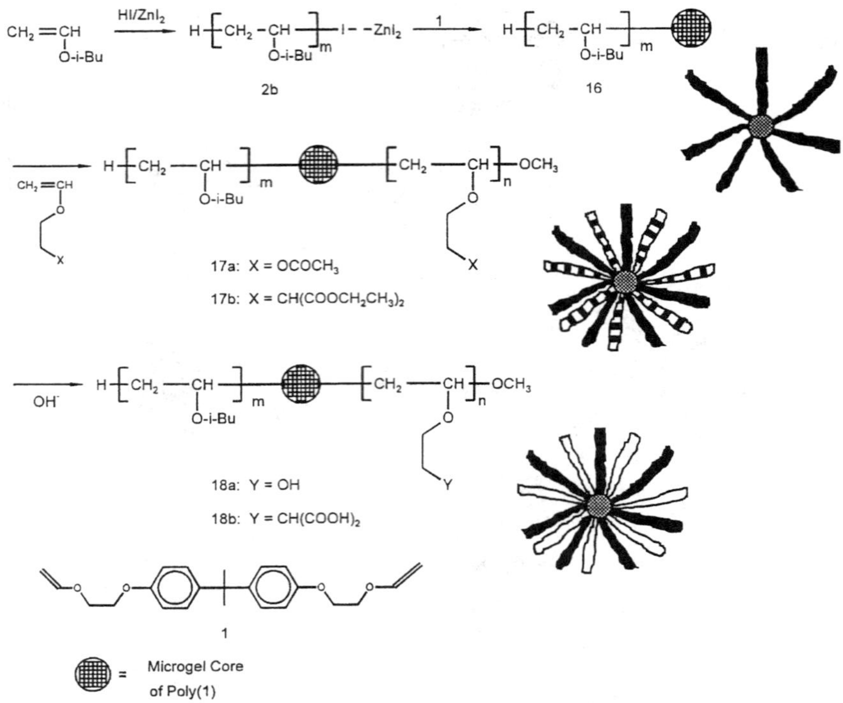

Scheme 5

chains of **19** would lead to a star-shaped precursor polymer **20** and core-functionalized amphiphilic star polymer **21** with hydrophobic poly(IBVE) arms and a hydrophilic microgel core.

The core functionalization in **21** may lead to the higher accumulation of polar hydroxyl groups in a core region that should be smaller in size than the arm moiety of the star-shaped polymers with functionalized arms. Another important feature of **21** is that an outer hydrophilic shell of **21** can effectively surround the hydrophilic microgel core with many hydroxyl groups. Therefore, the core-functionalized star polymers are amphiphilic but are expected to possess properties that differ from those of the star block and the heteroarm star amphiphiles.

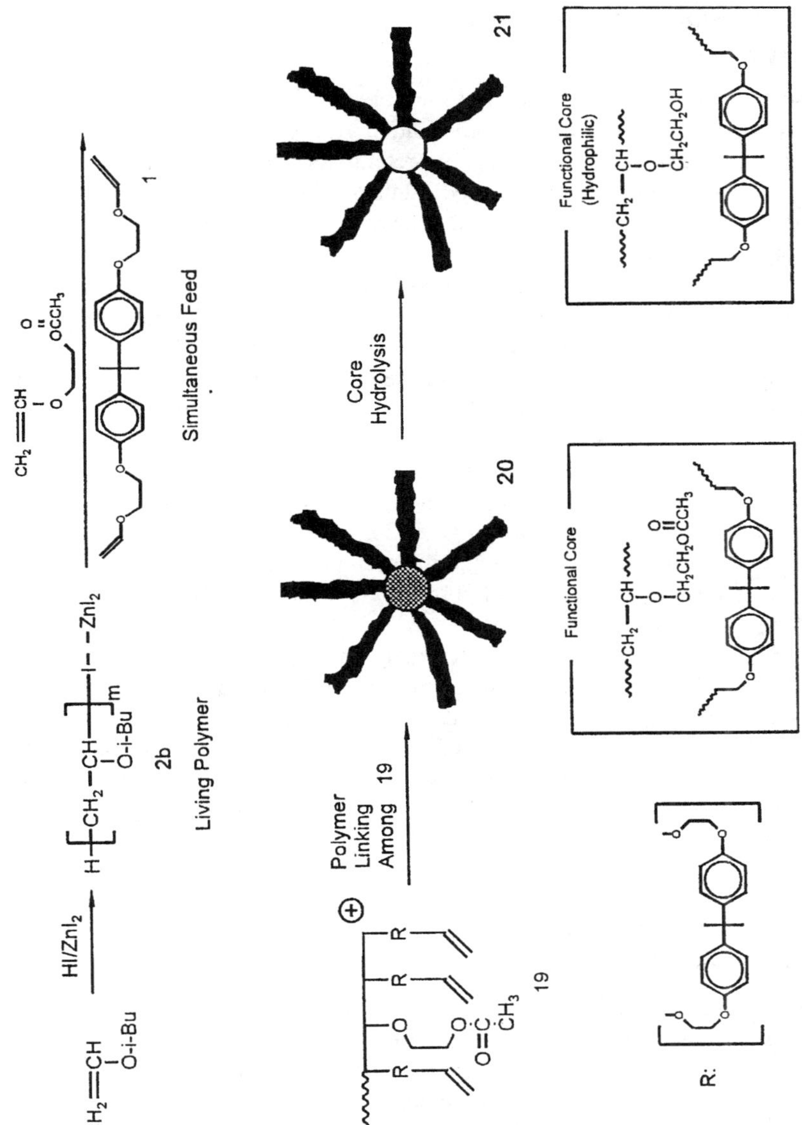
HI/ZnI2
2b
Living Polymer
Simultaneous Feed
1
Polymer Linking Among 19
19
20
Core Hydrolysis
21
Functional Core
Functional Core (Hydrophilic)
R:

Scheme 6

D. Amphiphilic Graft Polymers of Vinyl Ethers with Hydroxyl Groups

Another method for the synthesis of multibranched polymers is the homo- or copolymerization of a macromonomer into a graft polymer. For a review, see Rempp and Franta [10]. In particular, we refer to the synthesis of "amphiphilic" graft polymers with a predetermined number of arms (branches) via living cationic polymerization of macromonomers. Although a number of macromonomers have been prepared, there are no examples of polymerization of "block" macromonomers [10], which consist of more than two different segments. Therefore, it was of interest to study the synthesis of amphiphilic graft polymers with a controlled branch number. As shown in Scheme 7, 2-acetoxyethyl vinyl ether and isobutyl vinyl ether are sequentially polymerized with the HI/ZnI_2 initiating system. The resulting living block polymer **2a** is terminated with diethyl 2-(vinyloxy)ethylmalonate anion **22** in toluene at –15°C to give a macromonomer **23,** which bears a cationically polymerizable vinyl ether terminal [11]. The subsequent homopolymerization of **23** with HI/ZnI_2 in toluene at –15°C affords a graft polymer **24.** Alkaline hydrolysis of the ester groups in **24** then leads to the amphiphilic graft polymer **25** where each branch consists of a hydrophilic polyalcohol [poly(2-hydroxyethyl vinyl ether)] and a hydrophobic poly(IBVE) segment.

Such amphiphilic graft polymers may have almost uniform number of branches, provided that living polymerization of **23** is available. In addition, unlike the various types of star-shaped polymers previously reported [7,12,13], graft polymer **24** has no hydrophobic microgel core that may possess some dimensions. Therefore, the amphiphilic graft polymers are expected to possess properties and functions differing from those of the corresponding star block amphiphiles.

E. Three-Arm Star Polymers By Living Cationic Polymerization

Three-arm star polymers are among advanced polymer material that may find uses, for example, as crosslinking agents, ionomers, surface active agents, compatibilizers, and prepolymers for elastomer. For their usefulness and versality, they should satisfy at least two criteria: perfect end functionality (exactly three living ends per polymer molecule) and controlled molecular weight and narrow molecular weight distribution of each arm chain. These criteria may be achieved most readily by living polymerization with a trifunctional initiator. Some trifunctional initiators have been developed by Kennedy and co-workers, who

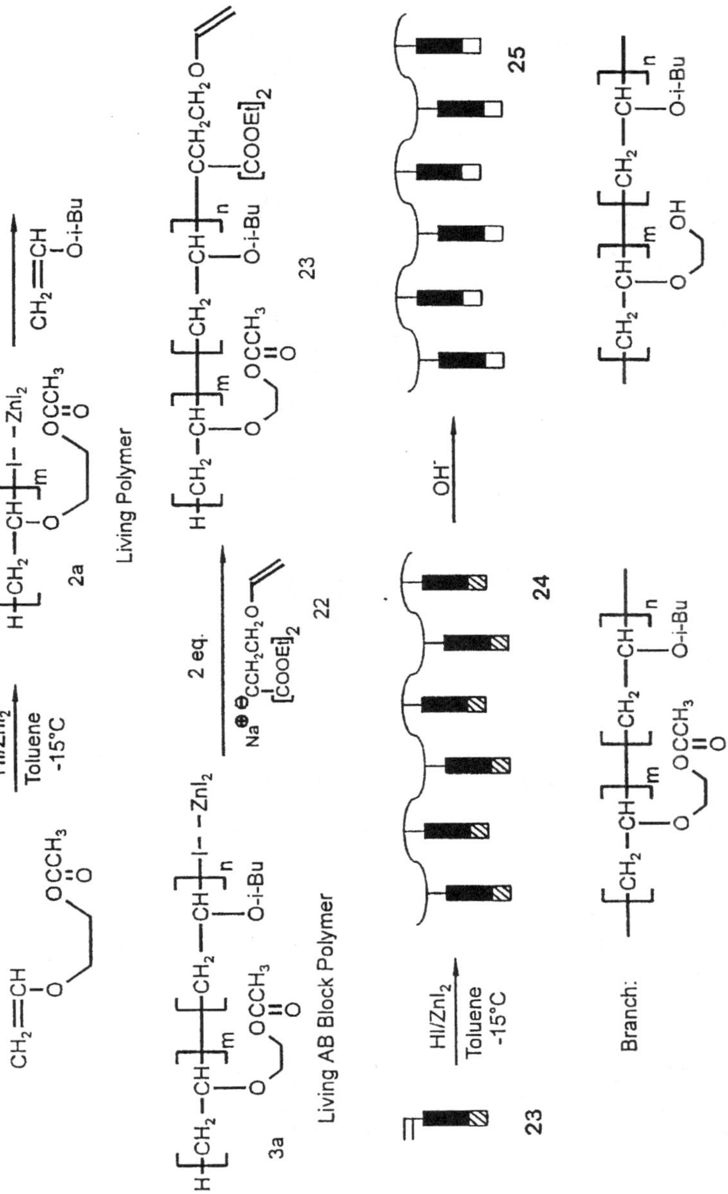
HI/ZnI$_2$
Toluene
-15°C
2a
Living Polymer
3a
Living AB Block Polymer
2 eq.
22
23
OH$^-$
24
25
Branch:

Scheme 7

obtained three-armed star polyisobutylenes via the inifer method [14] and living cationic polymerization [15–19]. However, these methods have been limited to nonpolar hydrocarbon monomers, and three-arm polymers with functional pendant groups have not been obtained yet. Higashimura and Deffieux have prepared three-arm poly(vinyl ether)s from perfectly trifunctional initiators [20]. An important objective is the synthesis of three-arm star polymers with functional pendant groups, particularly from vinyl ethers, to which a variety of polar functions can readily be introduced without adversely affecting their living cationic polymerization [21]. Living cationic polymerization of vinyl ethers may be initiated with a carboxylate adduct in conjunction with ethylaluminum dichloride and an excess amount of weak Lewis base (Scheme 8) [22].

The adduct is obtained from a vinyl ether and CF_3COOH, and its trifluoracetate group initiates cationic polymerization of vinyl ethers with $EtAlCl_2$, in the presence of such a weak Lewis base as 1,4-dioxane, and generates living polymer, where the growing carbocation is stabilized by the added base.

Based on this methodology, this first study employs trifunctional compounds **26** and **27,** which were prepared from CF_3COOH and trifunctional vinyl ethers, as trifunctional initiators for living cationic polymerization of isobutyl vinyl ether (Scheme 9).

Another recent finding is that living cationic polymerizations of *p*-methoxy- and *p-tert*-butoxystyrenes can be initiated not only with the classical HI/ZnI_2 system, but also with a cationogen derived from vinyl ether and hydrogen iodide in the presence of zinc iodide as activator [23].

The living polymerizations by the cationogen/ZnI_2 systems follow the nearly identical pathways with those originally initiated by the HI/ZnI_2 counterpart, and it is important that the vinyl ether–type carbocation can polymer-

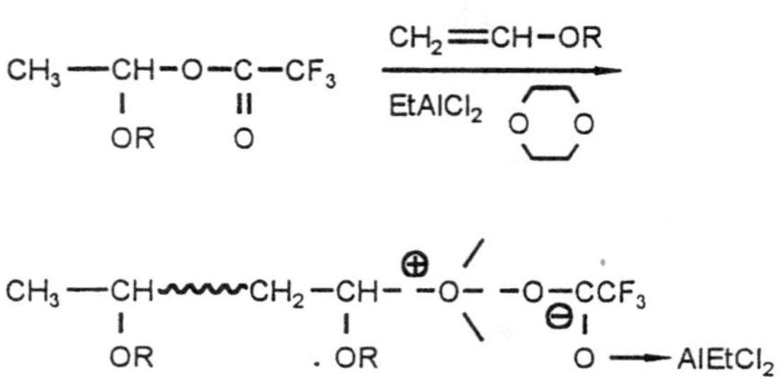

Scheme 8

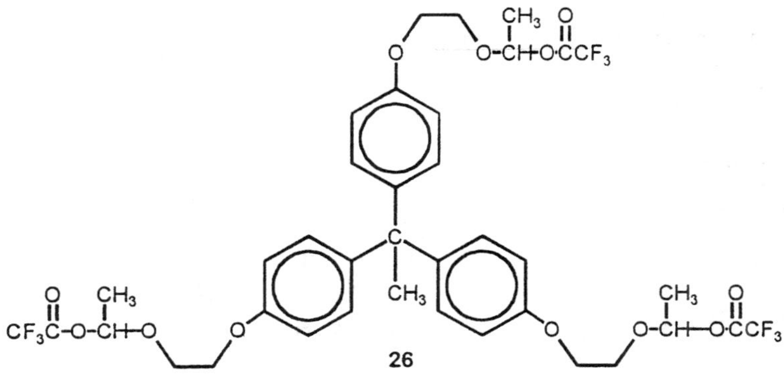

26

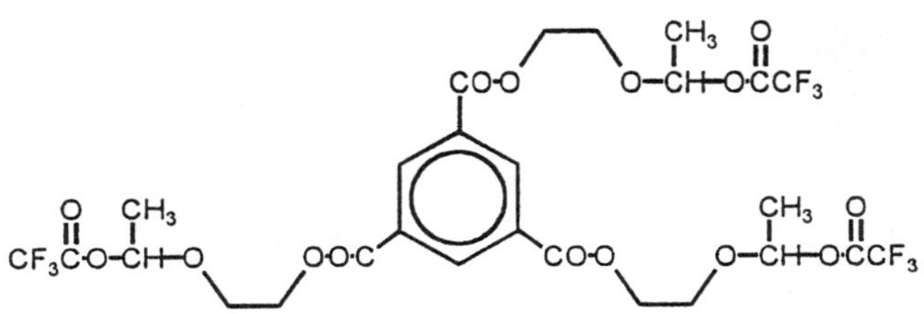

27

Scheme 9

ize *p*-methoxystyrenes (*p*MOS), which are less reactive than vinyl ethers (Scheme 10).

This chemistry was adapted for the design of a three-arm star polymer. Trifunctional initiator **28** was prepared by the treatment of a trifunctional vinyl ether **29** with 3 equivalents of hydrogen iodide. *p*MOS was polymerized with **28**/ZnI_2 to give three-arm living poly(*p*MOS) (Scheme 11) [24].

II. STAR-SHAPED POLYMERS CONTAINING POLYISOBUTYLENE

A. Three-Arm Star PIBs and Three-Arm Star Blocks

The story of star polyisobutylene (PIB) began with the synthesis of three-arm star PIBs via the inifer method [14]. Over the years the process has been refined

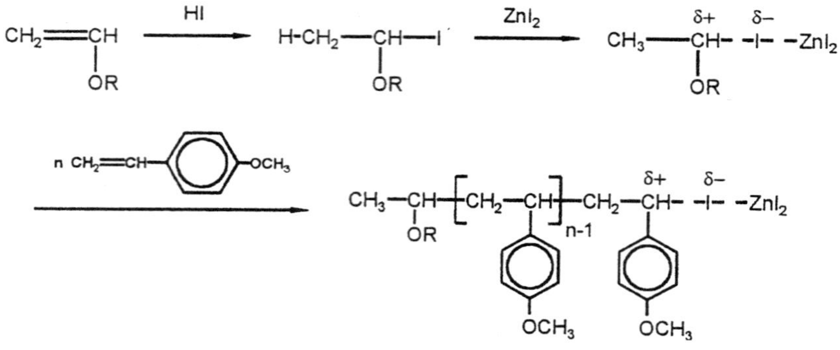

Scheme 10

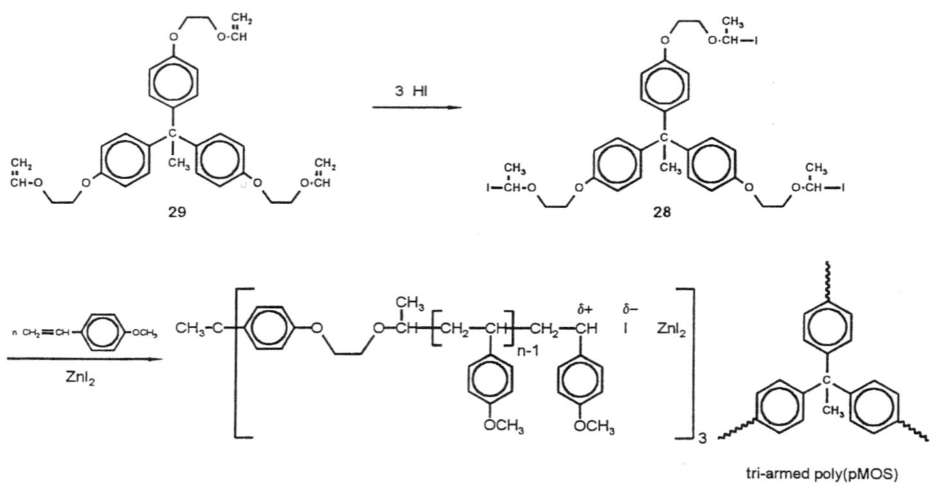

Scheme 11

with the discovery of the cationic living polymerization of isobutylene by tertiary ethers and ester-based initiating systems [15,25]. The first synthesis of a three-arm star PIB via living polymerization was carried out by using a trifunctional initiator such as 1,3,5-tris(2-methoxypropane) benzene/Lewis acid initiating system [15] (Scheme 12). The *tert*-chlorine end groups can be quantitatively converted to prepare trifunctional (isopropylidine functions) PIB as well. Terminally functional (telechelic) prepolymers are of great scientific and commercial importance [25]. The syntheses of three-arm star PIBs capped by

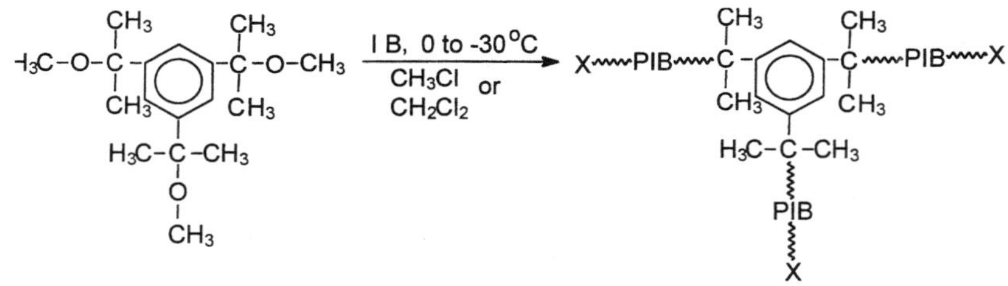

Scheme 12

three *tert*-chlorine end groups have been described by the use of forced termination of living carbocationic polymerization. Subsequently, three-arm star allyl-telechelic PIB [19] was prepared by a direct two-step one-pot process—i.e., carrying out the polymerization of IB and the functionalization of the PIB formed sequentially in the same reactor (Scheme 13). The allyl-terminated PIBs were also converted to primary alcohol groups and epoxide groups after further modification via hydroboration and oxidation.

PIB-toughened poly(methyl methacrylate) (PMMA) networks have been reported in which the PIB domains are covalently bound to a PMMA matrix [26]. The synthesis involved a macromonomer approach where a combination of radical and cationic polymerization techniques were used. Methacrylate tritelechelic PIBs were prepared by living cationic polymerization followed by end functionalization and used in free-radical solution copolymerization with various amounts of MMA. Scheme 14 represents the general reaction pathways for the synthesis of network.

By a different approach [18] a three-arm star block copolymer of isobutylene and tetrahydrofuran has also been synthesized. The synthesis in-

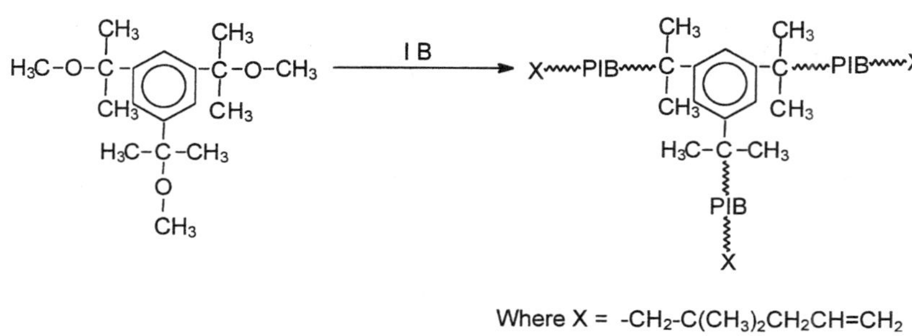

Scheme 13

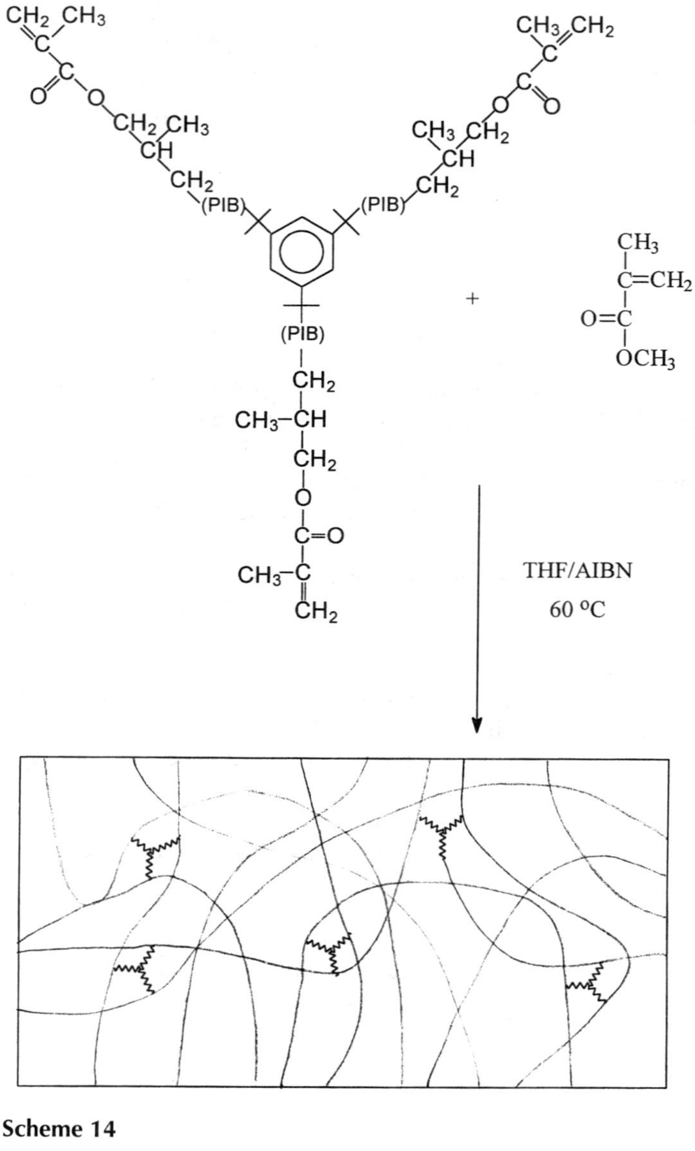
CH2 CH3
C
C
O O
CH2 CH3
CH
CH2
(PIB)
CH3 CH2
C
C
O O
CH3 CH2
CH
CH2
(PIB)
(PIB)
CH2
CH3–CH
CH2
O
C=O
CH3–C
CH2
+
CH3
C=CH2
O=C
OCH3
THF/AIBN
60 °C

Scheme 14

volved the preparation of triflate (—OSO_2CF_3) terminated three-arm star telechelic PIBs carrying exactly three primary triflate end groups followed by the quantitative conversion of primary hydroxyl end groups with triflic anhydride. These triflate-telechelic linear and three-arm star PIBs were employed as macroinitiators to induce the polymerization of tetrahydrofuran leading to three-arm star block copolymers of PIB and polytetrahydrofuran (PTHF). Blocking efficiencies (B_{eff}) up to about 90% were obtained by this method. Triflate-(—OSO_2CF_3)-terminated PIBs are useful intermediates since this moiety is an excellent leaving group. Also, primary triflates are efficient initiators for the ring-opening polymerization of THF. The following reaction Scheme 15 describes the reaction pathways for the synthesis of three-arm star block polymers.

Kennedy et al. [27] have also demonstrated the preparation of three-arm star radial thermoplastic elastomer (TPE) comprising rubbery polyisobutylene center blocks connected to glassy poly(*p*-chlorostyrene) outer blocks by sequential monomer addition technique. The radial block copolymers exhibited excellent TPE characteristics. The TPE exhibits two Tg characteristics of glassy P*p*ClSt (129°C) and rubbery PIB (–70°C) segments. Transmission electron microscopy (TEM) of a triblock polymer containing about 38% P*p*ClSt suggests cylindrical P*p*ClSt domains 40–70 nm in length and 25–35 nm in diameter embedded in a PIB matrix.

B. Multi-Arm Star PIB

Very recently, Kennedy and his group [28] have described the first synthesis of a multi-arm radial-star PIB by the arm-first, core-last method. The synthesis occurred by the addition of excess divinylbenzene (DVB) linking agent to a living PIB arm by the "arm-first" method under specific conditions. The weight average molecular weight in excess of 1 million with M_w/M_n about 3 has been reported. The number average number of arms emanating from the core was N_n,arm = 56. One of the key features of the work is that it provides a synthetic route to polyolefin-based, cationically synthesized star polymers that have essentially no residual unsaturation. Although earlier work concerning three- and four-arm star PIBs by direct initiation has been reported [25], this research has certainly been a step forward in the arena of star polymers via cationic polymerization. The synthesis of PIB-based star-branched polymers was achieved via living cationic polymerization using the aliphatic initiating system consisting of 2-chloro-2,4,4-trimethylpentane (TMPCL), $TiCl_4$, and triethylamine (TEA, electron donor) and a core-forming monomer such as divinyl benzene (DVB). The general reaction to describe the synthesis of star-branched PIB is presented in Scheme 16. The exact time of the addition of the linking agent is

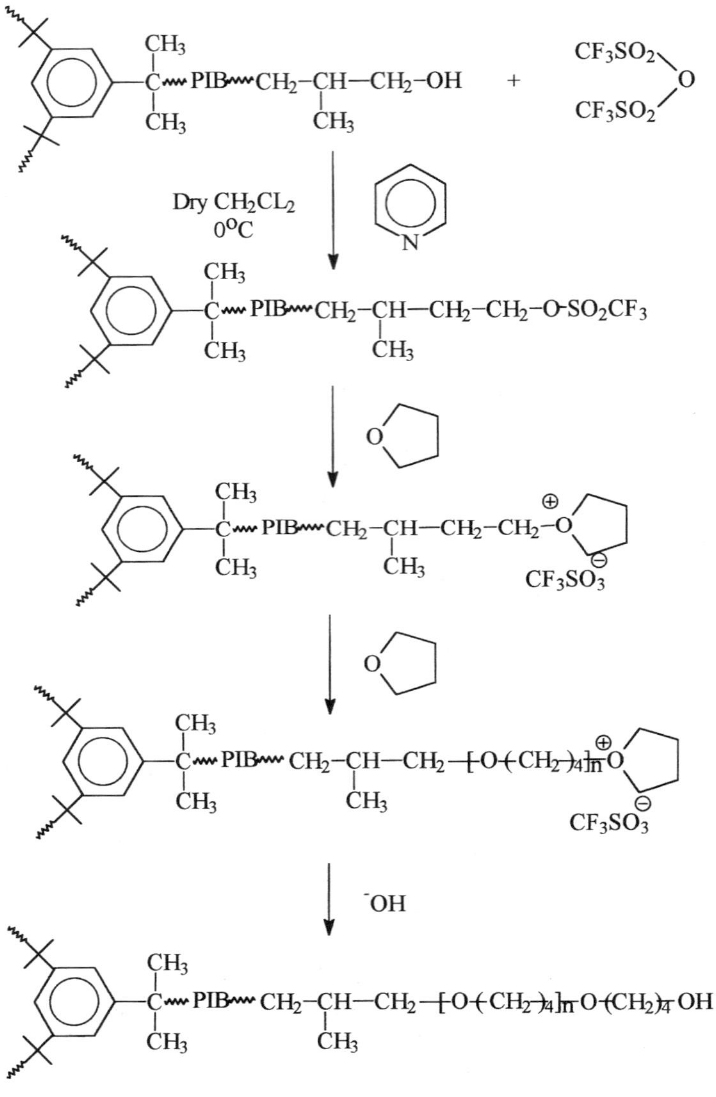
CH3
C PIB CH2-CH—CH2-OH
CH3 CH3
+
CF3SO2
O
CF3SO2
Dry CH2CL2
0°C
N
CH3
C PIB CH2-CH—CH2-CH2-O-SO2CF3
CH3 CH3
O
CH3
C PIB CH2-CH—CH2-CH2-O
CH3 CH3
CF3SO3
O
CH3
C PIB CH2-CH—CH2
O-(CH2)4 n O
CH3 CH3
CF3SO3
-OH
CH3
C PIB CH2-CH—CH2
O-(CH2)4 n O-(CH2)4-OH
CH3 CH3

Scheme 15

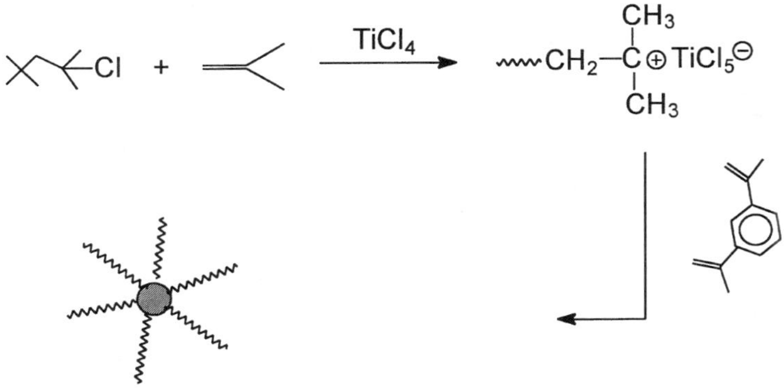

Scheme 16

of significance: DVB addition at lower than about 95% IB conversion leads to undesirable IB/DVB copolymers and ill-defined low molecular weight product, whereas DVB addition after 100% IB conversion may result in the loss of activity of PIB^+ growing chain. Although the crossover reaction from PIB^+ to $\rightarrow$ PIB-DVB^+ is extremely rapid, the overall formation of the star polymer (i.e., core growth, intermolecular linking to multi-arm stars) is rather slow. Kennedy's group [29] has characterized these star structures by a variety of methods. The presence of radial structure in a star PIB containing a PDVB core was proven directly by core destruction as represented in Scheme 17. The number of arms was calculated by comparing the molecular weight data of the star molecule with the arms. An insight into the mechanism of the formation of multi-arm star PIBs emanating from polydivinylbenzene (PDVE) cores was reported by Marsalko et al. [30]. It has been reported that star–star coupling occurs in specific conditions during the synthesis of star-branched PIBs. The re-

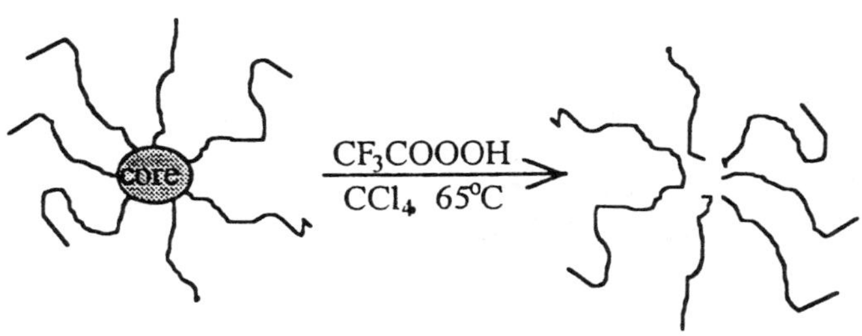

Scheme 17

port [30] described the factors that influence intermolecular star–star coupling to higher-order stars and the mechanism of star–star coupling and core contraction.

A similar method for the preparation of star-branched PIB was also reported by Storey et al. [31], who used a different core-forming monomer such as 1,3-di-isopropenylbenzene. The reactions were very similar to the work of Kennedy et al. [28]. Functional star-branched PIBs were prepared by living cationic polymerization via haloboration initiation [32] using an approach similar to Kennedy et al. IB was polymerized using BCl_3 in polar solvents (CH_2Cl_2) with a proton trap (DTBP, 2,6-*ditert*-butylpyridine) to produce low molecular weight PIBs(M_n = 3000–4000) carrying Cl_2B- head group and a *tert*-chloro end group. These PIBs were utilized for linking reaction with DVB in the presence of $TiCl_4$ in CH_3Cl / hexane solvent. Scheme 18 describes the synthesis of star PIB.

In a recent work Kennedy et al. [33] have described the synthesis of star-shaped PIBs by linking olefin-terminated PIBs with multifunctional cyclosiloxanes via hydrosilation. The synthesis has been achieved by hydrosilating olefin-capped (e.g., allyl- or isopropenyl-terminated) PIB with siloxanes carrying a plurality of Si–H groups. They have also extended their work [34] to produce "higher-order" star polymers having multiple, well-defined arms of PIB emanating from a condensed core of cyclosiloxane. Higher-order stars

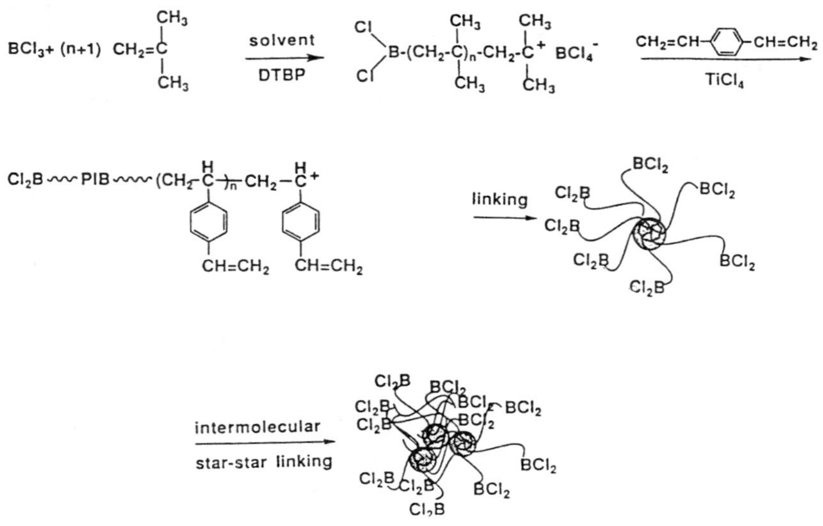

Scheme 18

might be envisioned as clusters of first-order star polymers linked together by their siloxane cores. The synthesis involves hydrosilation of allyl-terminated PIB prearms by small methylcycloxiloxanes (e.g., D_4^H, D_6^H) and competitive moisture-mediated core–core coupling. Well-defined 21–28 arm stars have been prepared via a controlled polymerization process. The intrinsic viscosities of stars are much lower than those of linear PIBs of the same molecular weight over the 30–100°C range. The synthetic pathway is represented in Scheme 19.

C. Multi-Arm Star Block Copolymers

After the demonstration of the synthesis of multi-arm star PIBs, Kennedy's group [35] extended their work to design multi-arm star block copolymers containing polyisobutylene. (Pst-*b*-PIB)$_n$-PDVB star blocks can be prepared by at least two methods, specifically by a one-pot (direct method) or two-pot (indirect) method. The advantage of the one-pot method is simplicity, where star blocks are formed by sequential addition of styrene and isobutylene and DVB to a living charge. However, in a two-pot process the diblock copolymer Pst-*b*-PIB was isolated and used as a macroinitiator in the presence of $TiCl_4$ to initiate the linking step by the addition of DVB. It indicates that the two-pot method is less simple (includes an additional isolation step of the Pst-*b*-PIB-Cl^t prearm). It has the distinct advantages of making use of a well-defined prearm in the synthesis, allowing for rapid and efficient linking at relatively high temperatures. The work by Kennedy's group [35] suggests that the physical properties of the final star blocks formed by the one-pot method are somewhat inferior to those obtained by the two-pot route. Scheme 20 represents the synthetic strategy.

The synthesis of Pst-*b*-PIB star blocks emanating from cycloxiloxane cores has also been reported [37]. The synthetic stragey involves the preparation of allyl-terminated Pst-*b*-PIB prearms (Pst-*b*-PIB-CH_2-CH=CH_2), followed by reaction with a single hexamethylhexacyclosiloxane (D_6^H) core and higher order star blocks with multiple D_6^H cores ([Si-H]/[C=C] > 2). The preparation of higher-order star block copolymers is certainly a very slow process (it took almost a week for about 90% prearm conversion to higher-order star blocks). The synthetic pathway for the preparation of the prearm as well as star-block copolymers is shown in Scheme 21.

D. Eight-Arm Star PIBs and Block Copolymers

Recently, Kennedy and co-workers [37] have reported the synthesis of well-defined star polymers having well-defined arms of PIB emanating from a calixarene derivative initiator. Calixarene derivative initiators in conjunction with

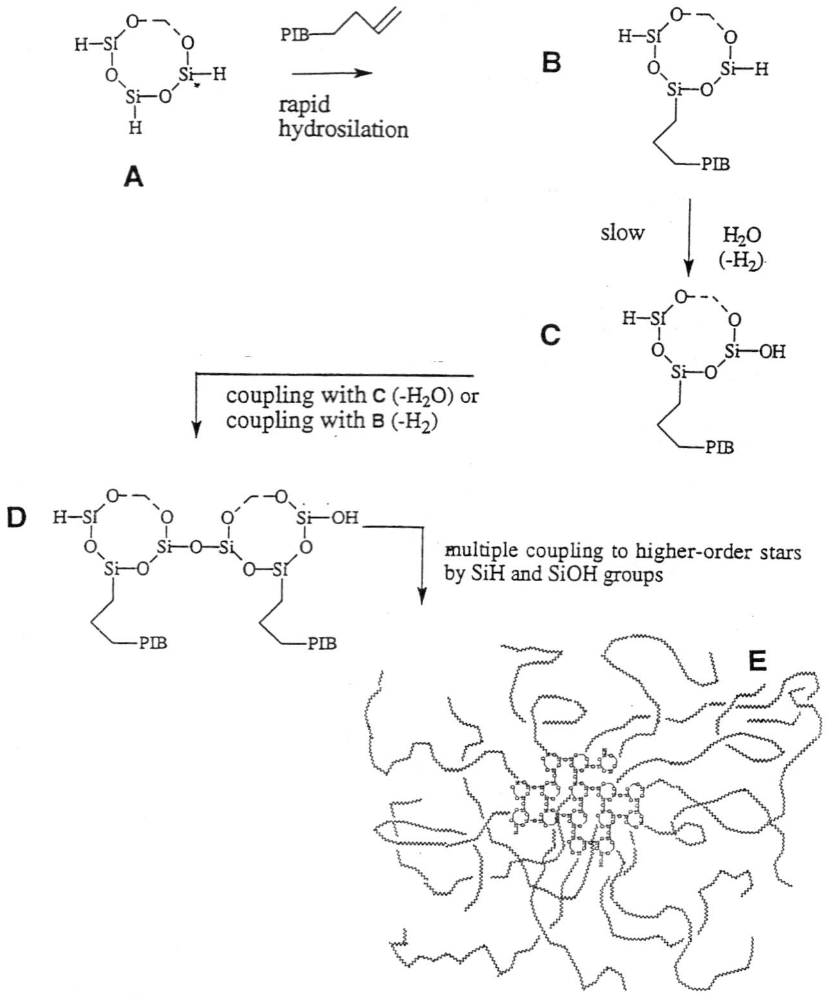
PIB
rapid
hydrosilation
A
B
slow
H_2O
$(-H_2)$
C
coupling with C ($-H_2O$) or
coupling with B ($-H_2$)
D
multiple coupling to higher-order stars
by SiH and SiOH groups
E

Scheme 19

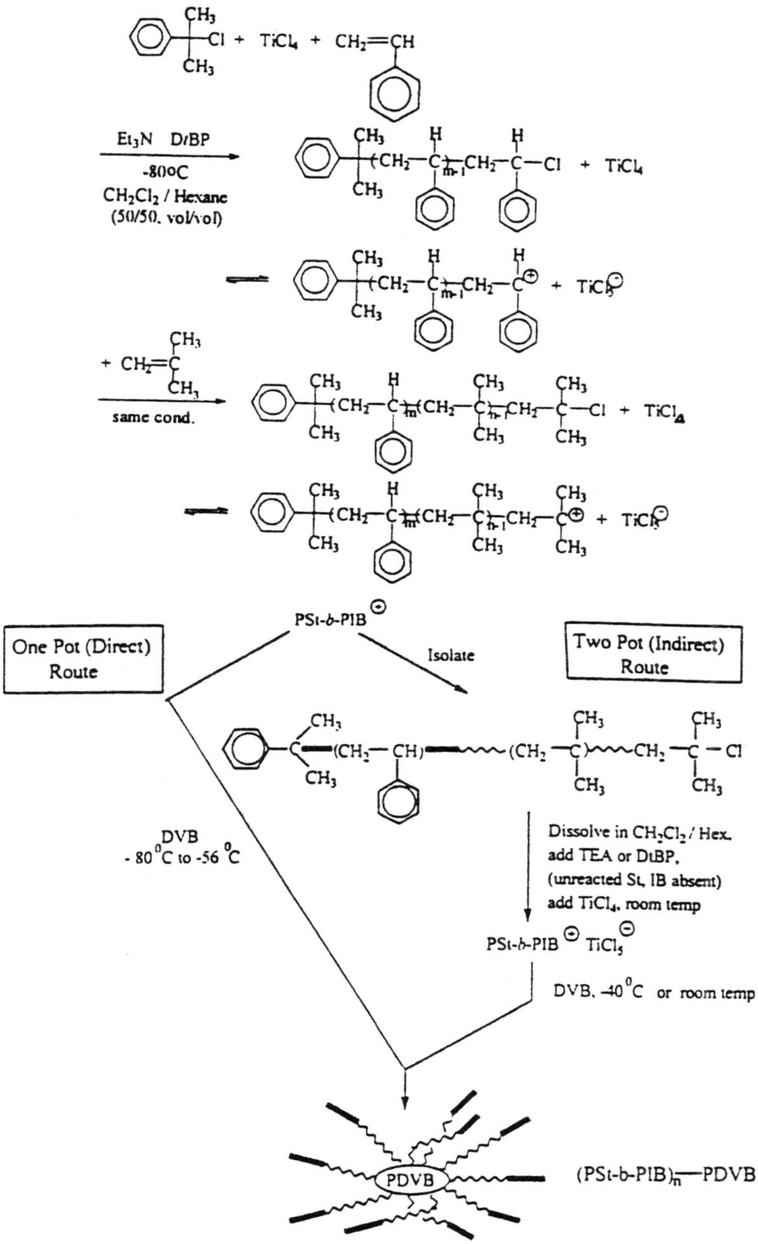
Et3N DtBP
-80oC
CH2Cl2 / Hexane
(50/50, vol/vol)
same cond.
PSt-b-PIB
One Pot (Direct) Route
Isolate
Two Pot (Indirect) Route
DVB
- 80 °C to -56 °C
Dissolve in CH2Cl2 / Hex.
add TEA or DtBP,
(unreacted St, IB absent)
add TiCl4, room temp
DVB, -40 °C or room temp
PDVB
(PSt-b-PIB)n—PDVB

Scheme 20

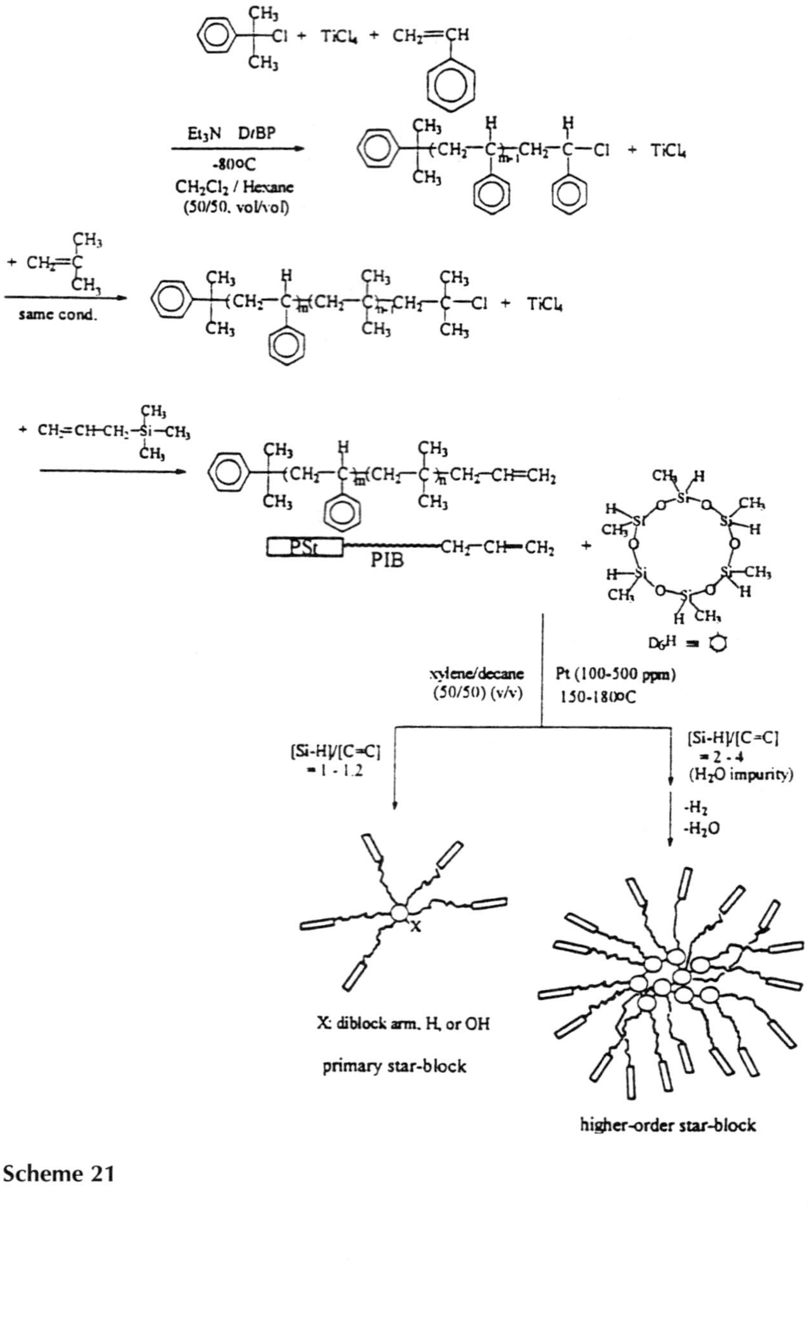
Et3N D/BP
-80oC
CH2Cl2 / Hexane
(50/50, vol/vol)
same cond.
PSt
PIB
D6H
xylene/decane
(50/50) (v/v)
Pt (100-500 ppm)
150-180oC
[Si-H]/[C=C]
= 1 - 1.2
[Si-H]/[C=C]
= 2 - 4
(H2O impurity)
-H2
-H2O
X: diblock arm, H, or OH
primary star-block
higher-order star-block

Scheme 21

BCl_3–$TiCl_4$ induce the living carbocationic polymerization of isobutylene or a similar cationically polymerizable monomer to form star polymers or block copolymers. The resultant star polymers have a well-defined core as well as well-defined arms. These stars also have functional groups at the end of each arm of the star polymer upon termination of the polymerization reaction. Calix[*n*]arenes ($n = 4, 6, 8$) are cyclic condensation products of a *p*-substituted phenol and formaldehyde. Jacob et al. [37] have demonstrated the preparation of octa-arm (^{t}Cl functional) PIB stars by octafunctional initiators derived from calix(8)arenes, i.e., calix(8)arene *tert*-methyl ether initiator in conjunction with BCl_3–$TiCl_4$ coinitiators. The preparative strategy called for the precision synthesis of calix(8)arene derivatives carrying eight initiating sites for the living polymerization of IB and its use to induce the living polymerization of IB to desirable arm lengths. The *tert*-Cl terminal groups of the eight arms are sites for further derivatization. For example, the allyl-telechelic octa-arm PIB stars [38] can be synthesized by terminating the growing PIB chain by allyltrimethylsilane (ATMS). The relative concentrations of BCl_3 and $TiCl_4$ are critical for the synthesis of well-defined eight-arm stars. The polymerization of IB with this octafunctional initiator was achieved by a two-stage procedure. In stage I a small amount of IB was polymerized in the presence of BCl_3 in $CHCl_3$ and other additives followed by the addition of calculated amount of hexanes and $TiCl_4$ and the remainder of the IB in the second stage. After the required time (i.e., $\sim$100% IB conversion) ATMS was added to achieve the allyl functionality. Scheme 22 makes it easier to visualize the structures involved and the key steps of this process.

In another example Kennedy's group [39] also described the synthesis of novel star PIBs, octa(polyisobutylene-dimethylsiloxy)octasilsesquioxane, consisting of eight polyisobutylene arms emanating from an octa(dimethylsiloxy)octasilsesquioxane (T_8^D) core. The synthesis was accomplished by the hydrosilation of allyl-terminated polyisobutylene prearms with T_8^D core. This work was similar to earlier work by the same group [33,34]. Earlier work [34] has shown that the synthesis of stars (octaisobutylene–octasilsesquioxane) consisting of eight PIB arms radiating from an octasilsesquioxane (T_8) core by hydrosilation of allyl-terminated PIB with octahydridooctasilsesquioxane (T_8^H) was incomplete even under forcing conditions, and that core–core coupling was a side reaction. The problem might be due to the steric hindrance around the SiH groups on the T_8^H. The steric congestion can be eliminated by moving the SiH group one O–SiH bond away from the relatively rigid T_8 skeleton and thus T_8^Dcore was synthesized [39]. Surely, by using the T_8^D core the steric hindrance problem was eliminated. Well-defined octa-arm stars were produced using Karstedt's catalyst bis(divinyltetramethyldisiloxane)plat-

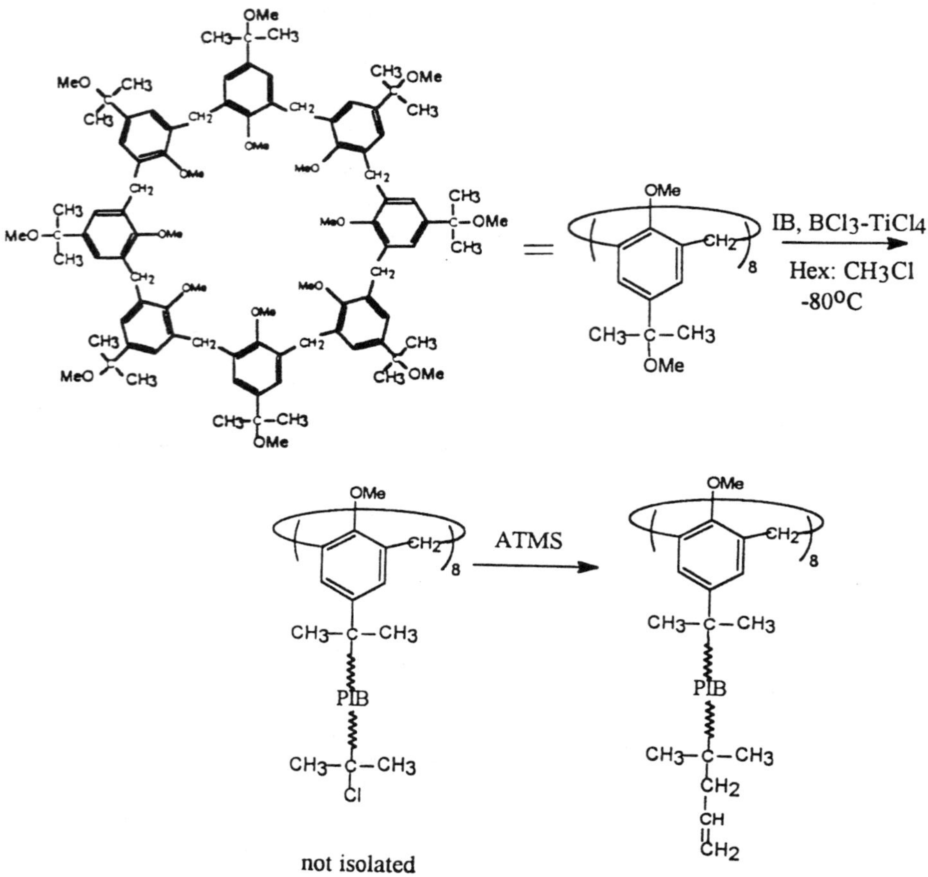

Scheme 22

inum(0) with a T_8^D core. The conventional hydrosilation catalyst $H_2PtCl_6{\cdot}H_2O$ produces side products by core–core coupling. The rate of hydrosilation is faster than the rate of side reactions using karstedt's catalyst.

Kennedy's group also extended its work on the synthesis of octa-arm star block copolymers [40]. The synthetic strategy is very similar to the method described above for the synthesis of allyl group–terminated octa-arm star PIB. The synthesis involved the use of an octafunctional calix(8)arene–based initiator to initiate the polymerization of isobutylene to desirable molecular weights; then styrene or *p*-chlorostyrene was polymerized to obtain the sought stars,

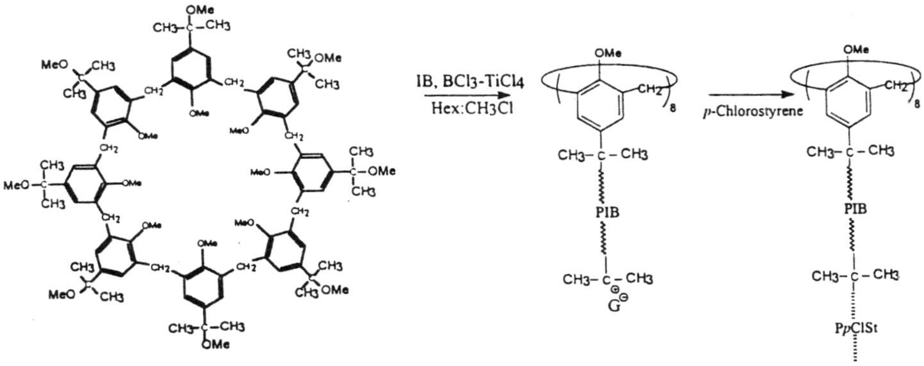

1

Scheme 23

such as eight poly(styrene-*b*-isobutylene) or poly(p-chlorostyrene-*b*-isobutylene) arms emanating from a calix(8)arene core (Scheme 23). The star blocks exhibited an excellent combination of thermoplastic elastomeric properties.

E. Six-Arm Star Polymers Containing PIB

Previous sections cover the preparation of three, four, or eight-arm stars containing polyisobutylene. Gnanou and his co-workers [41] have recently reported the synthesis of hexa-armed polystyrene stars using a hexafunctional initiator via carbocationic polymerization. Earlier, Chang et al. [42] successfully introduced six benzyl halide functions on phosphazene rings and utilized the latter compound as a plurifunctional initiator for oxazoline polymerization. Hexa-armed poly(oxazoline)s with different arm sizes have been made available in this way. The synthesis of hexa-armed star polystyrene was made possible [41] by the clean synthesis of a novel hexafunctional initiator $C_6[(CH_2)_2$-*p*-$C_6H_4CH(Cl)Me]_6$. This initiator was prepared via Fe(η-$C_5H_5)^+$ mediated perbenzylation of hexamethylbenzene, followed by regiospecific acetylation, reduction and chlorination of outer phenyl rings. Polystyrene stars with narrow molecular weight distribution and a precise functionality of 6 were prepared. A general reaction pathway for the preparation of hexa-armed star polystyrene is presented in Scheme 24. Hexa-armed star block copolymers of styrene and isobutylene have been synthesized [43] by the cationic polymerization of isobutylene using a low molecular weight hexa-armed polystyrene as a macroinitiator. Although it was possible to synthesize such block copolymers, the

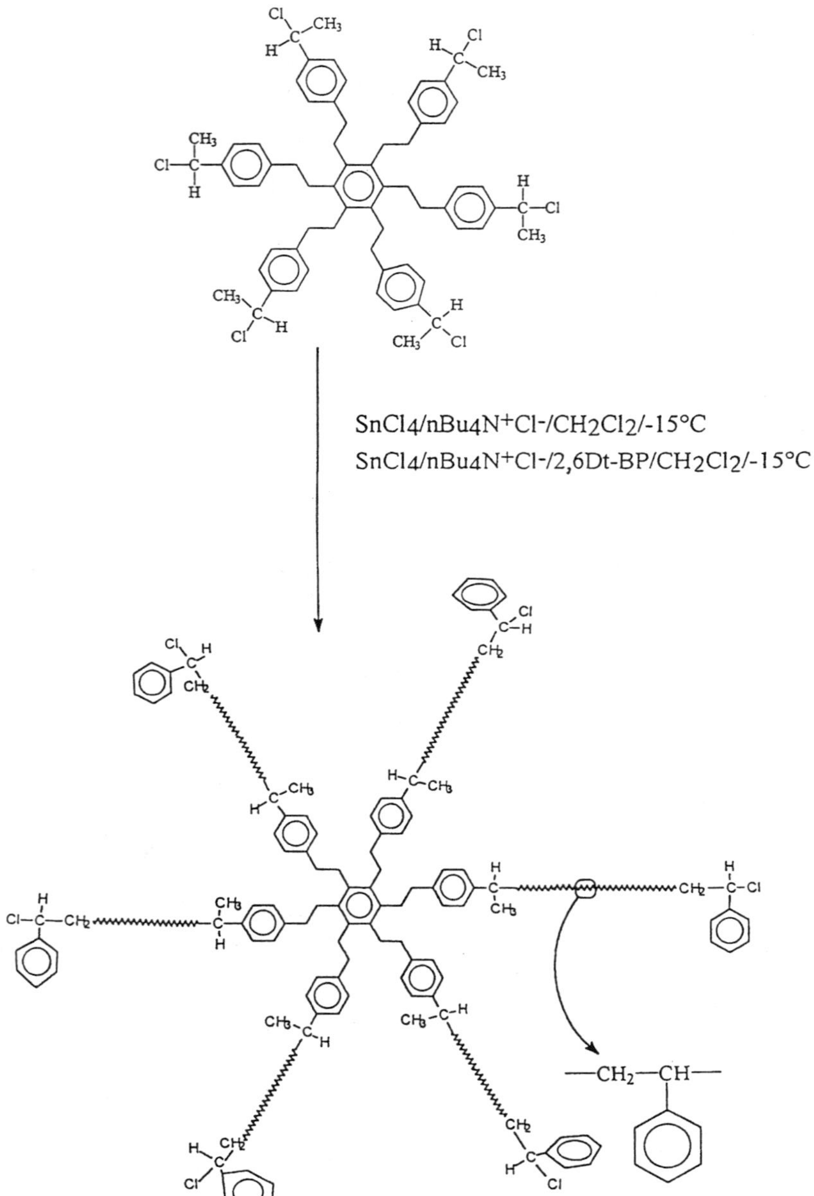

Scheme 24

polymerization was very complex and care must be taken to avoid side reactions.

REFERENCES

1. (a) J. G. Zilliox, P. Rempp, and J. Parrod, *J. Polym. Sci., Part C* **22,**145 (1968); (b) D. J. Worsfold, J. G. Zilliox, and P. Rempp, *Can. J. Chem.* **47,** 3379 (1969).
2. (a) D. N. Schulz and A. F. Halasa, *J. Polym. Sci., Polym. Chem. Ed.* **15,** 2401 (1977); (b) L. J. Fetters, W. W. Graessley, N. Hadijichristidis, A. D. Kiss, D. S. Pearson, and L. B. Younghouse, *Macromolecules* **21,** 1644 (1988).
3. (a) T. Higashimura, S. Aoshima, and M. Sawamoto, *Makromol. Chem., Macromol. Symp.* **13/14,** 457, 513 (1988); (b) T. Higashimura and M. Sawamoto, *Comprehensive Polymer Science.* Pergamon, London, 1989, Vol. 3, Chap. 42.
4. M. Minoda, M. Sawamoto, and T. Higashimura, *Macromolecules* **20,** 2045 (1987).
5. S. Kanaoka, M. Minoda, M. Sawamoto, and T. Higashimura, *J. Polym. Sci., Part A: Polym. Chem.* **28,** 1127 (1990).
6. M. Minoda, M. Sawamoto, and T. Higashimura, *Macromolecules* **23,** 1897 (1990).
7. S. Kanaoka, M. Sawamoto, and T. Higashimura, *Macromolecules* **24,** 5741 (1991).
8. S. Kanaoka, M. Sawamoto, and T. Higashimura, *Macromolecules* **24,** 2309 (Part I) (1991).
9. S. Kanaoka, M. Sawamoto, and T. Higashimura, *Macromolecules* **24,** 5741 (Part 2) (1991).
10. P. Rempp and E. Franta, *Adv. Polym. Sci.* **58,** 1 (1984).
11. M. Sawamoto, T. Enoki, and T. Higashimura, *Polym. Bull. (Berlin)* **16,** 117 (1986).
12. S. Kanaoka, T. Omura, M. Sawamoto, and T. Higashimura, *Macromolecules* **25,** 6407 (Part 3) (1992).
13. S. Kanaoka, M. Sawamoto, and T. Higashimura, *Macromolecules* **26,** 254 (Part 5) (1993).
14. J. P. Kennedy and E. Marechal, *Carbocationic Polymerization.* Wiley, New York, 1982.
15. M. K. Mishra, B. Wang, and J. P. Kennedy, *Polym. Bull. (Berlin)* **17,** 307 (1987).
16. G. Kaszas, J. E. Puskas, C. C. Chen, and J. P. Kennedy, *Polym. Bull. (Berlin)* **20,** 413 (1988).
17. C. C. Chen, G. Kaszas, J. E. Puskas, and J. P. Kennedy, *Polym. Bull. (Berlin)* **22,** 463 (1989).
18. A. Gadkari and J. P. Kennedy, *J. Appl. Polym. Sci., Appl. Polym. Symp.* **44,** 19 (1989).
19. B. Ivan and J. P. Kennedy, *J. Polym. Sci. A, Polym. Chem.* **28,** 89 (1990).
20. H. Shohi, M. Sawamoto, and I. Higashimura, *Makromol Chem.* **13,** 2027 (1992); M. Schappacher and A. Deffieux, *Macromolecules* **25,** 6744 (1992).
21. M. Sawamoto, S. Aoshima, and T. Higashimura, *Makromol. Chem., Macromol.*

Symp. **13/14,** 513 (1988); M. Sawamoto and T. Higashimura, *Makromol. Chem., Macromol. Symp.* **32,** 131 (1990).
22. Y. Kishimoto, S. Aoshima, and T. Higashimura, *Macromolecules* **22,** 3877 (1989).
23. K. Kojima, M. Sawamoto, and T. Higashimura, *Polym. Bull. (Berlin)* **19,** 7 (1988); idem, *Macromolecules* **23,** 948 (1990).
24. T. Higashimura, K. Kojima, and M. Sawamoto, *Makromol. Chem. Suppl.* **5,** 127 (1989).
25. J. P. Kennedy and B. Ivan, *Designed Polymers by Carbocationic Macromolecular Engineering: Theory and Practice,* Carl Hanser, Munich, 1991.
26. J. P. Kennedy and G. C. Richard, *Macromolecules* **26,** 567 (1993).
27. J. P. Kennedy and J. Kurian, *J. Polym. Sci., Polym. Chem.,* **28**(13), 3725 (1990).
28. T. M. Marasalko, I. Majoros, and J. P. Kennedy, *Polym. Bull,* **31,** 665 (1993); T. M. Marsalok, I Majoros, and J. P. Kennedy *Polym. Prepr.* **35,**(2), 504 (1994); T. M. Marsalok, I. Majoros, and J. P. Kennedy, *Macromol. Symp.* **95,** 39 (1995); J. P. Kennedy, I. Majoros, and T. Marsalko, US Patent 5,395,885 (1995).
29. T. M. Marsalok, I. Majoros, and J. P. Kennedy, *Polym. Mater. Sci. Eng.,* **73,** 181 (1995); T. M. Marsalok, I. Majoros, and J. P. Kennedy, *J. Macromol. Sci., Pure Appl. Chem.* **A34,** 775 (1997).
30. T. M. Marsalok, I. Majoros, and J. P. Kennedy, *Polym. Prepr.* **37**(1), 581 (1996); J. P. Kennedy, T. M. Marsalok, and I. Majoros, *Macromol. Symp.,* **107,** 319 (1996).
31. R. F. Storey and K. A. Shoemake, *Polym. Prep. Am. Chem. Soc.,* **35**(2), 578 (1994).
32. L. Wang, S. T. McKenna, and R. Faust, *Macromolecules* **28,** 4681 (1995).
33. J. P. Kennedy, N. Omura, and A. Lubnin, U.S. Patent 5,663,245 (1997).
34. N. Omura and J. P. Kennedy, *Macromolecules* **30,** 3204 (1997); N. Omura, A. V. Lubnin, and J. P. Kennedy, *ACS Symp. Ser.* **665,** 178 (1997).
35. S. Asthana, I. Majoros, and J. P. Kennedy, *Polym. Mater. Sci. Eng.* **77,** 187 (1997).
36. J. S. Shim, S. Asthana, N. Omura, and J. P. Kennedy, *Polym. Prepr.* (Am. Chem. Soc., Div. Polym. Chem.), **39**(1) 196 (1998).
37. S. Jacob, I. Majoros, and J. P. Kennedy, *Polym. Mater. Sci. Eng.* **76,** 298 (1997); S. Jacob, I. Majoros, and J. P. Kennedy, *Macromolecules* **29,** 8631 (1996).
38. S. Jacob, I. Majoros, and J. P. Kennedy, *Polym. Bull.* **40,** 127 (1998).
39. I. Majoros, T. M. Marsalko, and J. P. Kennedy, *Polym. Bull.* **38,** 15 (1997).
40. S. Jacob, I. Majoros, and J. P. Kennedy, *Polym. Prep.* (Am. Chem. Soc., Div. Polym. Chem.), **39**(1), 198 (1998); S. Jacob, I. Majoros, and J. P. Kennedy, *Polym. Mater. Sci. Eng.* **77,** 185 (1997).
41. E. Cloutet, J.-L. Fillaut, D. Astruc, and Y. Gnanou, in *Macromolecular Engineering: Recent Advances,* M. K. Mishra, O. Nuyken, S. Kobayashi, Y. Yagci, and B. Sar, eds., Plenum Press, New York, 1995, p. 47.
42. J. Chang, H. Ji, H. Han, S. Rhee, S. Cheong, and M. Yoon, *Macromolecules* **27,** 1376 (1994).
43. M. K. Mishra, *Macromol. Symp.* **107,** 243 (1996).

5

Star and Hyperbranched Polymers by Transition Metal Catalysis

Jun-ichi Kadokawa
Yamagata University, Yonezawa, Yamagata, Japan

Shiro Kobayashi
Kyoto University, Kyoto, Japan

I. INTRODUCTION

Star and hyperbranched (dendritic) polymers have attracted increasing attention in organic, supermolecular, and polymer chemistry as well as coordination chemistry because of their specific structures and characteristics [1–5]. For the preparation of these highly branched polymers, two kinds of synthetic methods have been developed: a one-step polymerization and a stepwise method. By the one-step method, various star and hyperbranched polymers with many structural and functional group variations have been prepared [6–10]. On the other hand, the stepwise method is quite useful, especially for the synthesis of dendritic polymers. Both divergent and convergent synthetic approaches have been employed [2,11].

Since the discovery of the Ziegler–Natta catalyst in olefin polymerizations, a number of polymerization reaction by transition metal catalysis have been developed [12–16]. The transition metal catalysis is not limited to the olefin polymerization, and extended to coordination and insertion polymerizations of various monomers [17,18]. Therefore, the reaction by transition metal catalysis is one of the most useful approaches in the polymer synthesis field.

This chapter deals with synthesis of star and hyperbranched polymers by transition metal catalysis. Highly branched polymers are also included. The topics to be discussed are as follows:

1. Phosphorus-containing star polymers by Ni-catalyzed ring-opening polymerization of phosphorus monomers.
2. One-step synthesis of dendritic polymers by Pd-catalyzed ring-opening polymerization of cyclic carbamates.
3. Soluble, highly branched rigid-rod polymers with long side chains by transition metal catalysis.

II. SYNTHESIS OF STAR POLYMER CONTAINING PHOSPHORUS ATOMS IN THE MAIN CHAIN BY NI(II) BROMIDE CATALYSIS

Star polymers having three, four, and six arms were prepared by the ring-opening polymerization of six-membered cyclic phosphonite (**1**) initiated with a catalyst system of $NiBr_2$ and polyhalobenzenes [19,20]. Various polyhalobenzenes (1,3,5-trihalobenzenes (**2a,b**), 1,2,4,5-tetrahalobenzenes (**3a–c**), and hexahalobenzenes (**4a–c**)) were employed as multifunctional initiators to give star-shaped polymers having three, four, and six arms, respectively (Scheme 1).

Polymerization of **1** using these initiator systems, followed by the reaction of the terminal group with pyridine gave star-shaped polymers **5–7** having a pyridinium group at chain ends (Table 1). The degree of polymerization (DP) was calculated from the integrated area of peak due to OCH_2 and peak due to *o*-protons of the pyridinium group in the ^1H NMR spectrum of the polymer. The DP values were relatively close to the values of the monomer/initiator feed ratio. The M_w/M_n values determined by gel permeation chromatography (GPC) were small. These results indicate that the polymerization initiated with these systems is of a living nature.

In relation to the preceding catalyst systems using the polyhalobenzenes, dihalobenzenes **8a–c** were used as bifunctional initiators for linear telechelic polymers (Scheme 2).

The polymerization of **1** initiated with these dihalobenzenes and $NiBr_2$ systems was carried out at 100 °C in CH_3CN followed by the reaction of the terminal group with pyridine to produce polymer **9** (Table 1). The DP determined by the ^1H NMR spectrum was relatively close to the monomer/initiator feed ratio. The M_w/M_n values obtained by GPC were small. These data indicate that the polymerizations also proceeded in a living manner.

In order to confirm that the polymerization was initiated from all carbons linked to the halogen atoms of polyhalobenzenes, the polymerizations initiated

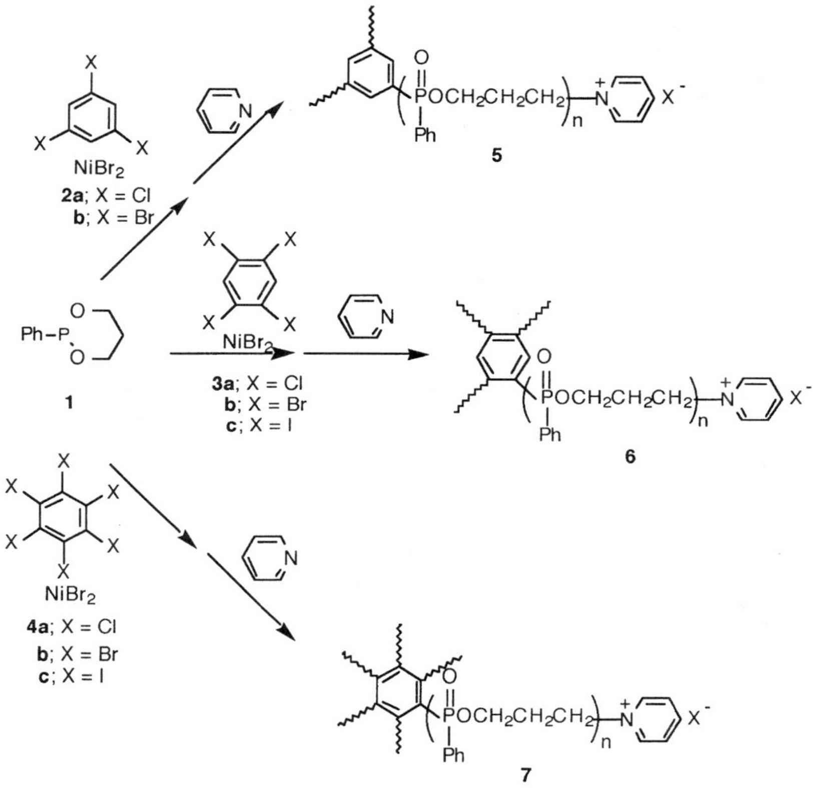

Scheme 1

with di-, tetra-, and hexaiodobenzenes (**8c, 3c,** and **4c,** respectively) and $NiBr_2$ systems were followed by ^{13}C NMR spectroscopy, in which the characteristic peaks due to C–I carbons of these polyiodobenzenes are observed at around δ 90–100. These values are very different from the chemical shifts of the peaks due to the aromatic carbons bearing phosphorus atoms of the monomer and the polymer (δ 128–140). When the monomer **1** was consumed, the peaks completely disappeared, implying that all initiating sites were consumed by initiation. In the ^{13}C NMR spectra of the reaction mixture initiated with the **3c**/$NiBr_2$ system, for example, peak b due to C–I carbons in Figure 1a (0 hours) completely disappears after heating at 100°C for 72 hours (Figure 1b).

Star polymers are known to be spherical in solution [6,7] Therefore, the

Table 1 Polymerization of **1** Initiated with Polyhalobenzenes/$NiBr_2$

Polyhalobenzene 2–4, 8	$\frac{[1]_0}{[polyhalobenzene]_0}$	$\frac{[1]_0}{[NiBr_2]_0}$	*Yield (%)*	M_n	*DP*	M_w/M_n
2a	18.7	56.4	81	4540	24.0	1.29
2b	33.2	57.6	86	8510	45.0	1.26
3a	28.4	76.8	83	7130	38.0	1.32
3b	43.3	80.0	92	9490	50.0	1.28
3c	43.4	76.1	84	9750	50.4	1.19
4a	49.0	112.7	81	10200	54.6	1.27
4b	39.4	109.3	100	9290	48.0	1.25
4c	59.5	36.4	80	9340	46.8	1.12
8a	10.5	36.4	96	2820	14.6	1.23
8b	19.9	36.8	95	4420	23.0	1.23
8c	19.3	38.4	97	4880	25.0	1.19

DP, degree of polymerization.

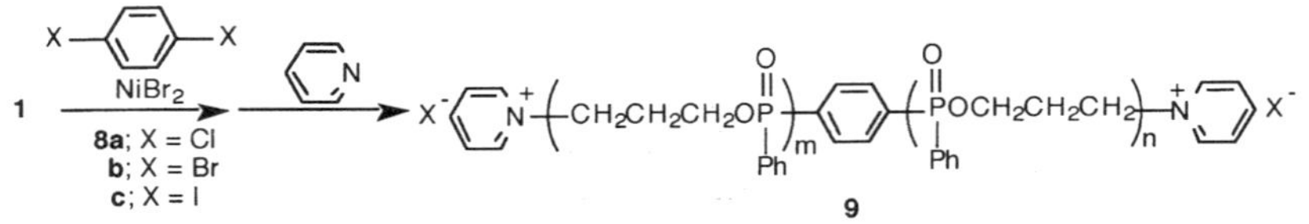

Scheme 2

molecular weight of a star polymer as shown by GPC is smaller than that of a linear polymer. Table 2 shows the GPC results of the star polymers **5–7** and the linear telechelic polymer **9.** The molecular weight of linear-type polymer **9** by GPC is almost the same as that obtained by ^{1}H NMR spectroscopy, whereas values of **5–7** were much smaller than those obtained by ^{1}H NMR spectroscopy. From these results, the spherical structure of **5–7** is definitely supported.

The background of the preceding polyhalobenzenes/$NiBr_2$ systems is based on the catalyst system consisting of monohalobenzenes and $NiBr_2$, which implied the polymerization of **1** to give a linear poly(phosphinate) **10** in a living manner (Scheme 3) [19,20]. The reaction was the first example of a living ring-opening polymerization of cyclic phosphorus(III) compounds.

In the polymerization mechanism (Scheme 4), the formation of a phos-

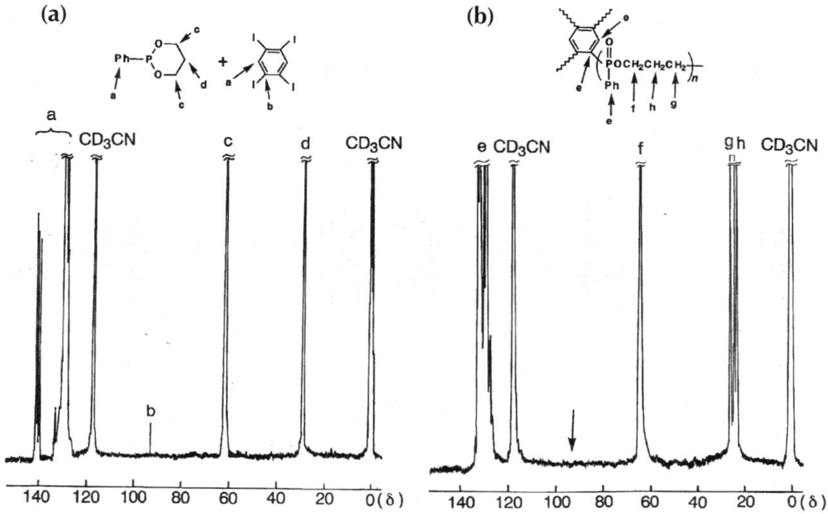

Figure 1 ^{13}C NMR spectra of the reaction mixture initiated with tetraiodobenzene in CD_3CN: (a) 0 hours (b) after heating at 100°C for 72 hours.

Table 2 Comparison of Molecular Weight of Star-Shaped and Linear Polymers Determined by ^1H NMR and Gel Permeation Chromatography

Polymer	$M_n(^1H\ NMR)$	$M_n(GPC)$	M_w/M_n
5	8510	3410	1.26
6	7130	4230	1.32
7	10200	4550	1.27
9	4420	4030	1.23

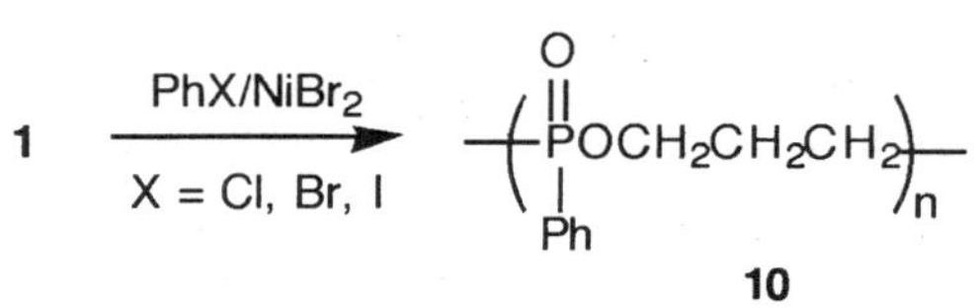

Scheme 3

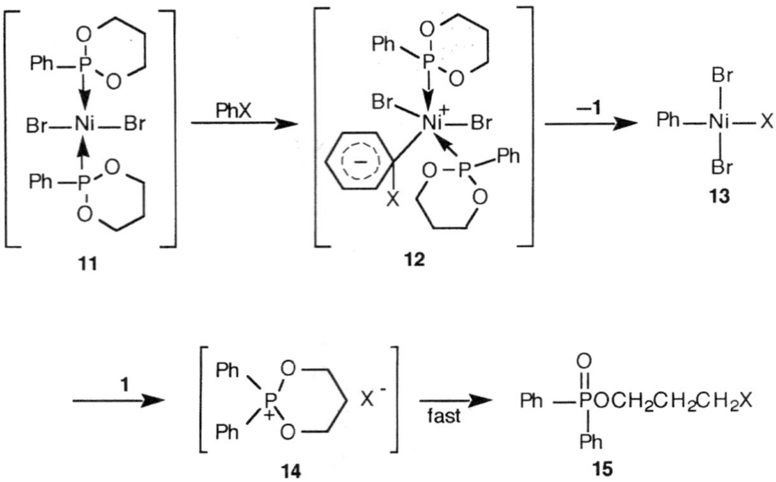

Scheme 4

phonium intermediate **14** is a key step in the present catalyst system and converts into a initiation species **15,** which implies the polymerization. The reaction of **1** and $NiBr_2$ forms a coordinated complex **11.** To this activated complex, PhX adds via an oxidative addition probably involving intermediate **12** to form the product **13** with cleaving of the phenyl–halogen bond. Then, **13** is able to give **14** readily.

In relation to the formation of intermediate **14** from **1** and halobenzene, the reaction of phosphorus(III) compounds with PhX in the presence of $NiBr_2$ as catalyst was reported to form phenyl-substituted phosphonium salts [21,22].

III. SYNTHESIS OF HYPERBRANCHED DENDRITIC POLYAMINE BY Pd-CATALYZED RING-OPENING POLYMERIZATION OF CYCLIC CARBAMATE

In order to prepare a hyperbranched dendritic polymer by chain polymerization, a new concept of polymerization, termed *multibranching polymerization* (MBP) was proposed, which provides a dendritic polymer involving the initiator as the core [23] The most characteristic aspect of MBP is *multiplication* of the propagating ends at every step of propagation according to Scheme 5:

On the basis of the preceding concept of MBP, Pd-catalyzed ring-opening polymerization of 5,5-dimethoxy-6-ethenylperhydro-1,3-oxazin-2-one

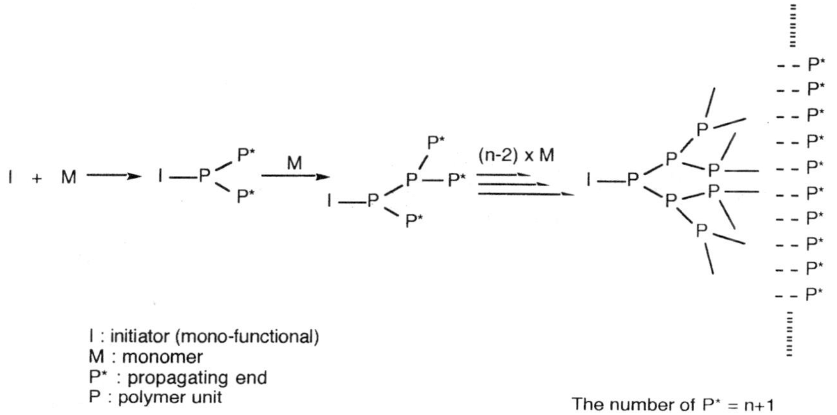

Scheme 5

(**16**) was carried out (Scheme 6) [23]. The key point of the monomer design is that **16** has an amidic proton that is a dormant propagating end. By using $Pd_2(dba)_3 \cdot CHCl_3$-2dppe as a catalyst, **16** was polymerized at room temperature in tetrahydrofuran (THF) solution to produce a dendritic polymer **17.** An idea for MBP of **16** was suggested by the exploration of Pd-catalyzed ring-opening polymerization of cyclic monomers involving π-allyl complex intermediates [24].

Table 3 summarizes the results of the polymerization employing benzylamine as an initiator. The control reaction without any initiators proceeded slowly to give an insoluble product, whose infrared spectrum was identical to that of the soluble polymer produced with the initiator. The insolubility can probably be ascribed to a physical character of this polymer. The soluble polymers produced with the initiator also became insoluble after they were stored for some days even in a nitrogen atmosphere.

The polymer structure was identified by the ^{13}C and ^{1}H NMR spectra, as compared with model compounds such as neopentylamine, *N*-allylneopentylamine, and *N,N*-diallylneopentylamine. These spectra indicated that the polymer has incomplete branching consisting of not only primary and tertiary but also secondary amino moieties; the former are a polymer end and a branching junction, respectively, and the latter is a nonbranching junction.

The following mechanism is proposed for the polymerization (Scheme 7); π-allylpalladium complex **18** is the key intermediate [25]. The initiator of a primary amine attacks the electrophilic site of **18** to produce diamine **19,** releasing CO_2 and regenerating a Pd(0) complex. Since both of the two amino groups of

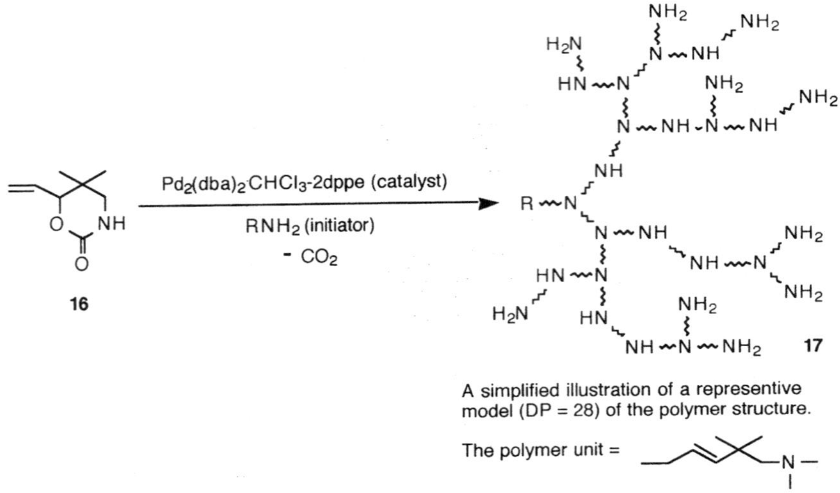

Scheme 6

Table 3 Pd-Catalyzed Ring-Opening Polymerization of **16**

Catalyst (mol%)	*Initiator (mol%)*	*Time (day)*	*Yield (%)*	M_n	*DP* VPO	*DP* NMR	*Degree of branching (%)*
0.5	10.1	2	60	1860	15.8	17.8	44
0.5	5.0	2	85	3190	27.8	28.6	44
0.5	2.5	2	Quant	5330	47.0	46.6	52
1.5	0	3	Quant	Insoluble			

19 have the ability to react with **18,** two kinds of triamines, **20a** and **20b,** are possible. Repetition of this reaction of **18** with primary and secondary amino groups, whose relative reactivity reflects the degree of branching, gives rise to production of hyperbranched dendritic polyamine incorporating the initiator, as core.

MBP of cyclic carbamate monomer, 5-methyleneperhydro-1,3-oxadine-2-one (**21**), using a polymeric initiator was also reported (Scheme 8). Poly(methyloxazoline) (**22**) having an amino group at a polymer end (DP = 20, functionality of amino group = 92%) was employed as the polymeric initiator [26]. The polymerization was carried out in the presence of $Pd_2(dba)_3 \cdot CHCl_3/8Ph_3P$

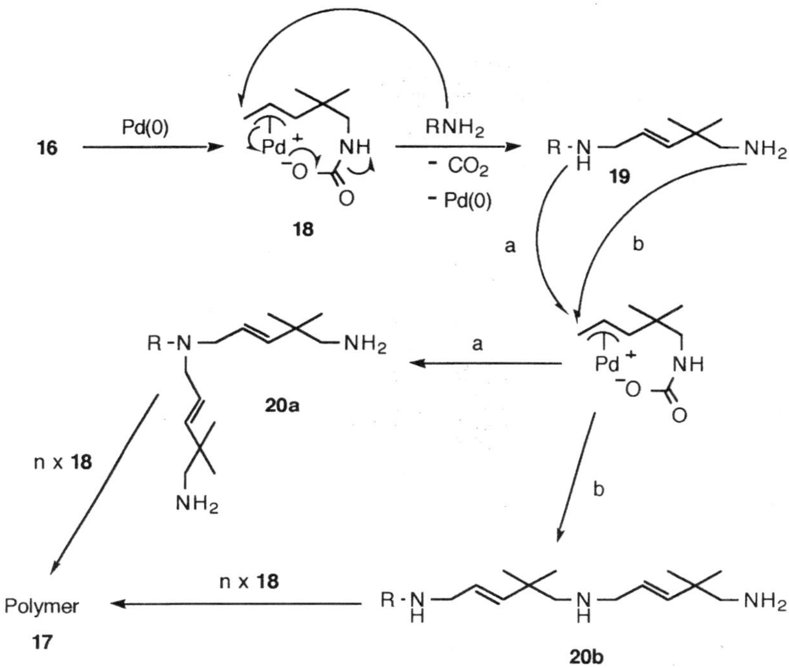

Scheme 7

(0.5 mol %) in CH_2Cl_2 solvent at room temperature. After the complete consumption of **21** was confirmed by IR analyses (48 h), an excess amount of *n*-butyl isocyanate was added to the reaction mixture to convert primary and secondary amino groups at propagation ends into urea groups. Then the mixture was poured into a large amount of diethyl ether to give a block polymer. The specific polymer structure **23** was confirmed, in which the hyperbranched dendritic polyamine is introduced from the poly(methyloxazoline) end of the linear structure.

IV. SYNTHESIS OF SOLUBLE, HIGHLY BRANCHED POLY(p-PHENYLENE)S HAVING LONG SIDE CHAINS BY TRANSITION METAL CATALYSIS

Poly(*p*-phenylene) (PPP) has been an important target for synthesis in polymer science for many years [27,28]. This interest is mainly due to the expectation

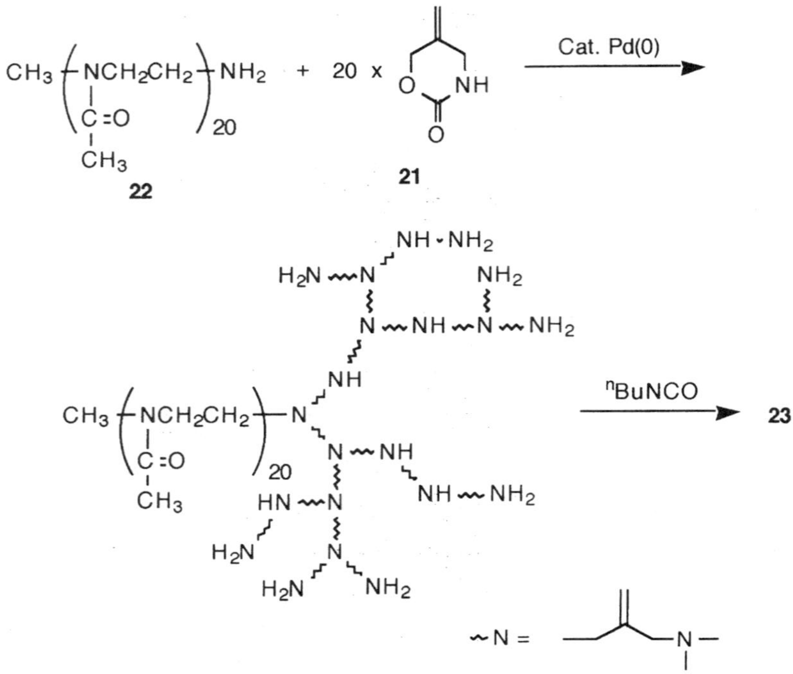

Scheme 8

of developing materials with good thermal and oxidative stability as well as electrical conductivity in the oxidized and reduced states. Several attempts, especially, a direct route from benzene, have been reported [29–31]. Only low molecular weight materials of irregular structure, however, have been yielded because of the insolubility of PPP. Although indirect methods using precursor polymers for the preparation of PPP have been investigated, the PPPs synthesized also have many structural defects [32–35]. Therefore, no method for the synthesis of structurally homogeneous and high molecular weight PPP is available. In order to produce a soluble PPP, aromatic monomers having long side chains were coupled using a transition-metal catalysis.

The first synthesis of soluble PPPs having long alkyl side chains was carried out with a Ni-catalyzed aryl/aryl coupling reaction of 1,4-dibromobenzenes (**24**) bearing long alkyl chains in the 2- and 5-positions (Scheme 9) [36].

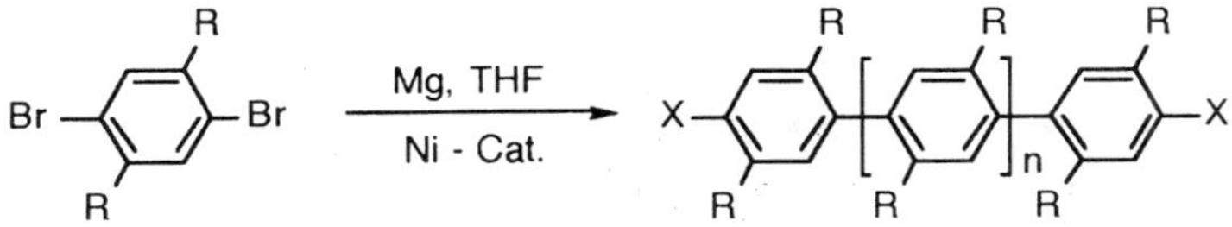

24

25

R = n-hexyl, n-octyl
X = H or Br

Scheme 9

This reaction is based on the synthesis of general, nonsubstituted PPP from 1,4-dibromobenzene in the presence of stoichiometric amounts of Mg under mild Ni-catalysis [31]. Ni-catalyzed polycondensation of **24** was carried out using Ni(bipy)Cl_2 (0.5–1.5 mol %, bipy = 2,2′-bipyridyl) in THF for 48 hours under reflux condition. The polymeric product was isolated as an acetone insoluble part in 40–60 wt % yields. As expected, the product polymer was soluble in various organic solvents (chloroform, methylene chloride, toluene). The structures of polymer **25** were established by the ^{1}H and ^{13}C NMR spectroscopies. In the HPLC trace of polymer an average DP of the order of 8 can be estimated, with largest species in the sample having a DP of about 20. Various Ni-complexes added as a catalyst were tested, Ni(bipy)Cl_2, Ni$(PPh_3)_2Cl_2$, Ni$(acac)_2$, and Ni(dppp)Cl_2 (acac = acetylacetonate, dppp = 1,3-bisdiphenylphosphinopropane). From the average DP data, a complex, Ni$(PPh_3)_2Cl_2$, was the most efficient catalyst for this process giving PPP with an average DP of 13.

Improved synthesis of PPP with significantly greater DP was reported via Pd-catalyzed coupling of 4-bromo-2,5-disubstituted benzeneboronic acid (**26**) [37,38] according to the Suzuki and Miller methods (Scheme 10) [39,40].

In the polymerization, monomers **26** having various-length alkyl chains were treated with 0.5 equivalent amounts of Pd$(PPh_3)_4$ in the heterogeneous system water/Na_2CO_3/toluene or benzene under reflux for 2 days. The structure of polymers **25** was determined as the proposed all-para-linkage of the repeating units by their ^{13}C NMR spectra. The absence of end group signals from the ^{1}H and ^{13}C NMR spectra of polymers **25** indicated the higher molecular weights of the produced polymers. The DP values of **25** were determined, using vapor pressure osmometry, to be around 30. As an extension of the preceding polymerization, nonsubstituted and *n*-hexyl- or *n*-dodecyl-substituted monomers were

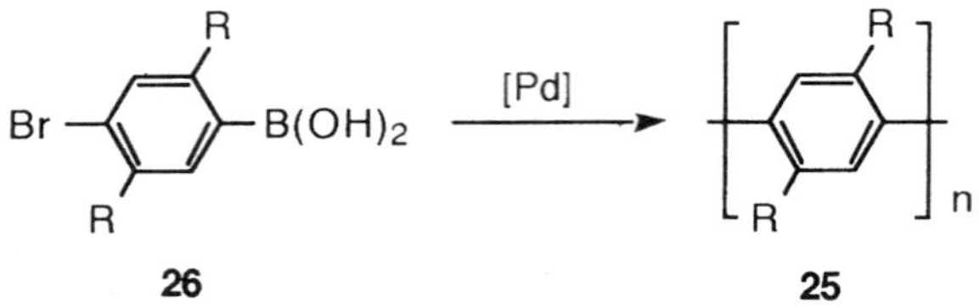

R = $(CH_2)_3CH_3$, $(CH_2)_5CH_3$, $(CH_2)_2C(CH_3)_3$, $(CH_2)_6CH_3$, $(CH_2)_7CH_3$, $(CH_2)_8CH_3$, $(CH_2)_{11}CH_3$, $(CH_2)_{15}CH_3$

Scheme 10

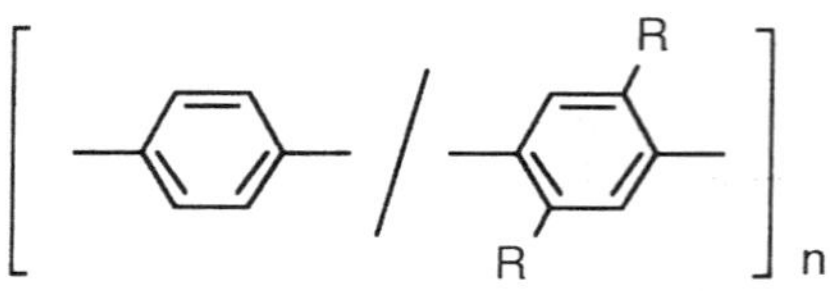

27

R = n-hexyl, n-dodecyl

Scheme 11

mixed in varying feed ratios and copolymerized under standard conditions to yield copolymers **27** (Scheme 11) [38].

The construction of PPP derivatives with well-defined blocks of nonsubstituted units of variable length linked to each other by only one alkyl-chain-substituted unit was carried out through copolymerization of functionalized telomers [38]. Two types of telomers, **31** and **32** ($m = 1, 2$), were synthesized according to Scheme 12.

Then, the Pd-catalyzed coupling reaction of diboronic acids **29** or **32** with dibromo compounds **28, 30,** or **31** gave telomer copolymer **33** with defined blocks of one to six nonsubstituted phenylene units (Scheme 13). A vapor pressure osmometry measurement of **33** ($y = 2$) gave $M_n = 21{,}000$. Thus, this polymer has more than 100 phenylene rings linked to each other exclusively in the para position.

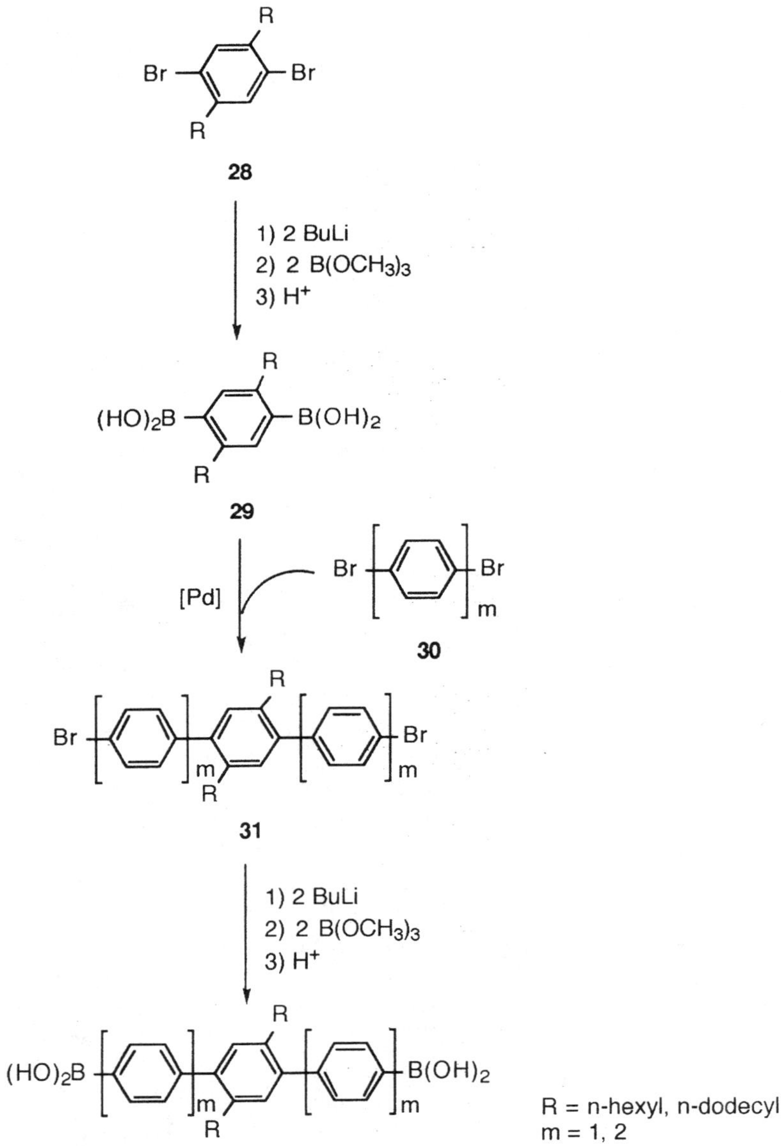
Br
Br
R
R
28
1) 2 BuLi
2) 2 $B(OCH_3)_3$
3) H^+
$(HO)_2B$
$B(OH)_2$
29
[Pd]
Br
Br
m
30
31
32
R = n-hexyl, n-dodecyl
m = 1, 2

Scheme 12

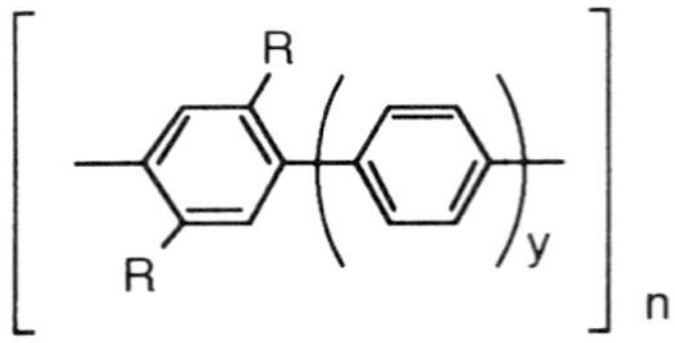

33

y = 1, 2, ... 6

Scheme 13

The synthesis of PPP derivatives with long alkyl chains and polar or ionic side groups was reported using a Pd-catalyzed coupling reaction (Scheme 14) [41]:

First, precursor PPP **35** having butoxymethyl side groups were prepared by coupling of monomers **34** and **29** in the presence of 0.5 mol % of $Pd(PPh_3)_4$. Polymer **35** with DPs of about 65 were obtained. Then, conversion of butoxymethyl groups in polymer **35** was carried out. In the first stage of the subsequent conversions, polymer **35** was transformed into the bromomethyl-substituted PPP derivative **36.** Then, polymer **36** was transformed into polymers **37** and **38.** Benzonitrile functionalities were introduced by refluxing polymer **36** with sodium *p*-cyanophenolate to lead to polymer **37.** PPP derivative **38** was synthesized by refluxing polymer **36** with sodium acetate. From the NMR analyses of the product polymer, the bromomethyl pendant groups of polymer **36** quantitatively reacted with the phenolate and the acetate. No absorptions indicating side reactions during the transformation processes were detected.

Furthermore, conversion of polymer **36** was extended to the introduction of carboxylic acid side groups to produce PPP derivative **40** (Scheme 15) [42]. Polymer **36** was reacted with *p*-hydroxybenzoic acid ethyl ester in the presence of sodium *tert*-butanolate and potassium iodide, followed by an alkaline hydrolysis of ethyl ester groups of product polymer **39** to give PPP derivative **40.** From the NMR analysis of polymer **39,** it was evident that almost 100% of the bromomethylene functionalities reacted to lead to the ester groups. For the alkaline hydrolysis of ester groups, when a homogeneous solution of polymer **39** in toluene was treated with 10 equivalents of potassium *tert*-butanolate and two equivalents of water, a nearly 100% ester cleavage was achieved.

Pd-catalyzed synthesis of PPPs **41** and **42** with 6-phenoxyhexyl side groups, followed by their three-step conversion sequence, gave the other PPP derivatives, **43** and **44,** having carboxylic acid groups attached to 6-phenoxyhexyl side chains (Scheme 16) [43].

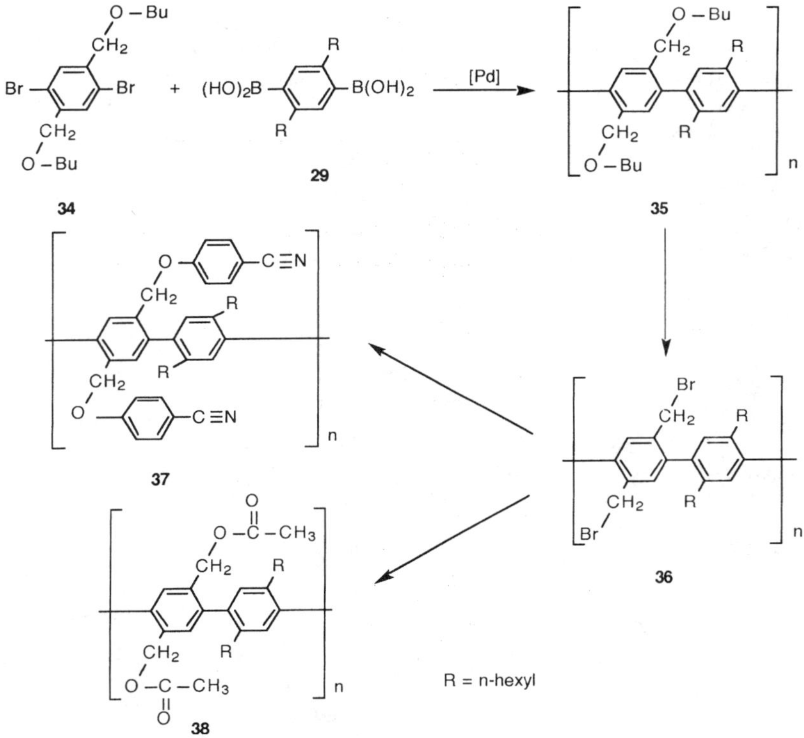

Scheme 14

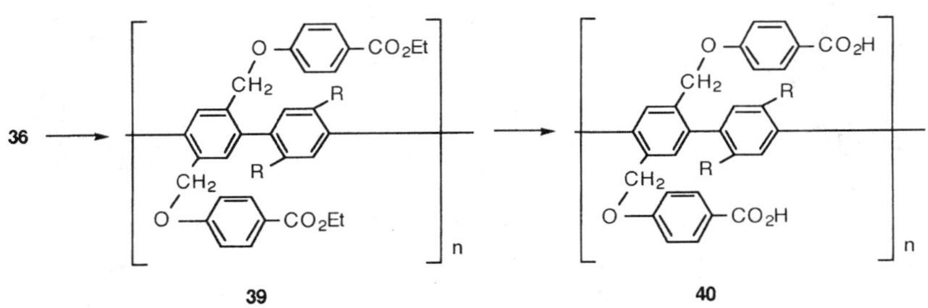

Scheme 15

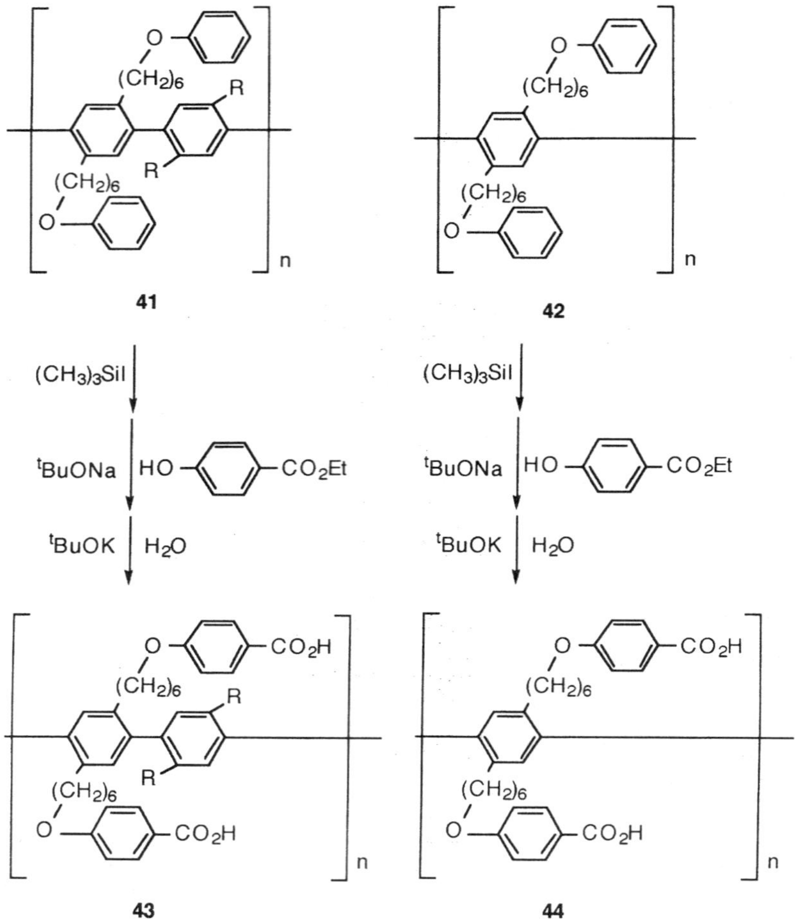

Scheme 16

Some highly branched PPP derivatives substituted with flexible long side chains have been used for adsorption studies on gold and copper [44,45]. The state of the monolayer is influenced not only by the rigidity of the polymer main chain, but also by factors such as the substrate and functional groups in the side chains. For some systems, uniform monolayers are obtained, in which the substrate is covered completely.

Pd-catalyzed polycondensation was extended to the synthesis of highly branched rigid-rod polyimide by incorporation of phenyl-substituted pyromellitic diimide units into the PPP main chain. The synthesis of the polyimide **46** obtained by Pd-catalyzed polycondensation of monomer **45** with bisboronic acid **29** was reported (Scheme 17) [46]. The polycondensation of **45** with **29** was carried out in the presence of catalytic amounts of $Pd(PPh_3)_4$ in the heterogeneous system water/Na_2CO_3/toluene. The product polyimide had good solubility due to the dodecyl side chains in the terphenylene units and two additional lateral phenyl substituents at each pyromellitic diimide unit. The polyimide structure of **46** was determined by the ^{1}H and ^{13}C NMR spectra. The M_n value determined by membrane osmometry was 72,300, corresponding to DP = 45.

This rigid-rod polyimide was also adsorbed from organic solutions on gold and copper [47]. The thickness of the layer was below 25 Å, and the substrates were covered to a high degree with the organic molecules. Coordination of the adsorbed molecules via carbonyl groups of the imide moiety onto metal atoms was rare; in most cases, electrostatic or van der Waals interactions are suggested to support the adsorption process.

Synthesis of rigid-rod polyimides **49** with dye-containing side groups via a precursor route was reported (Scheme 18) [48].

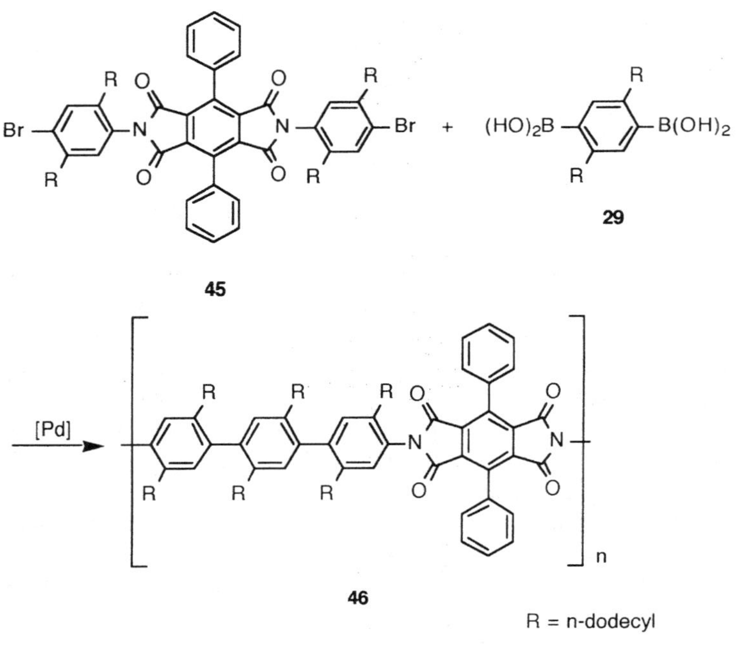

Scheme 17

First, rigid-rod precursor polyimides **48** were prepared via Pd-catalyzed terpolycondensation of monomers **45, 47,** and **29.** From these precursors, nucleophilic substitution of the phenoxy side groups with various hydroxy-azobenzene derivatives containing dye moieties was carried out to give rigid-rod polyimides **49** with dye-containing side groups. An almost quantitative substitution was achieved in the sterically less demanding dyes, whereas low conversion was observed for the bulky dyes. Chain degradation and side reactions did not occur during the substitution reaction.

Synthesis of soluble, highly branched ladder polymers with long side chains was achieved via bridging of functionalized PPP precursors (Scheme 19) [49].

The incorporation of substructures into ladder polymers provides molecular properties with well-defined configuration and conformation. The difficulties, however, in synthesizing ladder-type polymers have been reported [50,51] For the synthesis of soluble ladder-type PPP, first, PPP precursor **51** with functional ketone groups was prepared by Pd-catalyzed coupling reaction of dihexyl-substituted aromatic boronic acid **29** with didecyl-substituted aryl bromide **50** in 60–80% yields. The M_n values determined by GPC were 6100–9200 [52].

In order to synthesize the desired ladder-type polymer involving a ring-closure reaction, PPP precursor **51** was converted into a polyalcohol **52.** A complete conversion of the carbonyl groups in **52** occurred. The ring closure of **52** to the PPP ladder structure proceeded smoothly by the reaction of **52** with $BF_3{\cdot}OEt_2$. NMR analysis of the product polymer indicated that the ring-closure reaction occurred quantitatively to give ladder-type PPP **53.** The GPC profiles of **51, 52,** and **53** showed that there is no breakdown of polymeric chains during the reactions.

During the preceding synthetic route, however, the alkyl side chains were limited to hexyl and decyl groups, because of the lower solubility of PPP precursor **51.** Therefore, an alternative route to the ladder-type PPP has been worked out (Scheme 20) [53]. When the Pd-catalyzed coupling reaction of **29** with 2,5-dibromoterephtalic dialdehyde (**54**) was tested, it formed **55,** which was fully soluble. The polymer reaction of **55** with aromatic Grignard organic compounds took place giving rise to polyalcohol **52,** in which a great variation in the choice of the substituent R is possible (R = Ph, 4-$MeOC_6H_4$, 4-$Me_2NC_6H_4$). Then, the ring closure to the ladder-type PPP **53** proceeded completely.

An attempt of dehydrogenation of **53** was made to generate fully unsaturated ladder-type polymer **56** with 1,4-benzoquinone-bismethide structural units (Scheme 21) [54]. The synthesis of quinoid ladder-type polymer **56** pos-

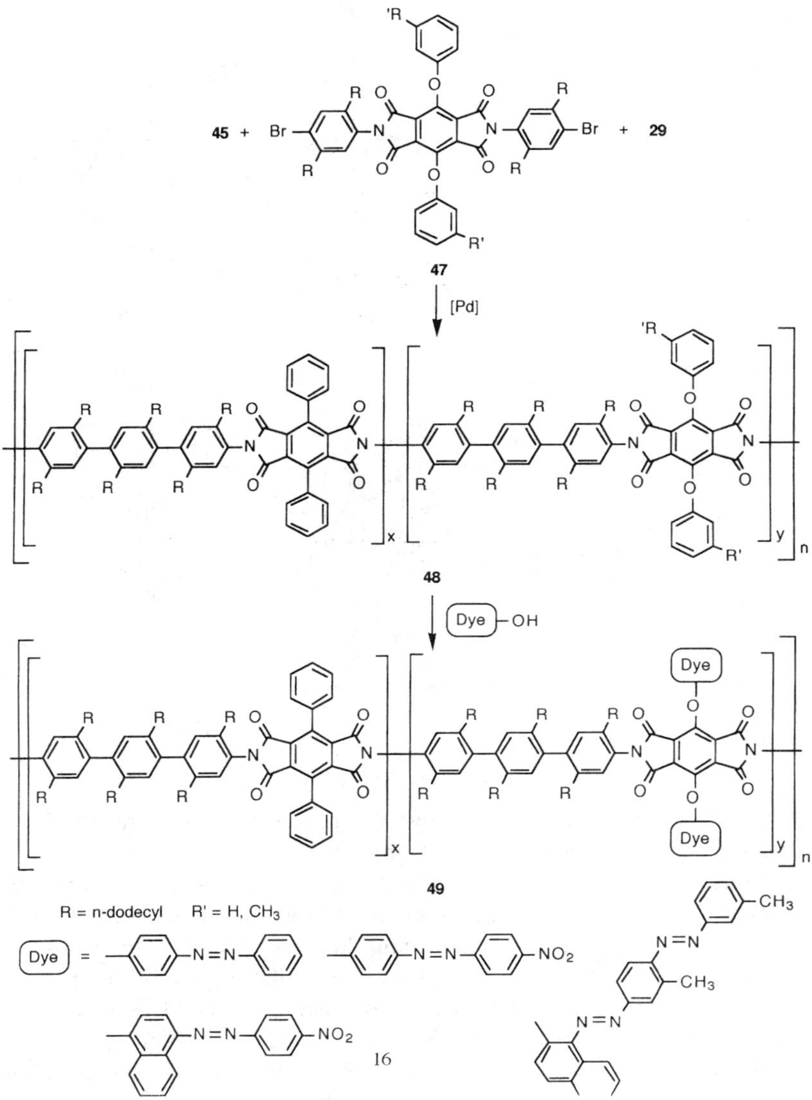

Scheme 18

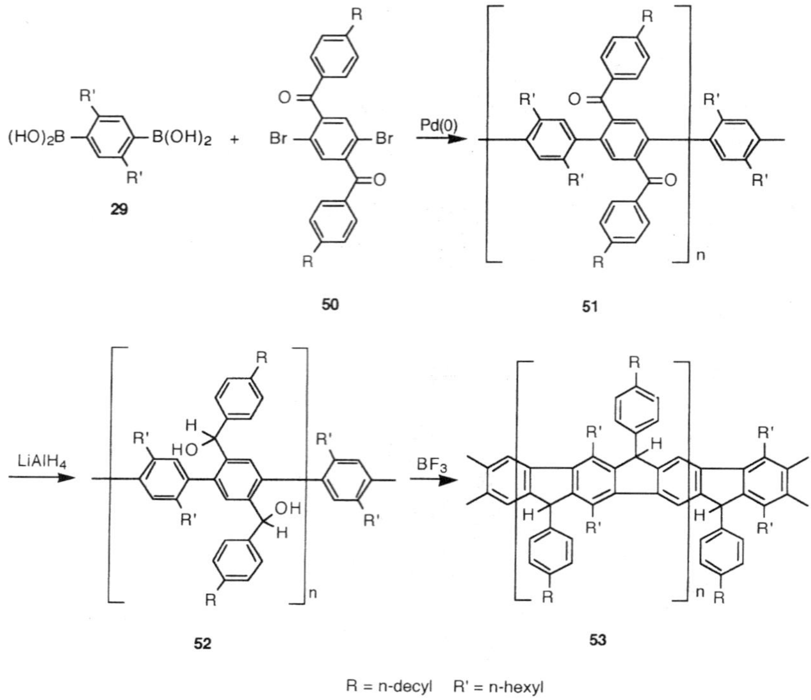

Scheme 19

sessing a degenerate ground state involves problems concerning the high chemical instability of such species [55,56].

To avoid these problems, the nonconjugated ladder-type PPP **58** with alternating 1,4- and 1,3-phenylene structural units was taken to subject to subsequent dehydrogenation experimentation (Scheme 22) [54]. Pd-catalyzed coupling reaction of **29** with alkyl-substituted 1,3-dibenzoyl-4,6-dibromobenzene derivative (**57**), instead of **50,** was carried out, followed by reduction and ring closure to give ladder-type PPP **58.** The conversion of **58** into a polymer with 1,4-benzoquinone-bismethide subunits **59** took place via dehydrogenation with DDQ. The ^{1}H NMR spectrum of the product polymer indicated 90% conversion of **58.** The electronic absorption spectrum of **59** (Figure 2) shows a new broad absorption in the vis-region (λ_{max} = 605 nm) assignable to the 1,4-benzoquinone-bismethide units generated. The maximum of the longest wavelength absorption (λ_{max} = 581 nm) of a model compound **60** lies in the same range as that of **59** (Scheme 23).

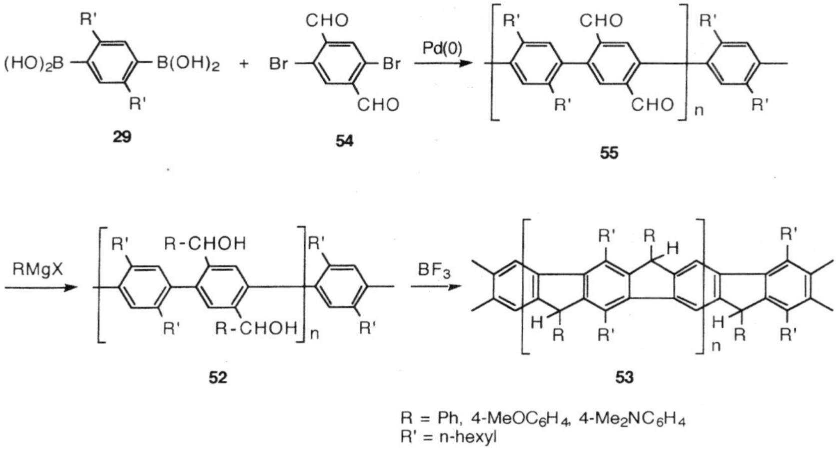

Scheme 20

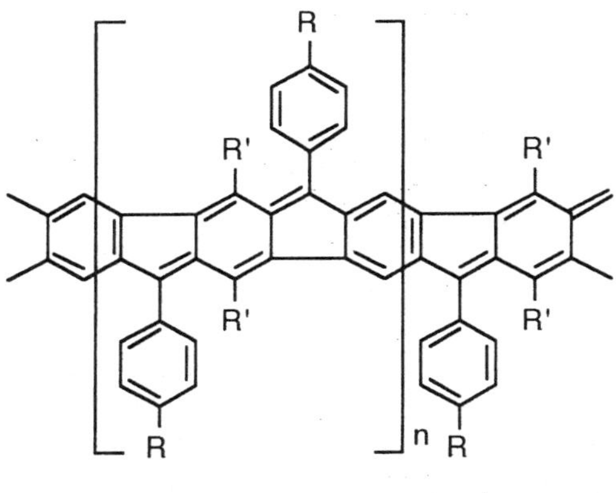

56

Scheme 21

A few investigations regarding the oxidative and reductive doping of PPP have been published [57,58]. The oxidative doping studies of the ladder-type PPP **53** via chemical and electrochemical methods were also reported (Scheme 24) [59]. The ladder-type PPP **53** was treated with antimony pentachloride or iron trichloride to give a deeply colored oxidation product **61.** The solutions of the reaction mixtures were ESR-active, supporting the generation

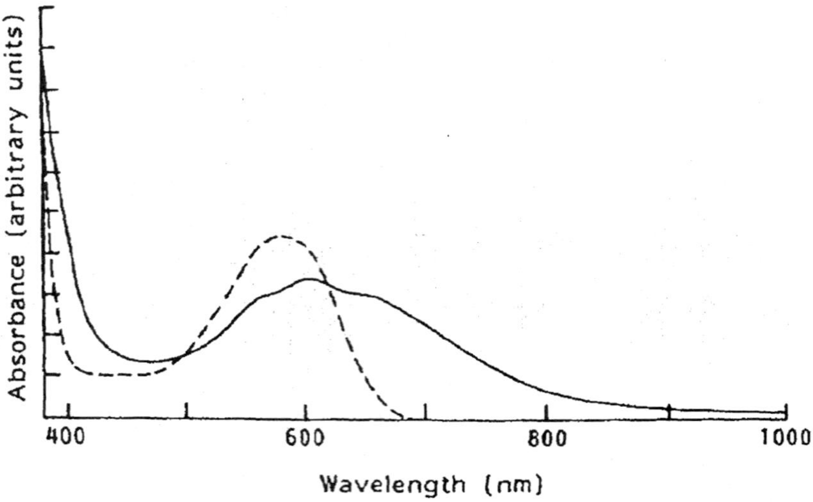

Figure 2 Vis absorption spectra of **59** (—) and **60** (····).

of radical cation species **61,** according to the electronic theory of doping [60,61]. Further oxidation of **61** gave dicationic state species **62** via a new level of oxidative doping.

When cyclovoltammetric oxidations of **53** were demonstrated in solution and in the solid state, a reversible electrochemical oxidation of **53** took place. The results of the cyclovoltammetric measurements were in good agreement with those of the chemical doping experiments.

Fully aromatic ladder-type PPP **65** was successfully synthesized, which is composed of all-carbon six-membered rings in the double-stranded main chain [62] (Scheme 25). Some attempts have already been reported to generate this type of polymer, but have failed or led to nondefined products [51,63]. Open-chain, linear PPP precursor **64** with alkoxybenzoyl substituents was prepared by a Ni-catalyzed reductive coupling method using 1,4-dibromo-2,5-bis(4-decyloxybenzoyl)benzene **(63)** according to the procedure described by Yamamoto et al. [64,65]. The colorless materials obtained possess M_n values of 4000–5000. The well-defined structure of **64** was established by NMR analyses. Ring closure to the desired ladder-type PPP **65** successfully occurred by adapting a carbonyl olefination procedure using B_2S_3 [66].

By using Ni-catalyzed coupling reaction, PPP with tetrahydropyrene repeating units **67** was also synthesized from 2,7-dibromo-4,9-di-*n*-octyl-4,5,9,10-tetrahydropyrene (**66**) (Scheme 26) [67]. This polymer structural

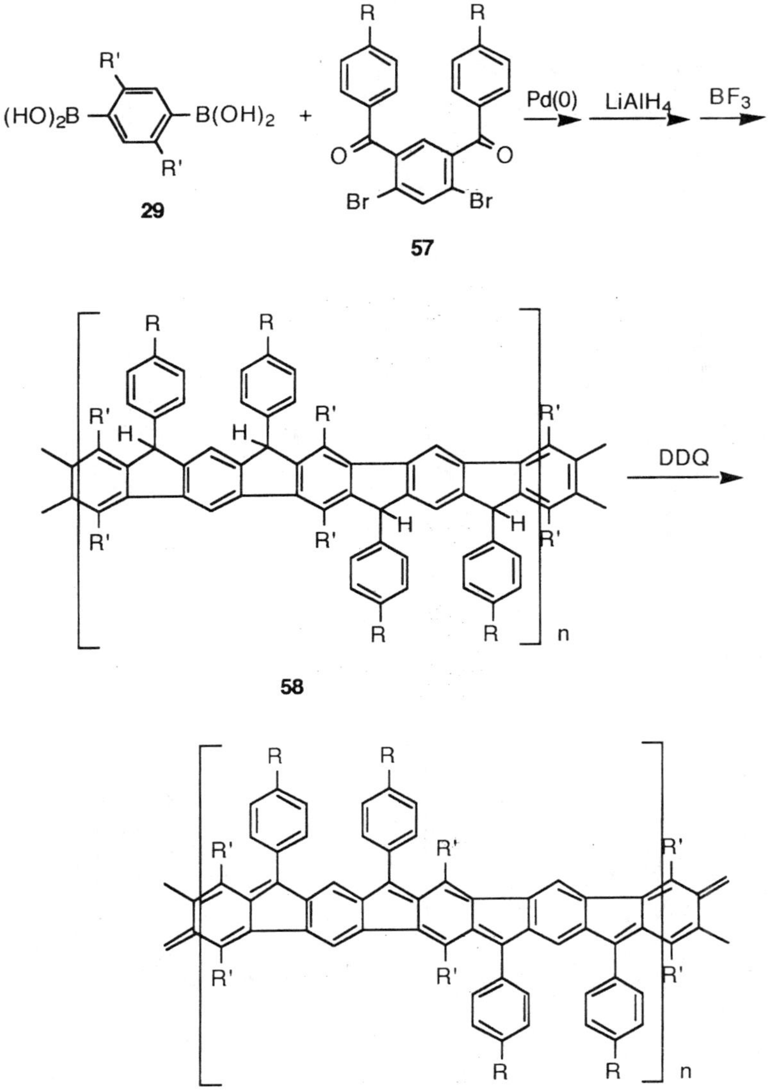

R = n-decyl   R' = n-hexyl

Scheme 22

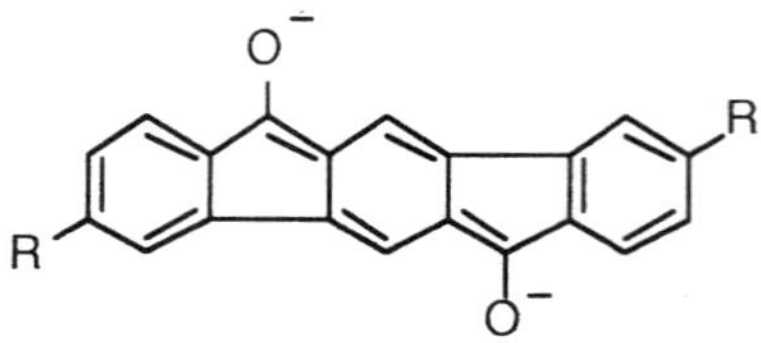

60

Scheme 23

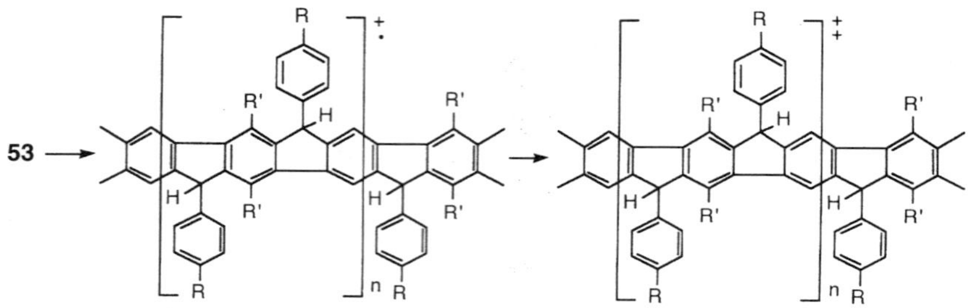

61 **62**

Scheme 24

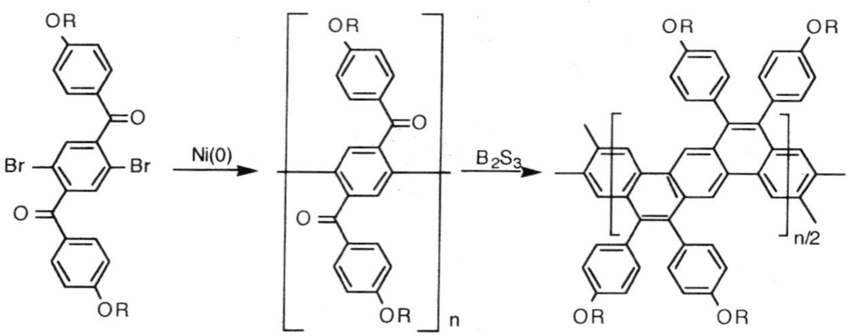

63 **64** **65**

R = n-decyl

Scheme 25

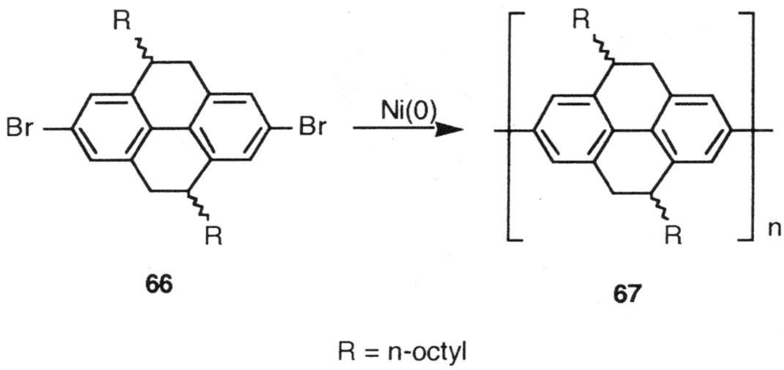

Scheme 26

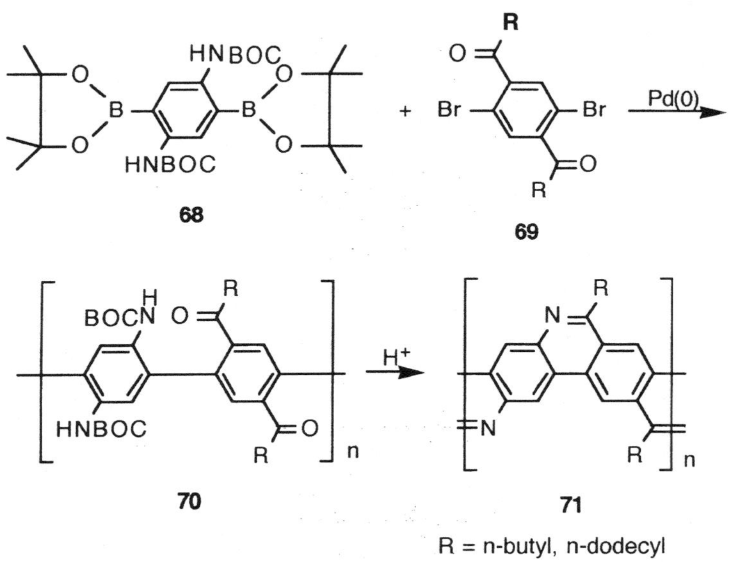

Scheme 27

design and the synthetic approach were based on the synthesis of 2,7-linked oligopyrenes [68]. The NMR spectra of the obtained polymer fully supported the structure **67.** The GPC analysis of **67** showed a monomodal distribution with M_n value of 17,400, corresponding to a degree of polymerization on the order of 40.

Nitrogen-containing PPP **70** was yielded by Pd-catalyzed coupling of **68** with **69** (Scheme 27). By treating **70** with trifluoroacetic acid, quantitative loss of BOC protecting groups and ring closure afforded ladder-type PPP **71.** [69].

REFERENCES

1. D. A. Tomalia, H. Baker, J. Dewald, M. Hall, G. Kallos, S. Martin, J. Roeck, J. Ryder, and P. Smith, *Polym. J.,* **17,** 117 (1985).
2. D. A. Tomalia, A. M. Naylor, and W. A. Goddard III, *Angew. Chem. Int. Ed. Engl.,* **29,** 138 (1990).
3. H. B. Mekelburger, W. Jaworek, and F. Vögtle, *Angew. Chem. Int. Ed. Engl.,* **31,** 1571 (1992).
4. D. A. Tomalia, *Aldrichim. Acta,* **26,** 91 (1993).
5. P. R. Dvornic and D. A. Tomalia, *Macromol. Symp.,* **88,** 123 (1994).
6. S. Bywater, *Adv. Polym. Sci.,* **30,** 89 (1979).
7. P. Rempp, E. Franta, and J. E. Herz, *Adv. Polym. Sci.,* **86,** 145 (1988).
8. O. Nuyken and S. Pask, *Encyclopedia of Polymer Science and Engineering,* 2nd Ed., Wiley, New York, 1989, Vol. 16, p. 494.
9. Y. K. Kim and O. W. Webster, *J. Am. Chem. Soc.,* **112,** 4592 (1990).
10. Y. H. Kim and O. W. Webster, *Macromolecules,* **25,** 5561 (1992).
11. C. Hawker and J. M. J. Frechet, *J. Am. Chem. Soc.,* **112,** 7638 (1990).
12. K. Ziegler, *Angew. Chem.,* **67,** 424 (1955).
13. K. Ziegler, E. Holkampf, H. Breil, and H. Martin, *Angew. Chem.,* **67,** 541 (1955).
14. K. Ziegler, *Angew. Chem.,* **67,** 548 (1955).
15. G. Natta, *J. Polym. Sci.,* **16,** 143 (1955).
16. G. Natta, P. Pino, P. Corradini, F. Danusso, E. Mantica, G. Mazzanti, and G. Moraglio, *J. Am. Chem. Soc.,* **77,** 1708 (1955).
17. E. J. Vandenberg, *Encyclopedia of Polymer Science and Engineering,* 2nd Ed., Wiley, New York, 1986, Vol. 4, p. 175.
18. P. Pino, U. Giannini and L. Porri, *Encyclopedia of Polymer Science and Engineering,* 2nd Ed., Wiley, New York, 1987, Vol. 8, p. 221.
19. S. Kobayashi and J. Kadokawa, *Acta Polym.,* **44,** 70 (1993).
20. J. Kadokawa and S. Kobayashi, *Macromol. Symp.,* **95,** 121 (1995).
21. V. Y. Hirusawa, M. Oku, and K. Yamamoto, *Bull. Chem. Soc. Jpn.,* **30,** 667 (1957).
22. H. Horner and U. M. Duda, *Tetrahedron Lett.,* **59,** 5177 (1970).
23. M. Suzuki, A. Ii, and T. Saegusa, *Macromolecules,* **25,** 7071 (1992).
24. M. Suzuki, S. Sawada, and T. Saegusa, *Macromolecules,* **22,** 1505 (1989).
25. I. Minami, Y. Ohashi, I. Shimizu, and J. Tsuji, *Tetrahedron Lett.,* **26,** 2449 (1985).
26. M. Suzuki, H. Ito, and T. Seagusa, *Polym. Prepr. Jpn.,* **42,** 376 (1993).
27. J. G. Speight, P. Kovacic, and F. W. Koch, *J. Macromol. Sci., Rev. Macromol. Chem.,* **C5,** 295 (1971).
28. P. Kovacic and M. B. Jones, *Chem. Rev.,* **87,** 357 (1987).
29. P. Kovacic and A. Kyriakis, *Tetrahedron Lett.,* 467 (1962).

30. P. Kovacic and A. Kyriakis, *J. Am. Chem. Soc.,* **85,** 454 (1963).
31. T. Yamamoto, Y. Hayashi, and A. Yamamoto, *Bull. Chem. Soc. Jpn.,* **51,** 2091 (1978).
32. C. S. Marvel and G. E. Hartzell, *J. Am. Chem. Soc.,* **41,** 448 (1959).
33. D. G. H. Ballard, A. Courtis, I. M. Shirley, and S. C. Taylor, *J. Chem. Soc., Chem. Commun.,* 954 (1983).
34. D. G. H. Ballard, A. Courtis, I. M. Shirley, and S. C. Taylor, *Macromolecules,* **21,** 294 (1988).
35. D. R. McKean and J. K. Stille, *Macromolecules,* **20,** 1787 (1987).
36. M. Rehahn, A. D. Schlüter, G. Wegner, and W. J. Feast, *Polymer,* **30,** 1054 (1989).
37. M. Rehahn, A. D. Schlüter, G. Wegner, and W. J. Feast, *Polymer,* **30,** 1060 (1989).
38. M. Rehahn, A. D. Schlüter, and G. Wegner, *Makromol. Chem.,* **191,** 1991 (1990).
39. N. Miyaura, T. Yanagi, and A. Suzuki, Synth. Commun., **11,** 513 (1981).
40. R. B. Miller and S. Dugar, *Organometallics,* **3,** 1261 (1984).
41. I. U. Rau and M. Rehahn, *Makromol. Chem.,* **194,** 2225 (1993).
42. I. U. Rau and M. Rehahn, *Polymer,* **34,** 2889 (1993).
43. I. U. Rau and M. Rehahn, *Acta Polym.,* **45,** 3 (1994).
44. U. B. Steiner, M. Rehahn, W. R. Caseri, and U. W. Suter, *Macromolecules,* **27,** 1983 (1994).
45. U. B. Steiner, W. R. Caseri, U. W. Suter, M. Rehahn, and I. U. Rau, *Langmuir,* **10,** 1164 (1994).
46. L. Schmitz, M. Rehahn, and M. Ballauff, *Polymer,* **34,** 646 (1993).
47. U. B. Steiner, W. R. Caseri, U. W. Suter, M. Rehahn, and L. Schmitz, *Langmuir,* **9,** 245 (1993).
48. L. Schmitz and M. Rehahn, *Macromolecules,* **26,** 4413 (1993).
49. U. Scherf and K. Müllen, *Synthesis,* 23 (1992).
50. W. J. Bailey, *ACS Polym. Mater. Sci. Eng.,* **60,** 400 (1989).
51. A. D. Schlüter, *Adv. Mater.,* **3,** 284 (1991).
52. U. Scherf and K. Müllen, *Makromol. Chem. Rapid Commun.,* **12,** 489 (1991).
53. U. Scherf and K. Müllen, *Macromolecules,* **25,** 3546 (1992).
54. U. Scherf and K. Müllen, *Polymer,* **33,** 2443 (1992).
55. J. M. Toussaint, B. T. Themans, J. M. Andre, and J. L. Bredas, *Synth. Met.,* **28,** C205 (1989).
56. J. L. Bredas, *Springer Ser. Solid State Sci.,* **63,** 166 (1985).
57. R. L. Elsenbaumer and L. W. Shacklette, *Handbook of Conducting Polymers,* T. A. Skotheim, Ed., Dekker, New York, 1986, Vol. 1, p. 213.
58. L. W. Shacklette, H. Eckardt, R. R. Chance, G. G. Miller, and R. H. Baughman, *J. Chem. Phys.,* **73,** 4098 (1986).
59. U. Scherf, A. Bohnen, and K. Müllen, *Makromol. Chem.,* **193,** 1127 (1992).
60. R. H. Friend, *Physics and Chemistry of Electrons and Ions in Condensed Matter,* J. V. Acrivos, Ed., Reidel, Dordrecht, 1984, p. 625.
61. J. L. Bredas, R. R. Chance, and R. Silbey, *Mol. Cryst. Liq. Cryst.,* **77,** 319 (1981).
62. K. Chmil and U. Scherf, *Makromol. Chem. Rapid Commun.,* **14,** 217 (1993).
63. M. Löffler and A. D. Schlüter, *GIT Fachz. Lab.,* **1101** (1992).

64. T. Yamamoto, A. Morita, Y. Miyazaki, T. Maruyama, H. Wakayama, Z. Zhou, Y. Nakumura, T. Kanbara, S. Sasaki, and K. Kubota, *Macromolecules,* **25,** 1214 (1992).
65. T. Kanbara, N. Saito, T. Yamamoto, and K. Kubota, *Macromolecules,* **24,** 5886 (1991).
66. K. Steliou, P. Salama, and X. Yu, *J. Am. Chem. Soc.,* **114,** 1456 (1992).
67. M. Kreyenschmidt, F. Uckert, and K. Müllen, *Macromolecules,* **28,** 4577 (1995).
68. M. Kreyenschmidt, M. Baumgarten, N. Tyutyukov, and K. Müllen, *Angew. Chem. Int. Ed. Engl.,* **33,** 1957 (1994).
69. J. M. Tour and J. J. S. Lamba, *J. Am. Chem. Soc.,* **115,** 4935 (1993).

6
Star Polymers by Immobilizing Functional Block Copolymers

Koji Ishizu
Tokyo Institute of Technology, Meguro-ku, Tokyo, Japan

I. INTRODUCTION

The star-branched or radial polymers have structures formed by linking linear polymers with a core having a small molecular weight. Generally, the star polymer has a smaller hydrodynamic dimension than a linear polymer with the same molecular weight. The interest in star polymers arises from their compactness and their enhanced segment density. Daoud and Cotton [1] were the first to study the conformation and dimensions of star polymers using scaling ideas. The picture presented for a star polymer consists of three regions: a central core, a shell of semidilute density in which the arms have unperturbed chain conformation, and an outer shell in which the arms of the star assume a self-avoiding conformation. Stars with many arms (the critical number of arms is estimated to be of the order of 10^2) are expected to form a crystalline array near the overlap concentration (C^*) [2].

De la Cruz and Sanchez [3] have calculated, using a mean-field theory, the phase stability criteria and static structure factors for n-arm star diblock copolymers [$(AB)_n$ star]. According to their calculations, as n becomes large, the $(AB)_n$ star begins to develop a "core-and-shell" structure. The core is rich in A monomer and the shell is rich in B monomer even in the disordered state.

This self-segregation or self-micellization tends to create significant concentration fluctuations at the core–shell interface.

More recently, we have established a preparative method for core–shell-type copolymers by the crosslinking reaction of diblock copolymer films having microphase-separated spherical cores with crosslinking reagents in the solid state [4]. The core–shell type and $(AB)_n$ star copolymers have a primary structure similar to that of star polymer. Thus, the core–shell type and $(AB)_n$ star copolymers can be expected to have supramolecular ordering.

This chapter deals mainly with the synthetic methods of the core–shell type, star block copolymers, and supramolecular ordering in solution and solid state.

II. CORE–SHELL TYPE COPOLYMERS FORMED BY IMMOBILIZING DIBLOCK COPOLYMER FILMS

Block and graft copolymers with incompatible sequences exhibit characteristic morphological behavior and interesting properties, owing to micelle formation in a selective solvent and microdomain formation in the solid state. Many types of micelles formed by diblock copolymers in solution were reported. Spherical self-assembly formed in the solvent for one sequence and not for the other forms a so-called polymer micelle with core–corona morphology.

Theoretical treatments of the microphase separation of block copolymers are classified into two types: strong segregation theories [5–7] and weak segregation theories [8–12]. In strong segregation regimes, the thermodynamic approach to the problem of microdomain formation for block copolymers has been worked out by demonstrating that the morphology of the domain structure—such as equilibrium size, shape, and interfacial thickness—can be described in terms of a balance of physical factors. Four ordered microphases are well-known, which consist of alternating layers, cylinders on a hexagonal lattice, spheres on a body centered cubic (BCC) lattice, and a bicontinuous "double-diamond" structure [13–14]. The spherical, cylindrical, and lamellar structures were all stable in the strong segregation regime [15]. The stable phases are considered to be spherical microdomains for the volume fraction of either diblock, $f = \sim 0.17$.

If the structure of the self-assemblies can be immobilized by crosslinking of the spherical parts (the spherical microdomains in the solid state and the core in solution), the crosslinked products should form the core–shell type copolymers (microspheres) (see Figure 1). In fact, poly[styrene(S)-*b*-butadiene (Bu)-*b*-S] block copolymer micelles with cores of polybutadiene (PBu) blocks in dilute solution were stabilized by crosslinking of the chains in the micellar

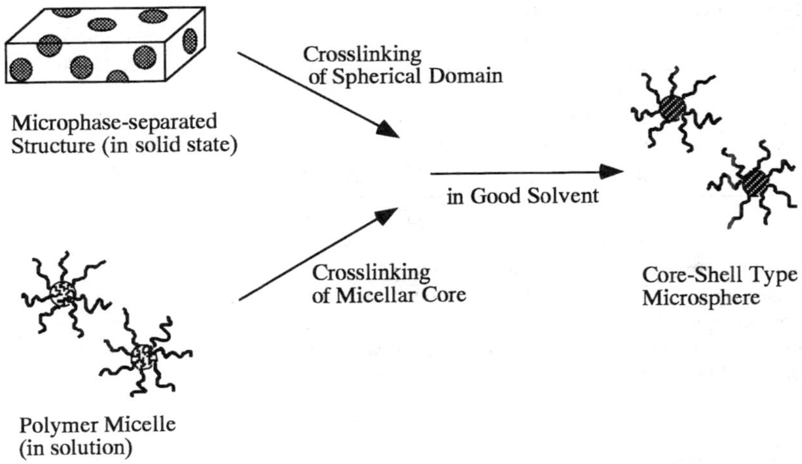

Figure 1 Schematic presentation of synthesis routes on core–shell type microspheres.

cores by UV irradiation in the presence of a photoinitiator and by fast electrons [16–18]. The stabilized micelles, examined by light scattering, sedimentation, and gel permeation chromatography (GPC), did not decompose upon heating or in good solvents for both blocks. Authors have also obtained such core–shell type microspheres by crosslinking of core domains in micelles formed in selective solvent for several block and graft copolymers [19–27]. It was concluded from these results that the reaction rate of intramicelle crosslinking and the lifetime of the polymer micelle were important for the crosslinking of the polymer micelle. When intramicelle crosslinking occurred before the breakup of the micelle, a core–shell type microsphere could be synthesized. Thus, the synthesis routes of these microspheres are being investigated very actively. On the other hand, the microphase-separated structures in bulk film are more stable in thermal equilibrium than the micelle in solution. So the crosslinking of spherical microdomains in the film is a superior preparative method for core–shell type microspheres than crosslinking of micellar cores.

A. Synthesis and Characterization of Core–Shell Type Copolymers

The well-defined poly[S-*b*-4-vinylpyridine (4VP)] diblock copolymer was prepared by the usual anionic polymerization using *n*-butyllithium (*n*-BuLi) as an initiator in tetrahydrofuran (THF) at –78°C [28]. Table 1 lists the characteristics

Table 1 Characteristics of Poly(styrene-*b*-4-vinylpyridine) Diblock Copolymer and Domain Spacing

	$10^{-4}\overline{M}_n$				
Specimen code	*Block copolymer*[a]	*P4VP block*[b]	*P4VP block*[c] *(wt %)*	*Shape of P4VP domain*	*Domain radius* $\overline{R}_{P4VP}$[d] *(nm)*
SV1	11.2	2.7	24	Sphere	17

[a]Measured by osmometry in 1,1,2-trichloroethane (TCE.)
[b]Estimated from the $\overline{M}_n$ data of the diblock copolymer and polystyren precursor.
[c]Determined from infrared spectra.
[d]Determined by means of electron microscopy of specimen cast from TCE. $\overline{R}_{P4VP}$ = average radius of P4VP sphere structure.

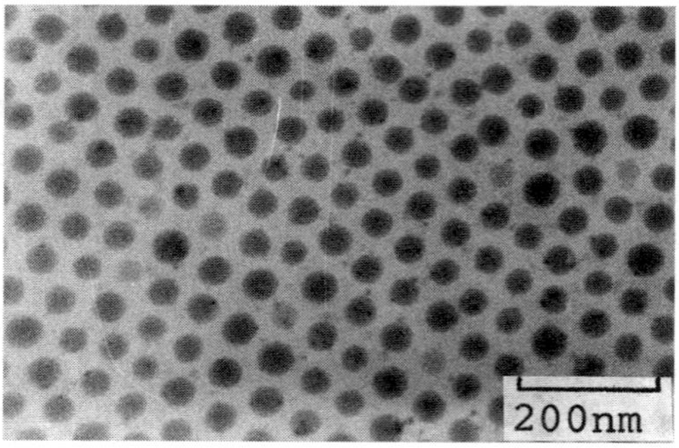

Figure 2 Transmission electron micrograph of SV1 diblock copolymer specimen cast from TCE. (From Ref. 28.)

of the "monodisperse" diblock copolymer SV1 and the microdomain spacing of the specimen cast from 1,1,2-trichloroethane (TCE). Figure 2 shows a transmission electron micrograph (TEM) of a diblock copolymer SV1 specimen cast from TCE. The dark portions are the selectively stained poly(4-vinylpyridine) (P4VP) blocks with osmium tetroxide (OsO_4). The microdomain structure in SV1 specimen (24 wt % P4VP) shows the texture of discrete P4VP spheres in a polystyrene (PS) matrix. This microphase-separated structure is more stable in thermal equilibrium than the micelle formed in selective solvent. The absolute value of the radius of the P4VP spheres, $\overline{R}_{P4VP}$ is far below the theoretical value (32 nm). The TCE used is a good solvent for PS and a somewhat poor one for

P4VP blocks, judging from the Hildebrand parameter. This morphology is considered to demonstrate the nonequilibrium state of the spherical domain system.

These segregated P4VP chains in a sphere were crosslinked by using quaternization with 1,4-dibromobutane (DBB) vapor at 60°C in the solid state. It was found from IR spectra that block copolymer films with P4VP spheres were almost quantitatively quaternized with DBB at the reaction time of 24 hours. Figure 3 shows a TEM micrograph of a crosslinked polymer microsphere of SV1 (SV1M) cast from 0.1 wt % TCE/nitrobenzene: 10/1 (v/v) solution [29]. This texture indicates a multimolecular micelle. In the preparation conditions of this cast specimen, multimolecular micelles are frozen with a structure like the core–shell type microspheres. The dark, gray, and white portions indicate the crosslinked P4VP cores, shell of PS chains, and carbon support, respectively. This micrograph shows clearly the structure of the core–shell type microspheres. It is clear that these microsphere particles have only a narrow size distribution.

Similar core–shell type microspheres could also be synthesized by crosslinking of the microphase-separated poly[S-*b*-2-vinylpyridine(2VP)] diblock copolymer film [30]. Two types of microspheres described above have positive charges in a core. On the other hand, the core–shell type microspheres having negative charges in a core were prepared by our groups [31]. That is to say, the crosslinking of segregated poly(methacrylic acid) (PMA) domains formed by poly[S-*b*-methacrylic acid (MA)] film was carried out by casting this block copolymer with *N, N*'-dicyclohexyl carbodiimide and hexamethylenediamine from dioxane solution. It was possible to prepare the microspheres composed of liquid rubber cores. Crosslinking of the microphase-separated

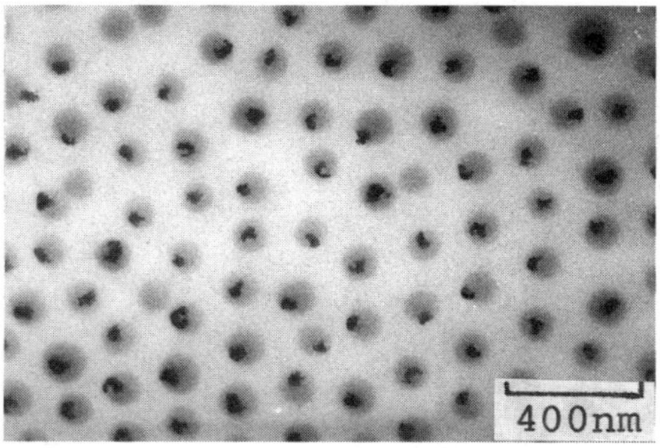

Figure 3 Transmission electron micrograph of core–shell type microsphere SV1-M. (From Ref. 28.)

polyisoprene (PI) chains in the core formed by poly[S-*b*-isoprene(I)] film was carried out by soaking in *n*-hexane solution of sulfur monochloride [32].

Authors have established a new architecture for the core–shell type microspheres with well-defined sizes of core radius and shell thickness [33,34]. Control of arm length can be achieved by cleavage of arm chains on core–shell type microspheres (method 1). In order to change arm number, removal of the arms from the microsphere is proposed. If the microsphere has both arms that can be removed and those that cannot, the arm number of the shell can be reduced by cutting off the removable arms. A schematic diagram of the microspheres for both methods is shown in Figure 4. Method 1 [33], concerning the control of arm length, is as follows: The well-defined poly(S_1-I-S_2-*b*-2VP) diblock copolymers were prepared by sequential anionic addition. PS blocks contained a small amount of isoprene sequence units at the prescribed position

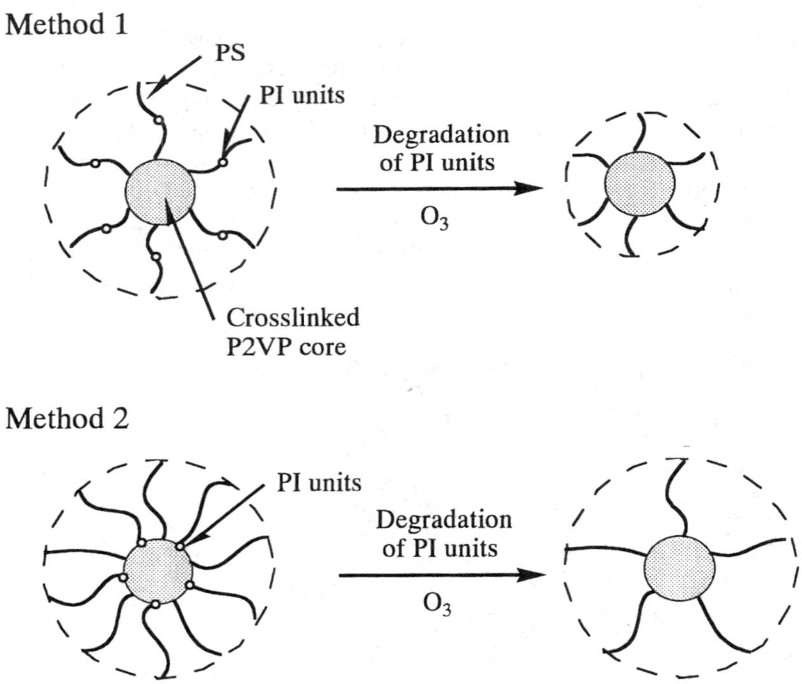

Figure 4 Schematic diagram illustrating the concept of controlling the arm length and arm number on core–shell type microspheres.

through three-stage addition of S_1, I, and S_2 monomers. To achieve rich 1,4-linkage of isoprene units, the polymerizations of S_1, I, and S_2 were carried out in benzene. After capping of poly(S_1-I-S_2) anion ends with 1,1-diphenyl ethylene, a THF solution of 2VP was added. The microstructure of the isoprene units was found to be 50% of the 1,4-structure from ^{1}H NMR spectrum analysis. After crosslinking of P4VP spherical domains, the degradation of PI units was achieved by ozonolysis.

Method 2, concerning the control of arm number, is described as follows [34]: Poly(S-*b*-2VP) diblock copolymer and poly(S-*b*-I-*b*-2VP) triblock copolymer were prepared by sequential anionic addition polymerization. Binary blends of these block copolymers were co-micellized in a P2VP microdomain homogeneously by varying the blend ratio. Subsequently, the P2VP microdomains of blend films were also crosslinked with 1,4-diiodobutane (DIB) vapor. Next, the PS arms from poly(S-*b*-I-*b*-2VP) triblock copolymer in the microsphere were cleaved by degrading the PI sequence by ozonolysis.

B. Solution Properties

We look at the interesting properties of core–shell type microspheres in solution. The turbidity curves of P4VP core/PS shell type microsphere SV1M and ionic block copolymer are shown in Figure 5 [28]. Where the corresponding ionic block copolymer was prepared from the quaternization of SV1 with methyl iodide, the ionic block copolymer of SV1 shows such behavior. It was observed that the appearance of new micelle was accompanied by a sharp maximum of turbidity (water fraction = 0.17). This behavior is explained as follows: First, micelles with a quaternized P4VP core and shell of solvated PS chains are formed in THF solution, and this micellar form changes into the reversed micelles with a PS core and a shell of ionized P4VP chains at a water fraction of about 0.17. These reversed micelles are stabilized until the water fraction reaches about 0.6, owing to the strong ionic strength of quaternized P4VP shell. However, the core–shell type microsphere SV1M prepared by crosslinking of the segregated P4VP chains in a core is very stable in a THF–water mixture. This result is explained by the formation of stabilized core–shell type microspheres with not only a crosslinked but also a quaternized P4VP core and a shell of PS chains. A similar result was obtained as shown in a ^{1}H NMR spectrum of PI core–PS shell type microspheres [32]. The spectrum of this microsphere showed only the peaks due to PS segments, but the microsphere was apparently soluble in $CDCl_3$, giving a clear solution. The NMR data indicated clearly that the crosslinked PI segments constituted a micellar core that behaved like a solid on the time scale of segment relaxation. The PS

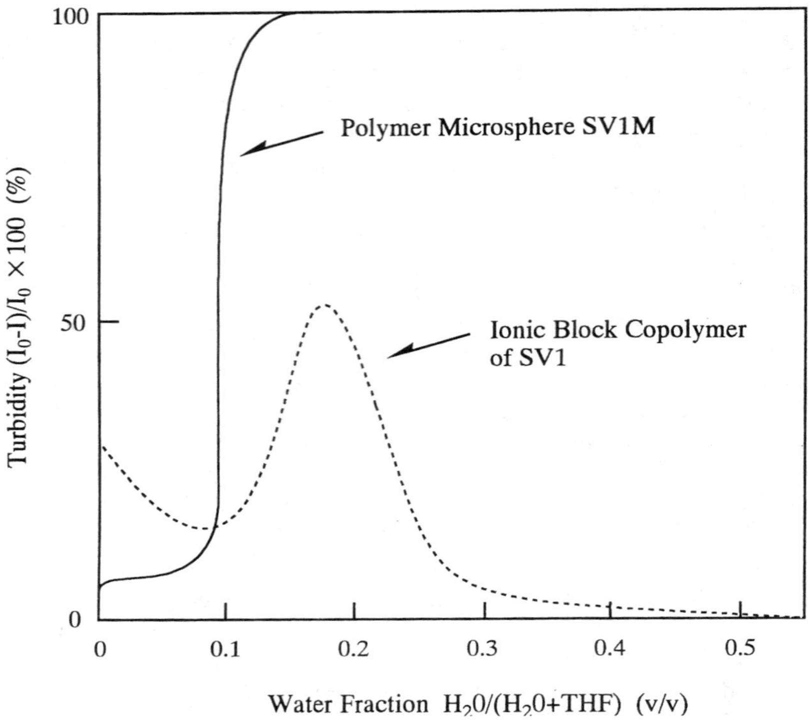

Figure 5 Turbidity curves of the core–shell type microsphere and ionic block copolymer. (From Ref. 28.)

segments in the shell were molecularly dispersed in $CDCl_3$ to show well-resolved NMR peaks. The solubility of the core–shell type microspheres depends strongly on the solubility of block chains constructing shell parts.

C. Hierarchical Structure Transformation in Lattice Formation

In a previous work [35], we studied the morphological behavior of core–shell type microspheres prepared from casting solution onto a carbon substrate. The particles of core–shell type microspheres were well aligned in a hexagonal array on the carbon substrate from two-dimensional observation. It was expected that in the core–shell type microspheres, such crystalline order in solution would appear close to the C^* concentration. However, the three-dimensional ordered structure in solution was not made clear by small-angle x-ray scattering (SAXS) because of the large diameter of the microsphere particle used.

We prepared core–shell type microspheres of a relatively small-diameter particle size [36]. Characteristics of starting poly(S-*b*-4VP) diblock copolymers are listed in Table 2. The segregated spherical P4VP domains were crosslinked using DBB vapor in the solid state. P4VP core/PS shell type microspheres (SV-M) freely dissolved in organic solvents such as benzene, THF, and chloroform. In order to estimate both the shape and the size of the microspheres, TEM observations of SV-M series, which had been stained with OsO_4 and shadowed with chromium (Cr), were carried out. Characteristics of the core–shell type microspheres are listed in Table 3. According to theoretical results of Witten et al. [2], a structure for the crystalline array should appear near

Table 2 Characteristics of Poly(S-*b*-4VP) Diblock Copolymers

	Block copolymer			*Domain size*	
Code	$10^{-4}\,\overline{M}_n{}^a$	$\overline{M}_w/\overline{M}_n{}^b$	*P4VP[c] block (mol %)*	$\overline{D}_n{}^d$ *(nm)*	$\overline{D}_w/\overline{D}_n{}^d$ *(nm)*
SV10	4.2	1.27	14	17	1.06
SV20	17	1.20	10	19	1.01

[a]Estimated from $\overline{M}_n$ of polystyrene precursor and composition of P4VP blocks.
[b]Determined by gel permeation chromatography.
[c]Determined by ^1H NMR.
[d]$\overline{D}_n$ indicates the diameter of P4VP spheres and was determined by transmission electron micrographs cast from TCE.

Table 3 Characteristics of Core–Shell Type Microspheres

		P4VP core		*Microsphere particle*	
Code	*DC[a] (mol %)*	$\overline{D}_n$ *(nm)*	$\overline{D}_w/\overline{D}_n{}^b$	$\overline{D}_n$ *(nm)*	$\overline{D}_w/\overline{D}_n{}^b$
SV10-M	35.4	16	1.01	36	1.01
SV20-M	32.8	20	1.01	40	1.03

[a]Crosslink density determined by Volhard titration.
[b]Size distribution ($\overline{D}_w/\overline{D}_n$) was estimated from transmission electron micrographs.

Table 4 Physical Values and C^* Concentration of Core–Shell Type Microspheres

	Microsphere			
Code	N^a	$10^{-8}\ \bar{M}_w(m)^b$	$\bar{D}_h$ (nm)[c]	C^{*d} (wt %)
SV10-M	1800	0.76	66	6.9
SV20-M	1500	2.6	86	6.8

[a]Arm number.
[b]Molecular weight of core–shell type microspheres
[c]Hydrodynamic diameter of microspheres determined by dynamic light scattering.
[d]Overlap concentration was calculated from the following equation [37]: $C^* = \bar{M}_w(\mathrm{m})/\bar{D}_h^3 N_A$, where N_A is the Avogadro number.

the C^*. The hydrodynamic diameter (D_h) of the microspheres must be known in order to estimate the C^*. These values were determined by dynamic light scattering (DLS) measurement. Table 4 lists the physical values and C^* of core–shell type microspheres.

We measured the SAXS of core–shell type microspheres near the C^*. The benzene solution of SV10-M microsphere had a bluish tint at 7.0 wt % (calculated C^* = 6.9 wt %). Figure 6 shows the SAXS pattern for the SV10-M in 7.0 wt % benzene solution. This SAXS pattern indicates that the microspheres are arranged with some kind of periodic structure in solution, owing to the appearance of several scattering patterns. Figure 7 shows the SAXS intensity distributions for this microsphere solution in the small-angle region, where the arrows show the scattering maxima. Table 5 lists the cubic packing (d_1/d_i) at the scattering angles relative to the angle of the first maximum according to Bragg's equation: $2d \sin \theta = \lambda n$ (where λ is one-half the scattering angle, λ = 1.5418 Å). It is shown that the first three peaks appear closely at the relative angular positions of 1: $\sqrt{2}$: $\sqrt{3}$ as shown in parentheses of Figure 7. These values correspond to the packing pattern of (110), (200), and (211) planes in a body-centered cubic (BCC) structure. Therefore, it is concluded that the SV-M microspheres are packed in a superlattice of BCC structure near the C^*.

The arrangement of SV-M microspheres in the bulk film was investigated from two- and three-dimensional aspects [36]. First, the two-dimensional arrangement of the microsphere is discussed. Figure 8 shows the TEM micrograph of a SV10-M specimen cast from 1.0 wt % TCE solution. These microspheres formed a monolayer film on the carbon substrate. In order to make

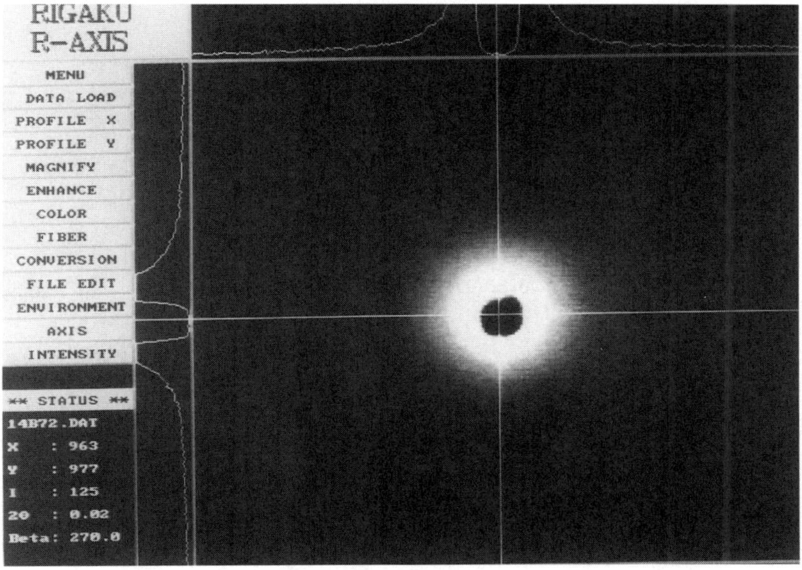

Figure 6 SAXS pattern for SV10-M in 7.0 wt % benzene solution. (From Ref. 36.)

clear the ordering of microspheres on the carbon substrate, the distribution of the distances $g(r)$ between the centers of the P4VP spherical microdomains was measured (Figure 9). Three clear peaks appear at distances of 31.2, 52.9, and 62.1 nm. When microspheres are completely packed in a hexagonal fashion, the distances between the centers of spheres are r, $\sqrt{3}r$, and $2r$ for the first, second, and third generation spheres, respectively (see Figure 10). The ratio of these distances for the peaks observed for SV10-M agrees well with the theoretical ratio of $1: \sqrt{3} : 2$. So, the microspheres are packed in a hexagonal arrangement on the carbon substrate.

Next, the three-dimensional arrangement of the microsphere film was studied by the SAXS measurement. Figures 11 and 12 show the SAXS pattern and SAXS intensity distributions in the small-angle region for SV10-M microsphere film. Table 6 lists the cubic packing (d_1/d_i) at the scattering angles relative to the angle of the first maximum. The maxima at $2\theta = 0.42$ and 0.66 are those occurring at the angular positions of $\sqrt{4/3}$ and $\sqrt{11/3}$ of the first-order maximum. It is therefore found that the SV-M microspheres are packed in a superlattice of a face-centered cubic (FCC) structure in the bulk film. This is the most efficient way to pack spheres. The packing structure of micro-

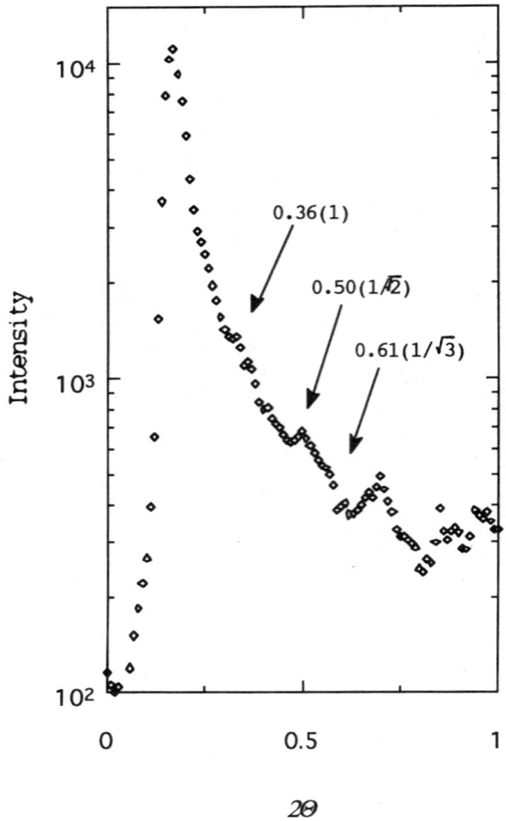

Figure 7 SAXS intensity distributions for SV10-M in 7.0 wt % benzene solution in the small-angle region; the arrows show the scattering maxima. The values in parentheses indicate the cubic packing (d_1/d_i). (From Ref. 36.)

Table 5 Small-Angle X-Ray Scattering Data for SV10-M in 7 wt % Benzene Solution

n	2θ	d_i (nm)	d_1/d_i
1	0.36	24.5	1
2	0.50	17.7	$\sqrt{2}$
3	0.61	14.5	$\sqrt{3}$

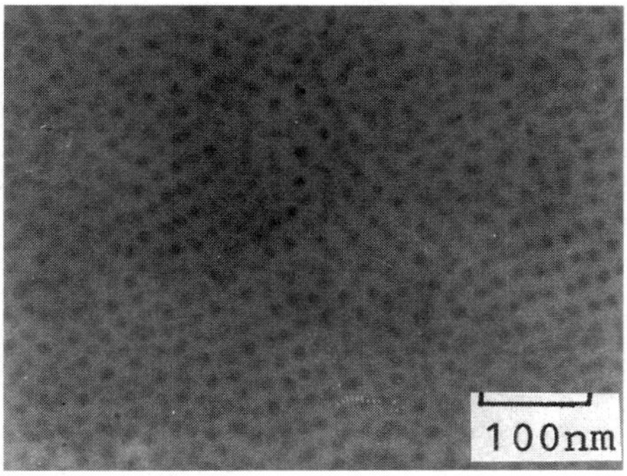

Figure 8 Transmission electron micrograph of SV10-M microsphere cast from 1.0 wt % TCE solution. (From Ref. 36.)

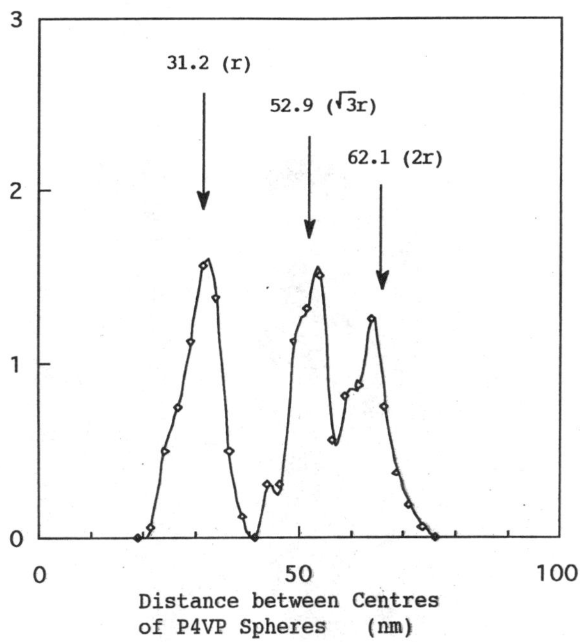

Figure 9 Distribution functions of the distance between the centers of P4VP spherical domains for SV10-M film. The values in parentheses indicate the ratio of distances between the centers of spheres for each generation. (From Ref. 36.)

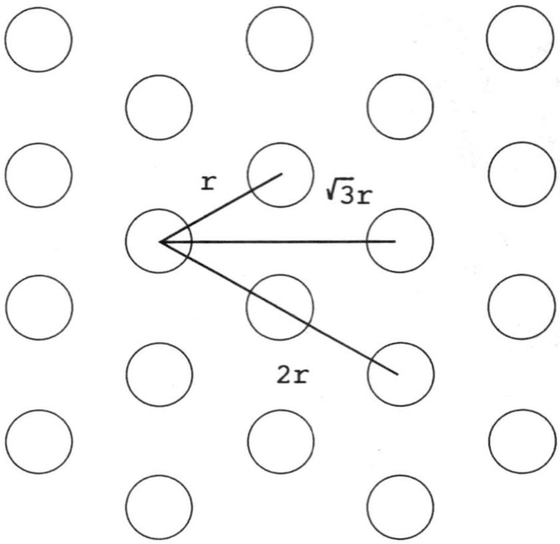

Figure 10 Schematic representation of microspheres showing a two-dimensional hexagonal packing arrangement.

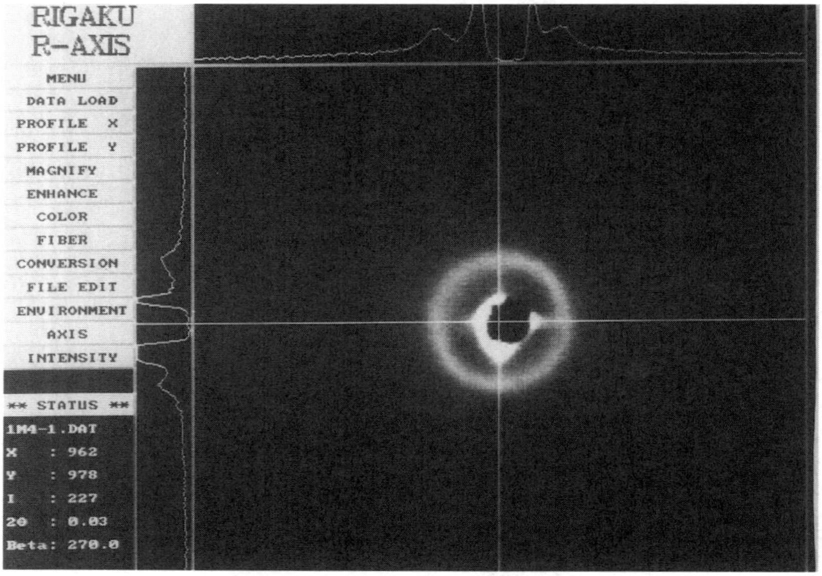

Figure 11 SAXS pattern for SV10-M microsphere film. (From Ref. 36.)

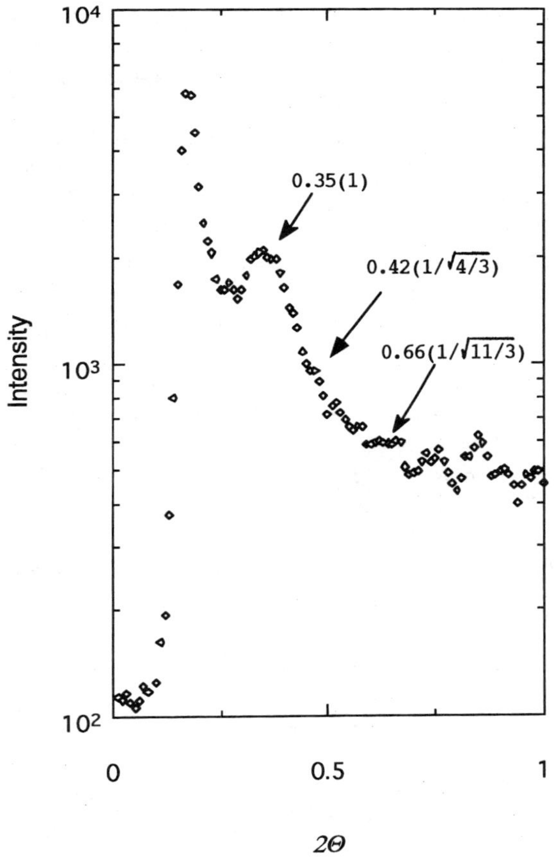

Figure 12 SAXS intensity distributions for SV10-M microsphere film in the small-angle region; the arrows show the scattering maxima. The values in parentheses indicate the cubic packing (d_1/d_i). (From Ref. 36.)

Table 6 Small-Angle X-Ray Scattering Data for SV10-M Film

n	2θ	d_i (nm)	d_1/d_i
1	0.35	25.2	1
2	0.42	21.0	$\sqrt{4/3}$
4	0.66	13.4	$\sqrt{11/3}$

spheres of large particle size could be determined by TEM observation using the tilt method [38].

In general, the spherical microdomains of diblock copolymers can be packed into one of the three cubic forms: simple cubic, FCC, and BCC. According to Ohta and Kawasaki [7], a BCC arrangement is only slightly more favored than a FCC arrangement. Chevalier et al. [39] have investigated the coalescence of latex particles through small-angle neutron scattering (SANS) and TEM. The latex particles arranged into polyhedral cells in dispersion, and these structures changed to colloidal crystals with an FCC lattice at the condensed solution (74 wt %). Thus, the transformation of a C^* solution into a continuous film for core–shell type microspheres seems very similar to that for latex particles.

Structural transformation of such core–shell type microspheres can be explained using a thermal blob model (Figure 13) proposed by Daoud and Cotton [1]. Explanation of each region was given in the introduction. In a core–shell type microsphere, the crosslinked P4VP domain can be regarded as the central core of region I. The PS arm of the shell can be seen as a succession of growing blobs in regions II (shell with semidilute segment density in which the arms have their unperturbed chain conformation) and III (self-avoiding

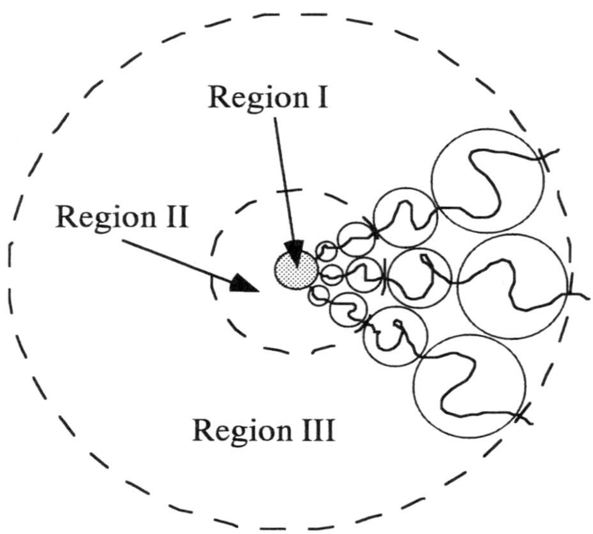

Figure 13 Thermal blob model for star polymer. (From Ref. 1.)

conformation). BCC ordering of these microspheres appears close to the C^*. The crosslinked P4VP cores do not interpenetrate with each other and the microspheres are compatible with their PS shells during solvent evaporation. As a result, a packing structure such as an FCC lattice appears in the dry film.

D. Three-Phase Separated Morphologies by Blending

Microphase separation with three phases can be obtained for an ABC triblock copolymer. The morphology of an ABC triblock copolymer has been suggested by Riess et al. [40]. An interesting morphology of microphase separation with three phases has been reported for ABC triblock copolymers [41–43]. On the other hand, we reported microphase separation with three phases in the blend system of AB and AC diblock copolymer micelles [44]. In that blend, two B and C spherical microdomains were dispersed at random in an A matrix.

The core–shell type microspheres can be considered the smallest units in the microphase-separated structure. The properties of the microsphere in the solvent area are governed not only by the core but also by the shell chains in a good solvent for the shell. Thus, the core–shell type microspheres can be thought of as one spherical molecule with multibranches in the solvent. Thus, core–shell type microspheres can be used as composite materials in blends for the introduction of spherical microdomains into a matrix. The effect of the ordering of microspheres on the morphology of the blend has not been investigated.

We investigated the morphologies of blends of well-ordered core–shell type microsphere with AB diblock copolymers, which form lamellar or spherical microdomains in a matrix [45,46]. The poly(S-*b*-4VP) with lamellar morphology or poly(S-*b*-4VP), poly(S-*b*-2VP), and poly(S-*b*-I) with spherical morphology, synthesized by anionic polymerization, were chosen as the blend materials. The results are listed in Table 7. The P4VP core–PS shell microsphere SV500-M used in this work easily forms an ordered structure on its own. Some characteristics of the microsphere SV500-V are shown in Table 8.

Figure 14 shows the TEM micrographs of S4V-100 and S4V-200 specimens cast from TCE and benzene/chloroform (1/1: v/v) mixture, respectively. The lamellar structure was observed for both specimens. The TEM micrographs of microsphere SV500-M are shown in Figure 15. The spherical shapes can be observed in the TEM of a sample cast from a dilute solution and shadowed with Cr (Figure 15a). When the specimen was cast from a 1 wt % polymer concentration, the microspheres formed a monolayer with hexagonal packing in two dimensions (Figure 15b).

To investigate the effect of C^* on the macrolattice formation of the mi-

Table 7 Characteristics of Diblock Copolymers

	$10^{-4}\ \bar{M}_n$			Morphology	
Sample code[a]	*PS*[b] *precursor*	*Block copolymer*	*PS block*[c] *(mol %)*	*Shape*	$\bar{D}^d$ *(nm)*
S4V-100	1.6	2.4	68.0	Lamella	10.8
S4V-200	5.6	7.6	73.0	Lamella	24.0
S4V-300	39.0	47.0	83.0	Sphere P4VP	68.0
S2V-100	2.0	2.8	70.0	Sphere P2VP	18.0
SI-100	2.3	3.2	72.5	Sphere PI	10.0
SV500	8.0	11.0	72.0	Sphere P4VP	48.4

[a]S4V and SV500, poly(S-*b*-4VP); S2V, poly(S-*b*-2VP); SI, poly(S-*b*-I).
[b]Determined by gel permeation chromatography (GPC).
[c]Determined by ^{1}H NMR.
[d]Domain size, diameter of spheres or thickness of lamellae. Casting solvent: TCE for S4V-100, S4V-300, S2V-100, and SV500; benzene/chloroform (1/1: v/v) mixture for S4V-200; benzene for SI-100.

Table 8 Characteristics of P4VP Core/PS Shell Type Microsphere

	Diameter (nm)					
Sample	*P4VP core*[a]	*Internal*[a]	*External*[b]	*Crosslink density*[c] *(mol %)*	N^d	C^{*e} *(wt%)*
SV500-M	49.7	77.8	180.5	20.8	1400	4.4

[a]Determined by TEM in the solid state.
[b]Measured by DLS in benzene solution at 20°C.
[c]Determined by Volhard titration.
[d]Aggregation number of block copolymer SV500 in a microsphere
[e]Overlap concentration in benzene calculated from the external diameter in benzene.

crosphere in solution, the microsphere and AB diblock copolymers were blended above and below C^* [45]. First, the microsphere SV500-M was blended with S4V-100 or S4V-200 below C^* (1 wt %). The weight fraction f of the AB diblock copolymer in the blend was 0.5. Figure 16 shows the two-dimensional TEM micrographs of binary blends of SV500-M with S4V-100 or S4V-200 cast onto the carbon substrate and stained with OsO_4. For both blends, spheres and lines of P4VP are observed. The diameter of the spherical microdomain was 48 nm for both samples. The widths of the dark lines were 11 and 22 nm for the blends with S4V-100 and S4V-200, respectively. The diam-

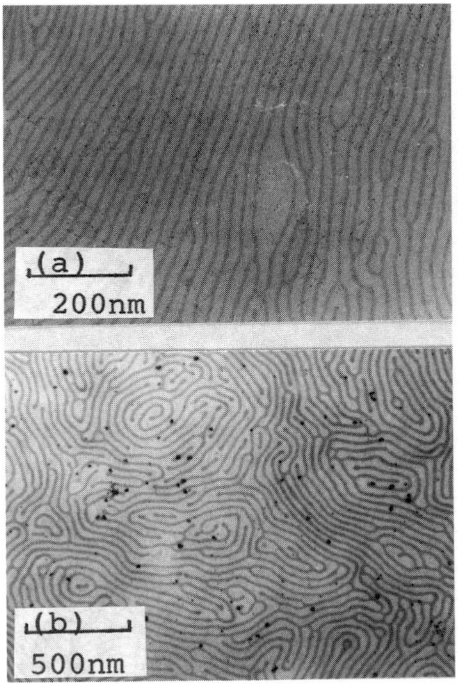

Figure 14 Transmission electron micrographs of poly(S-*b*-4VP) diblock copolymers stained with OsO_4: (a) S4V-100; (b) S4V-200. (From Ref. 45.)

eter of the spherical microdomain agrees well with the diameter of the P4VP core of the microsphere. The widths of the lines also agree well with the lamellar thickness of the block copolymers. It was concluded that the spherical microdomains were the P4VP cores of the microspheres and the P4VP lines were lamellar phases (bilayers) formed from the P4VP component of AB diblock copolymer in the blend.

Detailed observation of the blend of SV500-M with S4V-100 (Figure 16a) reveals an interesting and novel morphology. The P4VP cores of the microspheres form a regular structure, and the P4VP bilayer surrounds each microsphere with a honeycomb-like structure similar to a cell wall. Similar structures have been observed for ABC triblock copolymers [43]. However, the morphology of the ABC triblock copolymer has a cylindrical rather than spherical domain. On the other hand, in the blend of the microsphere with S4V-200

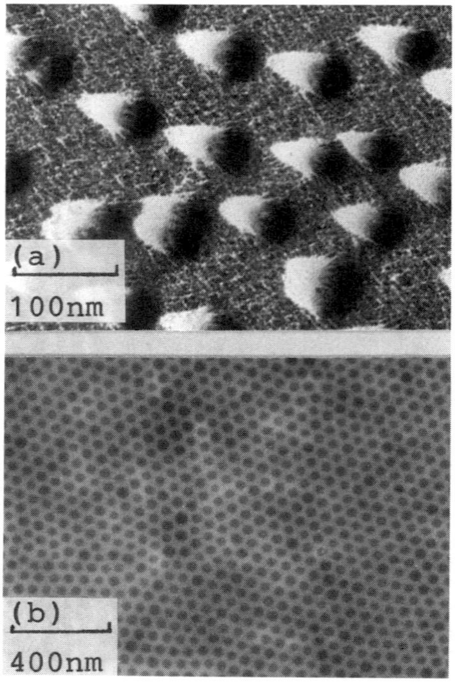

Figure 15 Transmission electron micrographs of microsphere SV500-M cast from benzene: (a) 0.05 wt % of polymer concentration and shadowed with Cr; (b) 1 wt % of polymer concentration. (From Ref. 45.)

(Figure 16b), no specific ordered structure can be observed. This suggests that the molecular weight of the diblock copolymer affects the morphology.

Next the novel morphology was investigated in detail by varying the weight fraction f of S4V-100 in the blend from 0.33 to 0.66 (Figure 17). For $f = 0.33$, the P4VP layer surrounded some microspheres in groups. The average number of microspheres in one domain surrounded by the P4VP layer (K) was 2.48. This indicates that the amount of S4V-100 was insufficient to surround each microsphere separately. For $f = 0.5$, as described above, most microspheres were surrounded by a P4VP layer similar to the cell wall. K was 1.08 for $f = 0.5$, and each microsphere was surrounded with a P4VP layer. For $f = 0.66$ (Figure 17c), all microspheres were surrounded with a P4VP layer of the S4V-100, and the P4VP spherical microdomains were locally ordered in hon-

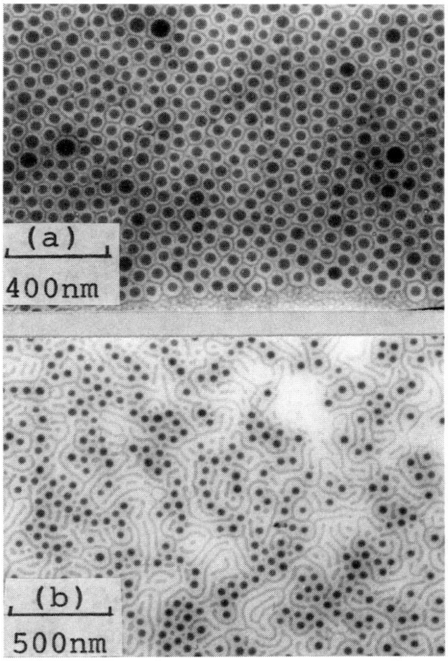

Figure 16 Transmission electron micrographs of binary blends of diblock copolymers with microsphere SV500-M at $f = 0.5$: (a) with S4V-100; (b) with S4V-200.

eycomb-like P4VP layers. However, wide and dark regions of P4VP were observed. These regions were horizontally oriented lamellar microdomains of S4V-100, resulting in a minimization of air–polymer surface tension [47]. It was confirmed that the S4V-100 was in excess of the microsphere SV500-M at $f = 0.66$.

In order to invesigate the packing state of the microspheres, the radial distribution function $g(r)$ of the P4VP spherical microdomain was measured and is shown in Figure 18. As shown in Figure 15b, the microsphere SV500-M itself is packed hexagonally. When all spheres are packed hexagonally, the peaks must appear at $r = b$, $\sqrt{3}b$, and $2b$ for the first, second, and third generations, respectively; where b indicates the distance from sphere n to sphere m at the first generation. Narrow peaks appear at r = 79.9, 139.4, and 159.7 nm, and the ratio of r for these peaks is $1 : \sqrt{3} : 2$ (Figure 18a).

At $f = 0.33$ (Figure 18b), the three peaks are wide. This indicates that the

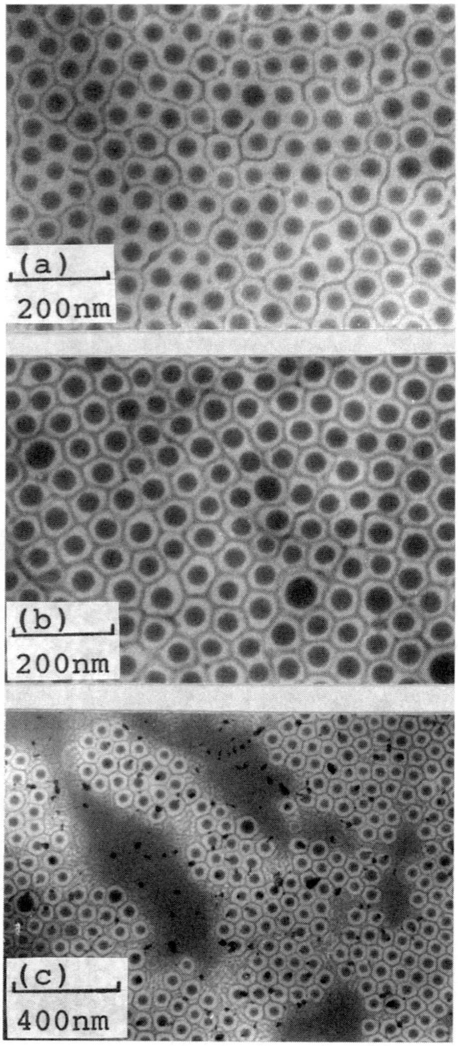

Figure 17 Transmission electron micrographs of binary blend SV500-M/S4V-100 (1.0 wt % of polymer concentration): (a) $f = 0.33$; (b) $f = 0.5$; (c) $f = 0.66$. (From Ref. 45.)

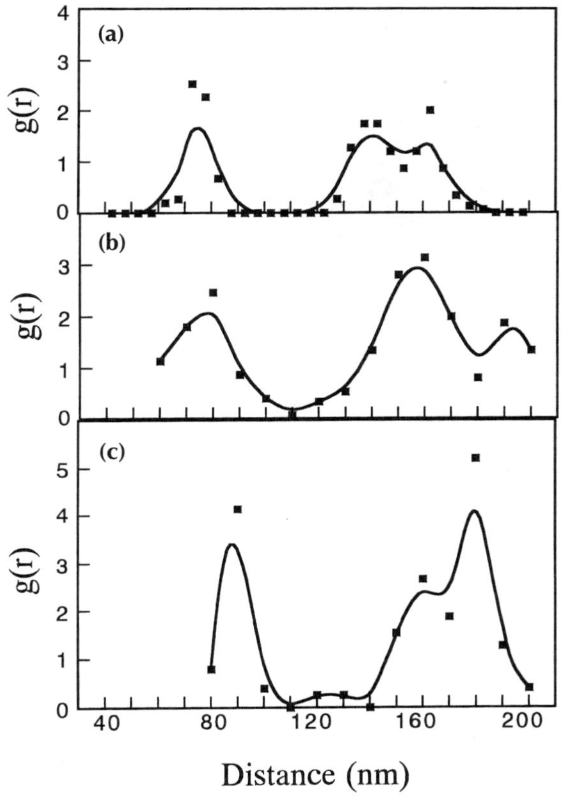

Figure 18 Distribution functions of microspheres of SV500-M/S4V-100 binary blend: (a) microsphere SV500-M; (b) blend at $f = 0.33$; (c) blend at $f = 0.5$. (From Ref. 45.)

packing structure of the microspheres was less ordered. For the blend at $f = 0.5$ (Figure 18c), three narrow peaks appear. The r values for the peaks are 88, 158, and 178, and the ratio of r for these peaks is 1 : 1.80 : 2.02. Thus, the microspheres were packed hexagonally, even though S4V-100 was blended into the system. It was concluded that novel morphology observed for SV500-M/S4V-100 blend with $f = 0.5$ had a completely ordered structure with hexagonal packing of the microspheres in two dimensions. Thomas et al. [48] and Birshtein and Zhulina [49] have proposed that the unit cell of the spherical microdomain

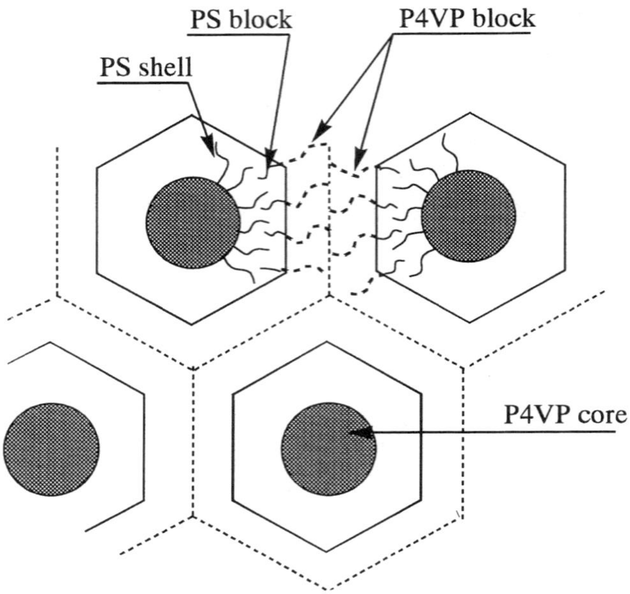

Figure 19 Schematic representation of two-dimensional chain conformations of diblock copolymer and microsphere in binary blend. (From Ref. 45.)

of the microphase-separated film is hexagonal in two dimensions. From these results the chain structure can be proposed, as shown in Figure 19 schematically. The cell of the microsphere has a hexagonal surface, and the AB diblock copolymer forms a bilayer between the microspheres. From the schematic rearrangement of the chains, the f value for a completely ordered morphology was calculated as 0.48. This value agrees well with the value $f = 0.5$, at which the blend showed a completely ordered morphology. On the other hand, when binary blend SV500-M/S4V-100 was cast from 5 wt %, that is, a concentration higher than C^* (4.4 wt %) at which ordering of the microspheres would occur, the ordered structure could not be obtained. This shows that the ordered superstructure of the blend was formed below the C^* concentration of the microsphere and maintained during the drying process.

We investigated controlling the morphology of microphase separation by blending the microsphere and block copolymer, which forms a spherical microdomain in a matrix on its own [46]. Figure 20 shows the TEM micrographs of diblock copolymer SI-100 and the blend film of SV500-M/SI-100. The mor-

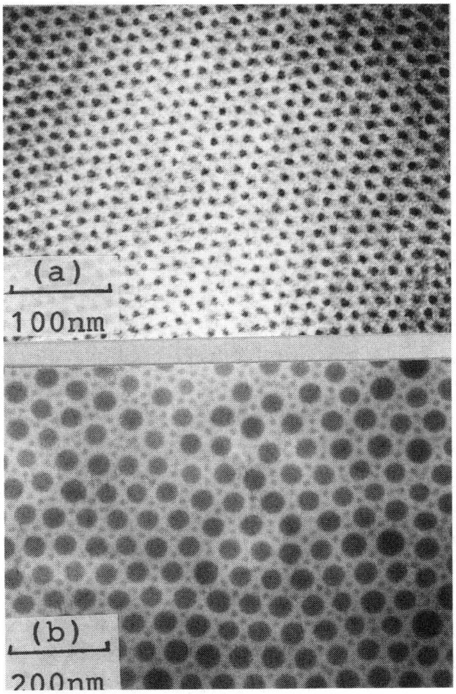

Figure 20 Transmission electron micrographs of diblock copolymer SI-100 and blend film of SV500-M/SI-100: (a) SI-100; (b) binary blend of SV500-M/SI-100. (From Ref. 46.)

phology of SI-100 shows the discrete PI spheres dispersed in a PS matrix. The polymer concentration of the blending solution was 1 wt %, that is, below C^*. An interesting morphology can be observed in Figure 20b. Six small spherical microdomains were arranged around each large spherical microdomain. The large spherical microdomains were hexagonally packed in two dimensions. The diameters of the spherical microdomains were 10 and 50 nm. These diameters agreed well with those of the PI spherical microdomain and P4VP core by themselves.

There are two possible explanations for ordered arrangements of the spherical microdomain when blended with AB diblock copolymer having low molecular weight. First, the short chain can be dispersed easily in the polymer matrix. Noolandi and Hong [50] calculated the conformation of the corona chain of polymer micelles in the blend of polymer micelle and homopolymer. When the molecular weight of homopolymer was smaller than that of the corona chain, the corona chain was expanded in the system and the homopolymer chain could be dispersed in the corona chain. The molecular weight of the block copolymer was much smaller than that of the PS shell; thus, the diblock copolymer dispersed between the microspheres and formed the spherical microdomain.

On the other hand, when considering the packing of the microspheres, the size factor should be taken into account. First, at equilibrium, microspheres should prefer to retain their symmetric, spherical shape, thus leaving gaps filled by block copolymer chains. The second explanation proposed is that the smaller microdomain, owing to the smaller molecular weight of the block copolymer, would be comfortable for the size problem (leaving gaps).

Quaternized P4VP core PS shell/poly(methacrylic acid) (PMA) core/PS shell type microsphere binary blend systems have a common PS sequence in the shell. Material with three-phase separated microdomains, such as both dispersed P4VP (positively charged region) and PMA (negatively charged region) spheres in a PS matrix, was obtained in this blend film [51].

III. STAR COPOLYMERS BY ORGANIZED POLYMERIZATION OF DIBLOCK MACROMONOMERS

It was mentioned in the preceding section that theoretical treatments of microphase separation of block copolymers were classified into two types for strong and weak segregation regimes. The phase behavior of AB diblock copolymer is determined by three experimentally controllable factors: the overall degree of polymerization N, the composition f (overall volume fraction of the A component), and the A–B segment–segment (Flory–Huggins) interaction parameter χ [8]. The first two factors are regulated through the polymerization stoichiometry and influence the translational and configurational entropies, whereas the magnitude of χ is determined by the selection of the A–B monomer pair. For symmetric (50/50 vol %) diblock copolymers, $(\chi N)_c = 10.5$, where subscript c denotes the order–disorder transition (ODT) [8]. More recently, the phase behaviors near the ODT of $(AB)_n$ and A_nB_n star block copolymers [3], ABA triblock copolymers [52], $(BC)_n$ multiblock copolymers [53,54], and ring block copolymers [55] have been derived using the mean-field principle. They exhibit different weak segregation behaviors at the ODT compared with that of diblock copolymers.

In this section the phase separation of $(AB)_n$ and A_nB_n star copolymers was discussed primarily. The preparation methods for both star copolymers are described in another book [4].

A. Organized Polymerization of Diblock Macromonomers

The schematic representation of $(AB)_n$ star and A_nB_n star copolymers are shown in Figure 21. Generally, the A_nB_n star copolymers were prepared by using a three-step anionic [56,57] or cationic [58] process. A living precursor polymer was made first, and used to initiate the polymerization of a small amount of a bis-unsaturated monomer, so as to build star molecules. Each of the resulting cores was linked to precursor chains that had contributed to its initiation. This living polymer was used subsequently to initiate the polymerization of another monomer, implying that new branches were growing out from the core.

On the other hand, the $(AB)_n$ star copolymers were prepared by the condensation of AB diblock anions using chlorosilane compound [59]. This star copolymer could also be synthesized by crosslinking diblock monocarbanionic [60,61] or monocarbocationic chains [62,63] with bis-unsaturated monomers.

We suggested a novel architecture for A_nB_n and $(AB)_n$ star copolymers by means of the organized polymerization of diblock macromonomers [64]. Block copolymers generally form the microphase-separated structure in the solid state and a micelle in solution. The concept of organized polymerization

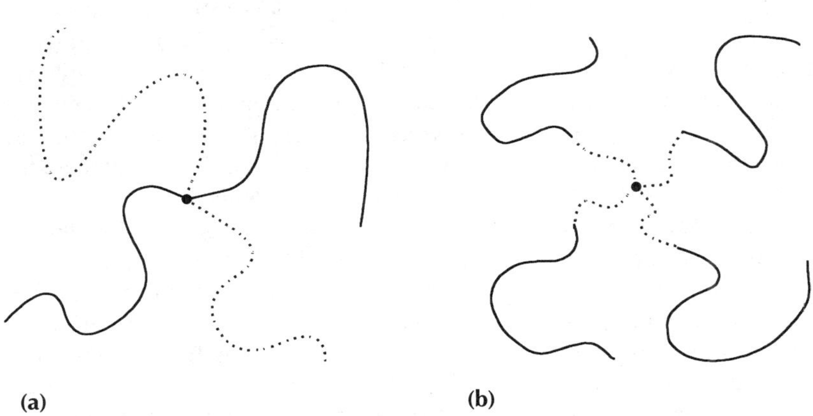

Figure 21 Schematic representation of star copolymers. (a) A_n B_n star block copolymer; (b) $(AB)_n$ star block copolymer.

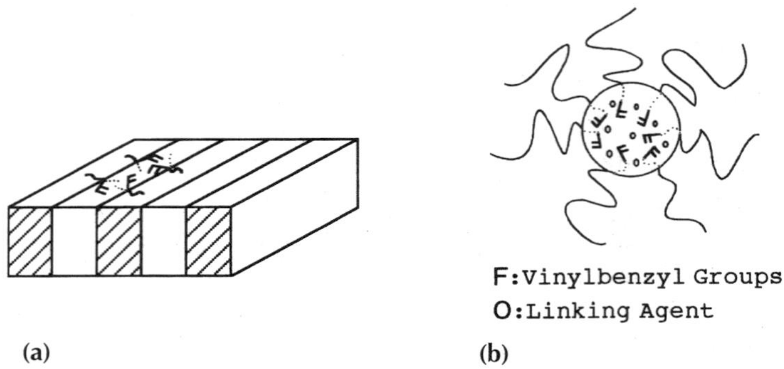

Figure 22 Schematic representations of conformation of AB diblock macromonomers in microphase-separated film and in micelle: (a) separated film; (b) micelle.

is based on utilizing a self-assembly of block copolymers as a reaction field. Schematic representations of the conformation of AB diblock macromonomers in microphase-separated film and in micelle are shown in Figure 22. Diblock macromonomers possessing central polymerizable groups are employed for the architecture of A_nB_n star copolymers. In the microphase-separated film (Figure 22a), the polymerizable groups at the position of the block junction should be oriented regularly at the domain interfaces. Some kind of polymerization of diblock macromonomer films is required in this environment to form the A_nB_n star copolymers. On the other hand, AB diblock macromonomers possessing terminal polymerizable groups are employed for the architecture for $(AB)_n$ star copolymers. The terminal polymerizable groups of AB diblock macromonomers are located with the condensed concentration within a micelle (Figure 22b). Thus, free radical polymerization of diblock macromonomers seems to proceed in an organized field, such as micellar domains, where the block junctions are located regularly at the interfaces of the micelles.

There have been attempts to prepare A_nB_n star copolymer using diblock macromonomers possessing central polymerizable groups. Berlinova and Panayotov [65] have prepared amphiphilic diblock macromonomers possessing a central unsaturated group. In the radical cis–trans isomerization of these diblock macromonomers, four- to eight-armed amphiphilic A_nB_n star copolymers were formed. We have reported the synthetic method for diblock macromonomers possessing central vinylbenzyl groups by anionic addition in a three-stage process using styrene, 1,4-divinylbenzene (DVB), and *t*-butyl methacrylate (BMA) monomers [66]. The morphology of poly(S-*b*-BMA) diblock macromonomer BP2 ($\overline{M}_n = 1.92 \times 10^4$, PS block 20.8 wt %, vinylbenzyl groups = 2.5 (number/1-polymer)) was a microstructure of PS spherical domains dispersed in a poly(*t*-butyl methacrylate) (PBMA) matrix. Photosensitive polymerization of diblock macromonomer BP2 film was carried out under various conditions [67]. A_nB_n star copolymer was formed in the polymerization system by adding both tetramethylthiuram disulfite (photosensitizer) and AIBN irradiated with a UV beam in a nitrogen atmosphere. This star copolymer exhibited an A_3B_3-type molecular structure.

The functional diblock macromonomers possessing central isoprene groups have also been synthesized by anionic addition in a three-stage process using styrene, isoprene, and 2VP monomers [68]. These functional diblock copolymers formed microphase-separated structures in the solid state. The P2VP blocks of the cast film were converted into P2VP·HCl salts by exposing them to HCl vapor. Crosslinking of the isoprene groups at the domain interfaces was carried out by soaking them in *n*-heptane/chloroform : 8/2 (v/v) mixture of S_2Cl_2 at room temperature, varying S_2Cl_2 concentration and reaction time. As a result, microgelation (addition condensation) of poly(S-*b*-2VP) diblock macromonomers formed A_nB_n star copolymers having the arm number $n = 3$–19.

Vinylbenzyl-terminated poly(S-*b*-I) diblock macromonomers were prepared by the direct coupling of corresponding diblock anions with a large excess of *p*-chloromethyl styrene (CMS) [69]. Free radical polymerization (in benzene initiated by AIBN) of these macromonomers formed the comb-shaped $(AB)_n$ copolymers. Both arm number and polydispersity $\overline{M}_w/\overline{M}_n$ increased gradually with an increase of macromonomer concentration $[M]$.

Subsequently, microgelation of these diblock macromonomers was carried out with the addition of ethylene glycol dimethacrylate (EGDM) into the preceding polymerization system [69]. The copolymerization of diblock macromonomers with EGDM led to microgelation in PI domains of micelles formed by macromonomers. The arm number of $(AB)_n$ star copolymers depended strongly on $[M]$ and the feed mole ratio of EGDM to $[M]$. In conclusion, $(AB)_n$ star copolymers ($n = 8$–30, $\overline{M}_w/\overline{M}_n = 1.1$–1.2) could be synthesized by means of organized polymerization of diblock macromonomers.

B. Phase Separation of Star Copolymers

According to calculation by de la Cruz and Sanchez [3], as the arm number n increases, the core of the $(AB)_n$ star copolymer (composition $f = \sim 0.2$) will naturally become richer in A monomers and the monomers deep in the core will be effectively screened from interacting with B monomers. Therefore, it is interesting to clarify the self-micellization of $(AB)_n$ star copolymers and their packing structures in solution and in the solid state.

We studied the self-micellization of $(AB)_n$ star copolymers as a parameter of arm number n by TEM observation of microphase-separated structure in the strong-segregation regime [70,71]. We also clarified the packing structure of $(AB)_n$ star copolymers in bulk film by SAXS measurements [71]. At present, the most rigorous way to prepare $(AB)_n$ star copolymers possessing more than 10 arms is to crosslink AB diblock monocarbanions with DVB. Then $(AB)_n$ star copolymers are prepared by crosslinking poly(S-*b*-I) diblock anions with a small amount of DVB (~55%, *m*-/*p*-isomer = 2) in benzene. The resulting solution is stirred at 20°C for 48 hours and the temperature is subsequently raised to 50°C. The $(AB)_n$ star copolymers are obtained by precipitation fractionation in the benzene–methanol system. Characteristics and domain sizes of $(AB)_n$ star copolymers are listed in Table 9. The morphology of poly(S-*b*-I) diblock arm precursor ($\overline{M}_n = 3.79 \times 10^4$, PI block 19.1 wt %) showed the texture of dispersed PI spheres (average radius $\overline{R}_{PI} = 5.8$ nm) in a PS matrix.

The microphase-separated structure of $(AB)_n$ star copolymers was observed in the solid state. The information obtained corresponds to the microphase separation in the strong segregation regime. All of the TEM micrographs of S2-1F–S2-3F star copolymers showed the morphology of dispersed PI spheres in a PS matrix. R_0 (in Table 9) is the calculated radius of segregated PI spheres, assuming that $(AB)_n$ star copolymer forms unimolecular micelles. Each observed value of $\overline{R}_{PI}$ was slightly larger than the calculated value of R_0. The measuring errors on the TEM micrographs appeared to be caused these dif-

Table 9 Domain Sizes of $(AB)_n$ Star Copolymers[a]

	$(AB)_n$ star copolymer		*Domain size (nm)*		
No.	*$\overline{M}_w \times 10^{-5}$ of PI blocks*	*Arm number*	$\overline{R}_{PI}$[b]	R_0[c]	$\overline{D}$[d]
S2-1F	1.06	1.56	6.0	3.6	20.6
S2-2F	1.42	20.9	7.1	3.9	23.8
S2-3F	1.79	26.3	7.5	4.2	26.0

[a]Composition of S2-F star copolymers, 19.1 wt % PI blocks.
[b]Average radius of PI spheres.
[c]Calculated radius of PI spheres, assuming that $(AB)_n$ star copolymer forms unimolecular micelle.
[d]Average diameter of microsphere particles.

ferences. In order to estimate both the shape and external size of star copolymers, TEM observations of the S2-F series (cast from dilute solution), which had been stained with Cr, were carried out. It was found from this shadowed sample that the star copolymer was a microsphere particle. The diameter $\overline{D}$ is also given in Table 9.

Figure 23 shows the observed domain size $\overline{R}_{PI}$ of star copolymers with the calculated domain sizes of diblock copolymers [72], as a function of the molecular weight of the PI blocks forming the spherical domains of poly(S-*b*-I) diblock copolymers. The domain size of the arm diblock copolymer is almost identical to that predicted from the equilibrium theory of Helfand and Wassermann [72]. The absolute values of PI spheres for $(AB)_n$ star copolymers are far below the theoretical line for diblock copolymers, and the 2/3 power law is not consistent in star copolymers. As predicted by de la Cruz and Sanchez [3], the $(AB)_n$ star copolymers ($n > 10$) begin to develop a core–shell type structure in the weak segregation regime. During solvent evaporation, the PI cores cannot interpenetrate with each other. The microphase-separated structure of star copolymers is formed with the dimension of a unimolecular micelle in the strong segregation regime.

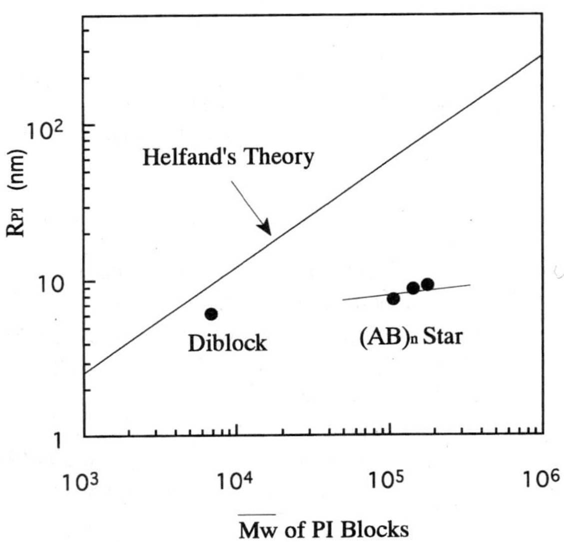

Figure 23 Plot of the average sphere size $\overline{R}_{PI}$ versus the molecular weight $\overline{M}_W$ of the PI blocks for $(AB)_n$ star copolymers and diblock copolymer. (From Ref. 71.)

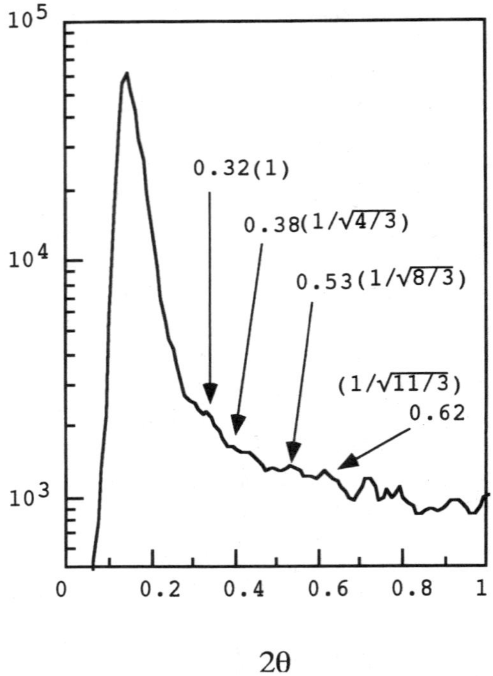

Figure 24 SAXS intensity distributions for S1-3F star copolymer film in the small-angle region; the arrows show the scattering maxima. The values in parentheses indicate cubic packing d_1/d_i. (From Ref. 71.)

Next, the structural arrangement of such star copolymers in the bulk film was studied by means of SAXS measurement [71]. Figure 24 shows the SAXS intensity distributions in the small-angle region for many-armed star copolymer S1-3F ($\overline{M}_w = 4.9 \times 10^6$, $n = 91.5$, 16.0 wt % PI blocks) film, where the arrows show the scattering maxima. It is shown that the first four peaks appear close together at the relative angular positions of $1 : \sqrt{4/3} : \sqrt{8/3} : \sqrt{11/3}$. These values correspond to packing pattern of (111), (200), (220), and (311) planes in a FCC structure. Star copolymers are packed in a superlattice of FCC structure in the bulk.

Subsequently, we measured the SAXS of $(AB)_n$ star copolymers near the C^*. Table 10 lists the characteristics and C^* of star copolymer S1-2F used in this work [73]. Below C^*, the star copolymers remain isolated, as any arrange-

Table 10 Characteristics and C^* of $(AB)_n$ Star Copolymer S1-2F

	Characteristics				
Code	$\bar{M}_w \times 10^{-5a}$	*PI block*[b] *(wt %)*	*Arm number*	$\bar{D}_h{}^c$ *(nm)*	C^* *(wt %)*
S1-2F	22.1	16.0	41.2	48.0	6.3

[a]Determined by GPC with small-angle light scattering and RI double detectors.
[b]Determined by ^{1}H NMR.
[c]Hydrodynamic diameter determined by DLS.

ment of star copolymers in solution is expected near or above C^*. The benzene solution of S1-2F star copolymer had a bluish tint at 6.0 wt % (calculated C^* = 6.2 wt %). Figure 25 shows the SAXS intensity distributions for S1-2F solution (6.0 wt % of polymer concentration) in the small-angle region, where the arrows show the scattering maxima. The values in parentheses indicate the cubic packing (d_1/d_i) at the scattering angles relative to the angle of the first maximum. It is found that the first four peaks appear closely at the relative angular positions of $1 : \sqrt{2} : \sqrt{3} : 2$. These values correspond to a packing pattern of (110), (200), (211), and (220) planes in a BCC structure.

Spatial packing of the cubic lattice in solution is considered as follows. The measured Bragg spacing d_1 is related to the cell edge a_c of the cubic lattice and the nearest-neighbor distance of the spheres D_0 (see Figure 26):

$$D_0 = (\sqrt{3}/2)a_c = (\sqrt{3/2})d_1 \qquad \text{for BCC}$$

Considering d_1 = 36.8 nm for S1-2F star copolymer, the D_0 and a_c were estimated to be 45.1 and 52.0 nm, respectively. The hydrodynamic diameter of S1-2F star copolymer particles ($\bar{D}_h$ = 48 nm) was almost the same value as D_0.

Structural transformation of such $(AB)_n$ star copolymers can be explained by the thermal blob model, as mentioned in Section II. C. Not only the core–shell type copolymers but also $(AB)_n$ star copolymers led to hierarchical structure transformation of cubic lattice in the film formation to minimize the thermodynamic energy.

More recently, Matsushita et al. [74] have reported $(AB)_n$ star copolymers (composition f = 0.5, n = 3–12) with lamellar structures. As a result, lamellar domain spacings of the star copolymers were essentially the same as those of the arm molecules, that is, linear diblock copolymers, irrespective of the arm number. It is necessary to accumulate information on the strong segre-

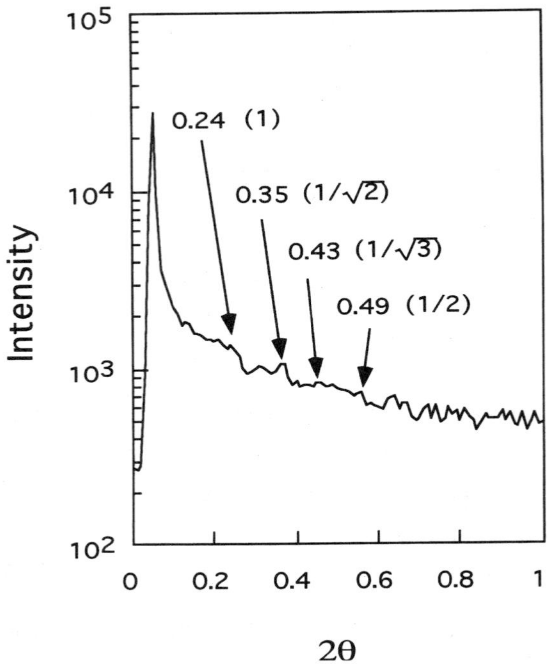

Figure 25 SAXS intensity distributions for S1-2F star copolymer in 6.0 wt % benzene solution in the small-angle region; the arrows show the scattering maxima. The values in parentheses indicate the cubic packing d_1/d_i. (From Ref. 73.)

gation regime for many-armed $(AB)_n$ star copolymers with cylindrical or lamellar structures.

On the other hand, reports on microphase separation of A_nB_n star copolymers have apparently not been published to date. We have recently synthesized A_nB_n starlike (comb-shaped) copolymers by anionic polymerization of binary vinylbenzyl-terminated PS and PI macromonomers and investigated the microphase-separated structures [75]. These A_nB_n star copolymers formed a clear microphase-separated structure. In this type of A_nB_n star copolymer, there are two entropic effects at work that oppose one another, according to de la Cruz and Sanchez's theory [3]. The first is the entropy of the melt. The entropy of the star copolymer is smaller than that of the corresponding diblock copolymer melt because of the additional constraints on the A–B junction point. However, for noncritical compositions, there will be a lowering of the transition entropy

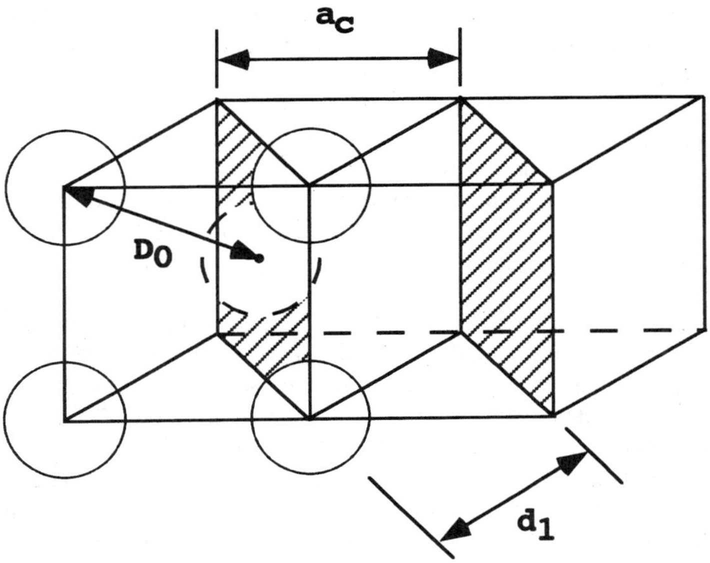

Figure 26 Schematic illustration of BCC lattice. (From Ref. 73.)

caused by the junction constraint. As a result, it is easier to phase-separate star copolymers than the corresponding diblock copolymers. It is very interesting to obtain clear microphase separation of A_nB_n star copolymers.

IV. MACROMONOMERS HAVING HYPERBRANCHED STRUCTURE

A. Discussion

Dendritic macromolecules are hyperbranched fractal-like structures that emanate from a central core and contain a large number of terminal groups. Two synthetic approaches have been reported for the preparation of these macromolecules: the divergent [76–78] and convergent growth approaches [79–82]. In both methods many synthetic steps are necessary to produce high molecular weight materials. To avoid synthetic problems, the macromonomer with hyperbranched dendritic moiety may be one of the most useful materials for the dendritic macromolecules.

Hawker and Frechet [83] have synthesized polyether macromonomer in which the macromolecular part is a hyperbranched dendritic moiety. Typical [G-4]-styrene macromonomer (**2**) was prepared from the reaction of hydroxymethyl-terminated [G-4]-OH polyether macromolecules (**1**) with *p*-chloromethyl styrene according to Scheme 1. Subsequently, several novel copolymers were prepared from copolymerization of the macromonomers with styrene by free radical polymerization.

Polyguanamine macromonomer (**4**) was synthesized by self-addition reaction of 2-amino-4-(*m*-amino-*N*-alkylanilino)-6-isopropenyl-1,3,5-triazine (**3**) in *N,N*-dimethyl-2-acetamide at 120°C for about 300 hours as shown in Scheme 2 [84].

Percec and co-workers [85] have synthesized macromonomers (**7**) having the tapered 3,4,5-tris(*p*-dodecyloxybenzyl)benzoate fragment of the structural unit. These macromonomers were prepared by the reaction of corresponding compounds (**5**) with methacryloyl chloride (**6**), as shown in Scheme 3.

Their corresponding polymethacrylates (**8**) were obtained by free radical polymerization. These taper-shaped monoesters of oligo(ethylene oxide) (EO) with 3,4,5-tris(*p*-dodecyloxybenzyloxy)benzoic acid and their corresponding polymers self-assembled into a tubular supramolecular architecture displaying an enantiotropic columnar hexagonal (Φ_h) mesophase. This interesting supramolecular architecture is mentioned in the next section.

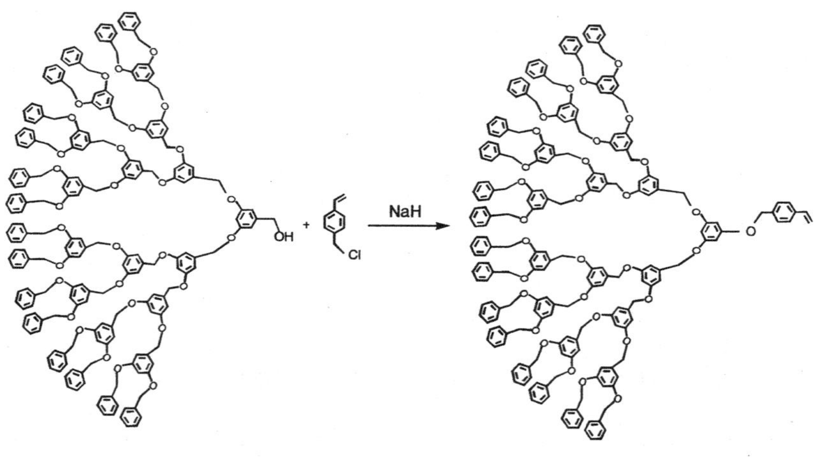

[G4]-OH

1

[G4]-styrene

2

Scheme 1

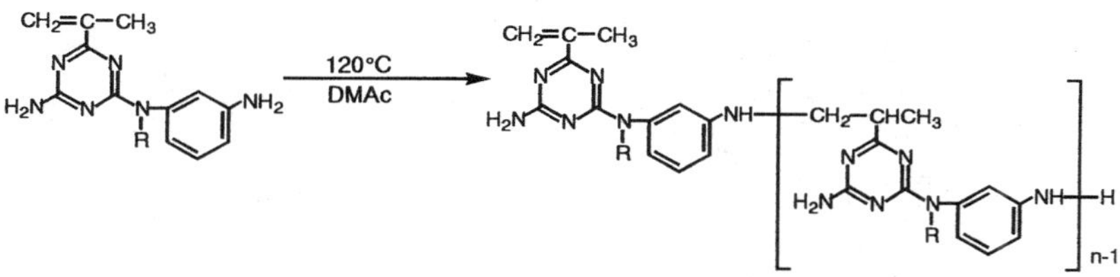

AN2,AN18

APAN2 : R=C_2H_5
APAN18: R=n-$C_{18}H_{37}$

3 4

Scheme 2

B. Self-Ordering of Star Polymer and Macromonomer Having Hyperbranched Structure

The phase behavior of the compounds (**5a–f**) and (**8a–d**) was determined by a combination of techniques consisting of differential scanning calorimetry (DSC), thermal optical polarized microscopy, wide- and small-angle X-ray scattering, and molecular modeling. As a result, characterization of this supramolecular architecture suggested a model in which the stratum of the column was formed by between 3.9 and 5.4 molecules of (**5**) or 4.5 and 5.9 repeat units of (**8**) with their oligo(oxyethylenic) segments melted and segregated in the center of the column and their melted alkyl tails radiating toward the columns periphery. These results indicate that endo-recognition by H-bonding of the oligo(oxyethylenic) receptor of (**5**) (most probably functioning by hydrophobic–hydrophobic interactions) provide the driving force for the self-assembly of this tubular supramolecular architectures (Figure 27). In the supramolecular architectures derived from (**8**), the H-bonding interaction is replaced by the poly(methacrylate) backbone, which leads to a Φ_h mesophase that undergoes isotropization at temperatures that are 40°C higher than those of (**5**). The channel penetrating the middle of the supramolecular cylinders derived from both (**5**) and (**8**) dissolved alkali–metal triflates, and the ionic interaction generated by the dissolved ion- pairs enhanced the thermal stability of their Φ_h phase.

Zimm and Stockmayer calculated the unperturbed dimensions of regular and randomly branched star polymers on the assumption that each chain segment adopts a random walk [86]. They showed that regular star polymers constitute the simplest architecture through which the segment density in a polymer coil can be increased at constant molecular weight. Daoud and Cotton [1] developed a scaling model that describes the radial distribution of the segment

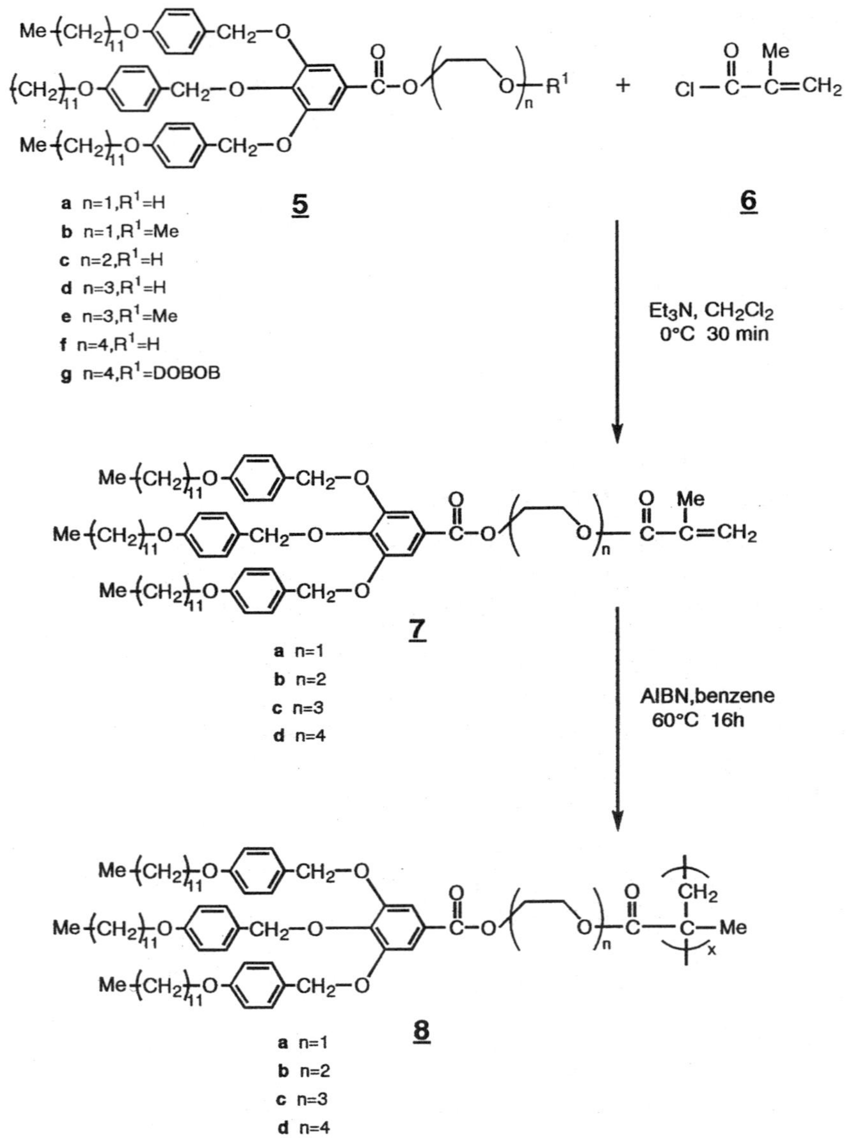

Me(CH2)11O
(CH2)11O
Me(CH2)11O
CH2—O
O
R1
+
O Me
Cl—C—C=CH2
a n=1,R1=H
b n=1,R1=Me
c n=2,R1=H
d n=3,R1=H
e n=3,R1=Me
f n=4,R1=H
g n=4,R1=DOBOB
5
6
Et3N, CH2Cl2
0°C 30 min
7
a n=1
b n=2
c n=3
d n=4
AIBN,benzene
60°C 16h
8
a n=1
b n=2
c n=3
d n=4

Scheme 3

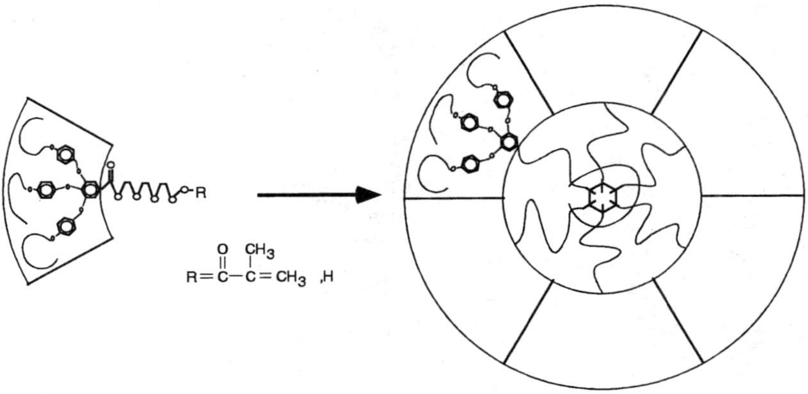

Figure 27 A top view of the schematic representation of the macromonomer **7c,** being polymerized to the self-assembled cylindrically shaped polymer **8c** in the Φ_h phase. (From Ref. 85.)

density more realistically in isolated stars with many arms. Because of their high segment density, regular stars with many arms are also predicted to have peculiar properties near the C^* in good solvents [2,87]. Roovers and co-workers [88,89] have reported the synthesis of regular star polybutadienes and their characterization by their dilute solution properties. The 64- and 128-arm stars have been prepared from the coupling of living poly(butadienyl)lithium with dendrimers having chlorosilane bonds at the surface. The dilute solution properties of the stars were determined in a good solvent (cyclohexane) and in a θ-solvent (dioxane) for polybutadiene. Measurements of the radius of gyration (R_G), second virial coefficient (A_2), translational diffusion coefficient (D_0), and intrinsic viscosity ([η]) indicated that the isolated stars behaved as hard spheres. The ratio of the hydrodynamic radius over R_G was slightly larger than $(5/3)^{1/2}$. It was found that $g = \langle R_G^2\rangle / \langle R_G^2\rangle_{lin} \propto f^{-0.73}$ in the good solvent and $f^{-0.64}$ in the θ-solvent, where $\langle R_G^2\rangle_{lin}$ and f indicate the mean square radius of gyration of the linear polymer and arm number, respectively. The former result is in good agreement with the Daoud–Cotton scaling model of stars. The latter dependence is stronger than the theory predicts: $g_\theta \propto f^{-0.5}$. The properties of these star polymers suggest that they behave much like hard-sphere particles in dilute solution.

More recently, Widawski et al. [90] have reported that the preparation of porous polymer membranes controlled both the size distribution and relative positions of pores. They have found a way to generate polymer films with an essentially monodisperse pore size, in which the pores are organized sponta-

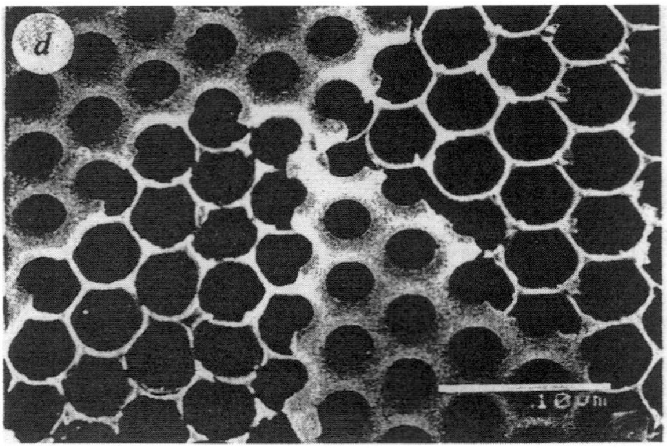

Figure 28 Scanning electron micrograph of a star polystyrene film showing the honeycomb morphology. (From Ref. 90.)

neously into periodic hexagonal arrays (Figure 28). The films were produced by evaporating solutions of star-shaped PS or poly[S-paraphenylene(PP)] block copolymers in carbon disulfide (CS_2) under a flow of moist gas. They are aggregated both in solution and in the solid state into quasi-spherical micelles with an insoluble polyparaphenylene (PPP) core surrounded by a PS shell (Figure 29). The mechanism of formation of the honeycomb morphology can be described as follows: Rapid evaporation increases the superficial concentration and induces a cooling of the solution surface, producing condensation of water. These three factors provoke a gelation process and phase separation, with the formation of a polymer film at the solution surface. Convection currents seen in the solution at the beginning of the evaporation process may be due to local variations of the superficial tension. These currents may favor a regular stacking of the spheres that result from the demixing process. With the return to room temperature, the liquid pressure would create holes in the superficial spheres, allowing the evaporation of solvent and water. Within the framework of this model, star-shaped polymers play a major role, firstly in decreasing the solution viscosity, and secondly in favoring the gelation and phase separation.

V. CONCLUSION

The interest in star polymers arises from their compactness and from their enhanced segment density. Essentially, both the core–shell type microspheres and

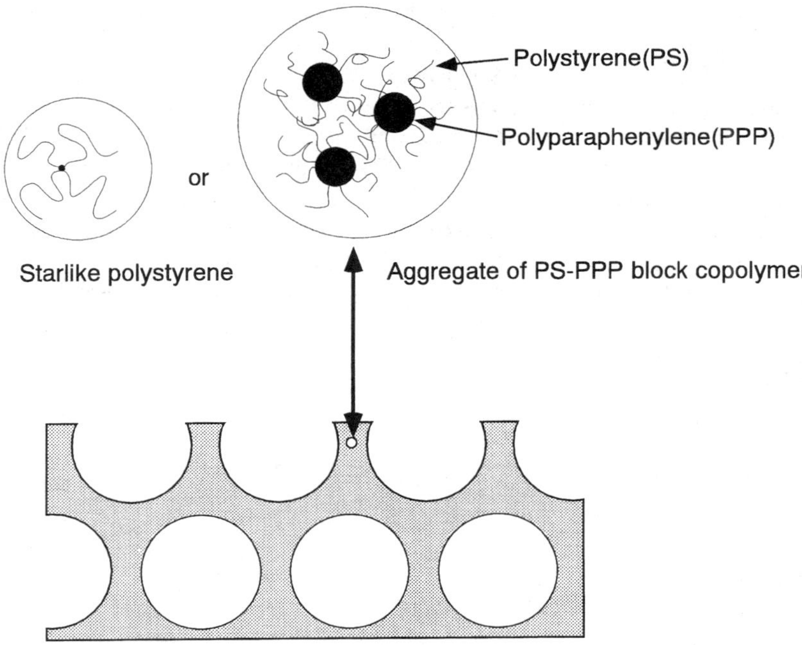

Figure 29 Schematic cross section of a regular microporous film prepared by evaporation of CS_2 solutions of star-shaped polystyrene or block copolymer. (From Ref. 90.)

$(AB)_n$ star copolymers (volume fraction of B block $f = \sim 0.2$) have a conformational structure similar to that of star polymers.

The core–shell type microspheres were obtained by crosslinking of the self-assemblies (spherical microdomains) in the microphase-separated film of diblock copolymers. The solubility of these microspheres depended strongly on the solubility of block chains constructing shell parts. The core–shell type microspheres formed a lattice with a BCC structure near the C^*. In bulk film, this structure changed to an FCC lattice. The crosslinked cores did not interpenetrate with each other and the microspheres were compatible with shells during solvent evaporation. Structural transformation of such microspheres could be explained by a thermal blob model. The core–shell type microspheres can be considered as the smallest units in the microphase-separated structure. These microspheres can be thought of as one spherical molecule with multibranchings in the solvent. The novel three-phase separated morphologies such as honeycomb structure were observed by microsphere/diblock copolymer binary blends.

The $(AB)_n$ and A_nB_n star copolymers were obtained by organized polymerization of diblock macromonomers. The $(AB)_n$ star copolymers (volume fraction of B block $f = \sim 0.2$) formed a lattice with a BCC structure near the C^*. In bulk film, this structure changed to an FCC lattice. Not only the core–shell type microspheres but also star copolymers led to hierarchical structure transformation of cubic lattice in the film formation to minimize the thermodynamic energy.

Thus, the hyperbranched polymers show supramolecular structures in solution and in the solid state. It will be necessary to create high functionalized materials using these polymers.

REFERENCES

1. M. Daoud and J. P. Cotton, *J. Phys. (Les Ulis,* Fr.) **43,** 531 (1982).
2. T. A. Witten, P. A. Pincus, and M. E. Cates, *Europhys. Lett.,* **2,** 137 (1986).
3. M. O. de la Cruz and I. C. Sanchez, *Macromolecules,* **19,** 2501 (1986).
4. K. Ishizu, in *Macromolecular Design: Concept and Practice,* M. K. Mishra, Ed., Polymer Frontiers International, New York, 1994, p. 39.
5. E. Helfand, *Macromolecules,* **8,** 552 (1975).
6. E. Helfand and Z. R. Wasserman, *Macromolecules,* **9,** 879 (1976).
7. T. Ohta and K. Kawasaki, *Macromolecules,* **19,** 2621 (1986).
8. L. Leibler, *Macromolecules,* **13,** 1602 (1980).
9. G. H. Fredrickson and E. Helfand, *J. Chem. Phys.,* **87,** 697 (1987).
10. G. H. Fredrickson, *Macromolecules,* **20,** 2535 (1987).
11. P. F. Green, T. M. Christensen, T. P. Russell, and R. Jerome, *J. Chem. Phys.,* **92,** 1478 (1990).
12. P. F. Green, T. M. Christensen, and T. P. Russell, *Macromolecules,* **24,** 252 (1991).
13. H. Hasegawa, H. Tanaka, K. Yamasaki, and T. Hashimoto, *Macromolecules,* **20,** 1651 (1987).
14. E. L. Thomas, D. B. Alward, D. L. Kinning, D. L. Martin, D. L. Handlin, Jr., and L. J. Fetters, *Macromolecules,* **19,** 2197 (1986).
15. F. S. Bates, *Science,* **251,** 845 (1991).
16. A. Prochazka, M. K. Baloch, and Z. Tuzar, *Makromol. Chem.,* **180,** 2521 (1979).
17. Z. Tuzar, B. Bednar, C. Konak, M. Kubin, S. Svobodova, and K. Prochazka, *Makromol. Chem.,* **183,** 399 (1982).
18. B. Bednar, J. Devaty, B. Koupalova, and J. Kralicek, *Polymer,* **25,** 1178 (1984).
19. M. H. Park, R. Saito, K. Ishizu, and T. Fukutomi, *Polym. Commun.,* **29,** 230 (1988).
20. M. H. Park, K. Ishizu, and T. Fukutomi, *Polymer,* **30,** 202 (1989).
21. R. Saito, K. Ishizu, T. Nose, and T. Fukutomi, *J. Polym. Sci., Polym. Chem. Ed.,* **28,** 1793 (1990).

22. R. Saito, K. Ishizu, and T. Fukutomi, *Polymer,* **31,** 679 (1990).
23. R. Saito, K. Ishizu, and T. Fukutomi, *Polymer,* **32,** 531 (1991).
24. R. Saito, K. Ishizu, and T. Fukutomi, *Polymer,* **32,** 2258 (1991).
25. R. Saito, K. Ishizu, and T. Fukutomi, *J. Appl. Polym. Sci.,* **43,** 1103 (1991).
26. R. Saito, K. Ishizu, and T. Fukutomi, *Polymer,* **33,** 1712 (1992).
27. R. Saito, K. Ishizu, and T. Fukutomi, *Trends Polym. Sci.,* **3,** 125 (1993).
28. K. Ishizu and T. Fukutomi, *J. Polym. Sci., Polym. Lett. Ed.,* **26,** 281 (1988).
29. K. Ishizu, *Polymer,* **30,** 793 (1989).
30. R. Saito, H. Kotsubo, and K. Ishizu, *Polymer,* **33,** 1073 (1992).
31. R. Saito, H. Kotsubo, and K. Ishizu, *Eur. Polym. J.,* **27,** 1153 (1991).
32. K. Ishizu and A. Onen, *J. Polym. Sci., Polym. Chem. Ed.,* **27,** 3721 (1989).
33. K. Ishizu, F. Naruse, and R. Saito, *Polymer,* **34,** 3929 (1993).
34. R. Saito, N. Kawachi, and K. Ishizu, *Polymer,* **35,** 867 (1994).
35. K. Ishizu, J. *Colloid Interface Sci.,* **156,** 299 (1993).
36. K. Ishizu, M. Sugita, H. Kotsubo, and R. Saito, *J. Colloid Interface Sci.,* **169,** 456 (1995).
37. W. W. Glaessley, *Adv. Polym. Sci.,* **16,** 1 (1974).
38. R. Saito, H. Kotsubo, and K. Ishizu, *Polymer,* **35,** 1747 (1994).
39. Y. Chevalier, C. Pichot, C. Graillat, M. Joanicot, K. Wong, J. Maquet, P. Lindner, and B. Cabane, *Colloid Polym. Sci.,* **270,** 806 (1992).
40. G. Riess, M. Schlienger, and G. Marti, *J. Macromol. Sci. Phys.,* **B17,** 355 (1980).
41. Y. Mogi, H. Kotsuji, Y. Kaneko, K. Mori, Y. Matsushita, and I. Noda, *Macromolecules,* **25,** 5408 (1992).
42. C. Auschra and R. Stadler, *Macromolecules,* **26,** 217 (1993).
43. S. P. Gido, D. W. Schward, E. L. Thomas, and M. C. Goncalves, *Macromolecules,* **26,** 2636 (1993).
44. K. Ishizu, A. Omote, and T. Fukutomi, *Polymer,* **31,** 2135 (1990).
45. R. Saito, H. Kotsubo, and K. Ishizu, *Polymer,* **35,** 1580 (1994).
46. R. Saito, H. Kotsubo, and K. Ishizu, *Polymer,* **35,** 2296 (1994).
47. K. Ishizu, Y. Yamada, and T. Fukutomi, *Polymer,* **31,** 2047 (1990).
48. E. L. Thomas, D. J. Kinning, D. B. Alward, and C. S. Henkee, *Macromolecules,* **20,** 2934 (1987).
49. T. M. Birshtein and E. B. Zhulina, *Polymer,* **27,** 1078 (1986).
50. J. Noolandi and K. M. Hong, *Macromolecules,* **15,** 482 (1982).
51. K. Ishizu and R. Saito, *Polym. Plast. Technol. Eng.,* **31,** 607 (1992).
52. M. D. Gehlsen, K. Almdal, and F. S. Bates, *Macromolecules,* **25,** 939 (1992).
53. H. Benoit and G. Hadziioannou, *Macromolecules,* **21,** 1449 (1986).
54. S. D. Smith, R. J. Spontak, M. M. Satkowski, A. Ashraf, A. K. Heape, and J. S. Lin, *Polymer,* **35,** 4527 (1994).
55. J. F. Marko, *Macromolecules,* **26,** 1442 (1993).
56. C. Tsitsilianis, P. Chaumont, and P. Rempp, *Makromol. Chem.,* **191,** 2319 (1990).
57. C. Tsitsilianis, S. Graff, and P. Rempp, *Eur. Polym. J.,* **27,** 243 (1991).
58. S. Kanaoka, T. Omura, M. Sawamoto, and T. Higashimura, *Macromolecules,* **25,** 6407 (1992).

59. L. K. Bi and L. J. Fetters, *Macromolecules,* **9,** 732 (1976).
60. L. K. Bi and L. J. Fetters, *Macromolecules,* **8,** 90 (1975).
61. D. B. Alward, D. J. Kinning, E. L. Thomas, and L. J. Fetters, *Macromolecules,* **19,** 215 (1986).
62. S. Kanaoka, M. Sawamoto, and T. Higashimura, *Macromolecules,* **24,** 5741 (1991).
63. S. Kanaoka, M. Sawamoto, and T. Higashimura, *Makromol. Chem.,* **194,** 2035 (1993).
64. K. Ishizu, *Polym. Plast. Technol. Eng.,* **32,** 511 (1993).
65. I. V. Berlinova and I. M. Panayotov, *Makromol. Chem.,* **190,** 1515 (1989).
66. K. Ishizu, S. Yukimasa, and R. Saito, *Polymer,* **33,** 1982 (1992).
67. K. Ishizu, S. Yukimasa, and R. Saito, *Polym. Commun.,* **32,** 386 (1991).
68. K. Ishizu and K. Kuwahara, *J. Polym. Sci., Polym. Chem. Ed.,* **31,** 661 (1993).
69. K. Ishizu, S. Yukimasa, and R. Saito, *J. Polym. Sci., Polym. Chem. Ed.,* **31,** 3073 (1993).
70. K. Ishizu and S. Yukimasa, *Polymer,* **34,** 3753 (1993).
71. K. Ishizu and S. Uchida, *Polymer,* **35,** 4712 (1994).
72. E. Helfand and Z. R. Wassermann, *Macromolecules,* **11,** 960 (1978).
73. K. Ishizu and T. Uchida, *J. Colloid Interface Sci.,* **175,** 293 (1995).
74. Y. Matsushita, T. Takasu, K. Yagi, K. Tomioka, and I. Noda, *Polymer,* **35,** 2862 (1994).
75. K. Ishizu and K. Kuwahara, *Polymer,* **35,** 4907 (1994).
76. D. A. Tomalia, H. Baker, J. Dewald, M. Hall, G. Kallos, S. Martin, J. Roeck, J. Ryder, and P. Smith, *Polym. J.,* **17,** 117 (1985).
77. D. A. Tomalia, H. Baker, J. Dewald, M. Hall, G. Kallos, S. Martin, J. Roeck, J. Ryder, and P. Smith, *Macromolecules,* **19,** 2466 (1986).
78. G. R. Newkome, Z. Yao, G. R. Baker, and V. K. Gupta, *J. Org. Chem.,* **50,** 2004 (1985).
79. C. J. Hawker and J. M. J. Frechet, *J. Chem. Soc., Chem. Commun.,* **15,** 1010 (1990).
80. C. J. Hawker and J. M. J. Frechet, *J. Am. Chem. Soc.,* **112,** 7638 (1990).
81. K. L. Wooley, C. J. Hawker, and J. M. J. Frechet, *J. Am. Chem. Soc.,* **113,** 4252 (1991).
82. C. J. Hawker, R. Lee, and J. M. J. Frechet, *J. Am. Chem. Soc.,* **113,** 4583 (1991).
83. C. J. Hawker and J. M. J. Frechet, *Polymer,* **33,** 1507 (1992).
84. H. Kunisada, Y. Yuki, S. Kondo, Y. Nishimori, and A. Masuyama, *Polym. J.,* **23,** 1455 (1991).
85. V. Percec, J. Heck, D. Tomas, F. Falkenberg, H. Blackwell, and G. Ungar, *J. Chem. Soc., Perkin Trans.* 1, **22,** 2799 (1993).
86. B. H. Zimm and W. H. Stockmayer, *J. Chem. Phys.,* **17,** 1301 (1949).
87. T. A. Witten and P. A. Pincus, *Macromolecules,* **19,** 2509 (1986).
88. L.-L. Zhou and J. Roovers, *Macromolecules,* **26,** 963 (1993).
89. J. Roovers, L.-L. Zhou, P. M. Toporowski, M. van der Zwan, H. Iatrou, and N. Hadjichristidis, *Macromolecules,* **26,** 4324 (1993).
90. G. Widawski, M. Rawiso, and B. Francois, *Nature,* **369,** 387 (1994).

7

Star Polymers via a Controlled Sol–Gel Process

Timothy E. Long
Virginia Polytechnic Institute and State University, Blacksburg, Virginia

Larry W. Kelts, Thomas H. Mourey, and Jeffrey A. Wesson
Eastman Kodak Company, Rochester, New York

I. INTRODUCTION

Living anionic polymerizations are especially suited for the preparation of oligomers and macromolecules with well-defined end-group functionality (macromonomers and telechelics), predictable molecular weight, and narrow molecular weight distribution. Because of the living nature of the propagating chain end intermediate, an appropriate electrophilic reactant can be added to the stable polymeric carbanion in order to place a specific functionality at the chain end. Reagents such as carbon dioxide [1] chlorosilanes [2], alkylene oxides [3], and benzyl chlorides [4] have been added to living carbanions to generate a terminal functional group. For example, the addition of $CO_2$to polystyryllithium leads to a lithium carboxylate anion and upon acidification generates a terminal carboxylic acid. A review by Morton [5] describes suitable additives and the corresponding functionalities formed.

Most efforts in end-group functionalization have been directed toward the preparation of difunctional oligomers for subsequent cocondensation to form a high molecular weight linear polymer or crosslinking with a multifunctional crosslinking agent to form an insoluble network [6,7]. However, monofunctional living chain ends have also served as key intermediates for radial star copolymer syntheses [8,9]. In addition, macromonomers are prepared by adding a molecule that contains both a site for the attack of a monofunctional polymeric carbanion and a site for subsequent free-radical copolymerization. Many novel graft copolymers have been prepared by utilizing the macromonomer technique [10–12]. Consequently, the ability to control the functionality ($f = 1$ or 2) allows for the preparation of a wide variety of well-defined macromolecular architectures.

Star polymers are generally synthesized by two popular arm-first strategies. The first process involves coupling of "living" anionic polymers with divinylbenzene (DVB) [13–17]. After complete polymerization of the linear arms, the sequential polymerization of DVB generates a network-like hub. The number of arms in the star polymer is controlled by varying the molar ratio of DVB to "living" anion, and star polymers containing as many as 56 weight average arms have been prepared [16–18]. Arms with predictable molecular weights and stars with narrow molecular weight distributions are possible by this technique. The second process has been pioneered by Fetters et al. [19,20]. This process, in a similar fashion to DVB coupling, involves direct reaction of the "living" polymeric arms. A molecule with a plurality of silyl chloride functionalities is added to an excess of the polymeric arm. In favorable cases, the number of arms in the final star polymer is equivalent to the number of silyl chloride functionalities in the coupling reagent. The preparations of these multifunctional coupling reagents generally involves multiple steps, and the compounds are difficult to isolate, characterize, and store for long periods of time. In addition, the excess homopolymer arm must be extracted in order to generate a well-defined star polymer. The condensation coupling of trialkoxysilyl-terminated macromonomers described here offers advantages over both procedures.

This work describes the direct reaction of living polymeric carbanions to form well-defined trialkoxysilyl functional organic polymers. The macromonomers ($f = 1$) were subsequently hydrolyzed and condensed to yield narrow molecular weight distribution star polymers. Although telechelic ($f = 2$) trialkoxysilyl functional oligomers have also been prepared and studied in detail, the product after hydrolysis and condensation is an insoluble network, not a soluble star polymer. These crosslinked condensation products are not amenable to structural characterization in solution.

The influence of the polymeric substituent alpha to the silicon atom on the subsequent sol–gel reaction kinetics has not been addressed in detail. Many workers have described the effect of organic substituents on the hydrolysis and condensation kinetics of conventional sol–gel monomers. For example, Schmidt and co-workers have studied the effects of organic substituent on low molecular weight monomers such as methyltriethoxysilane [21]. This report also addresses three different environments alpha to the silicon atom at the polymer chain end, and their influence on hydrolysis and condensation sol–gel reaction kinetics. The macromonomer structures are depicted in Figure 1.

II. MACROMONOMER SYNTHETIC TECHNIQUES

Two criteria must be met for the selection of a suitable functionalization reagent. First, an electrophilic site for direct deactivation of the polymeric carbanion must be present in the molecule. The extent of the functionalization reaction must be characterizable by a variety of complementary techniques (e.g., spectroscopic and chemical). This is not a trivial task since the concentration of the end group is generally quite low. Second, if the molecule contains the ultimate end-group functionality, then this group must be inert to a polymeric

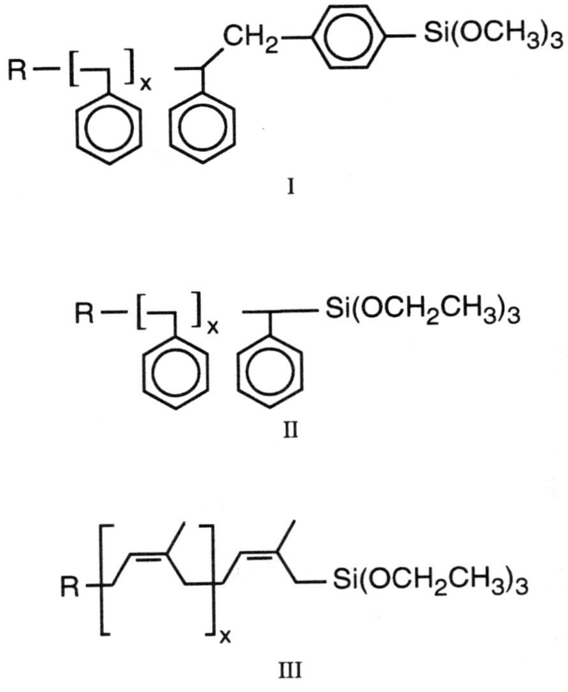

Figure 1 Trialkoxysilyl functionalized macromonomer structures.

carbanion at appropriate conditions for attack at the electrophilic site. Choromethylphenyl trimethoxysilane (CMPTMS) and triethoxychlorosilane (TECS) met both structural prerequisites and were commercially available. Details of the experimental techniques including monomer purification, polymerization processes, end-capping strategies, and characterization techniques were described earlier [22,23].

The end-capping of living polystyllithium with CMPTMS is depicted in Scheme 1. TECS was added in a similar fashion employing a 50% excess to generate triethoxysilyl end groups with an adjacent benzyl unit (Figure 1 II). The orange color that is associated with the polystyryllithium carbanion disappears immediately upon addition of end- capping reagent. ^{1}H NMR analysis indicates the presence of the phenyltrimethoxysilyl group (3.6 ppm). Presence of the initiator fragment residing at the other end of the polymer chain is verified by resonances between 0.6 and 1.2 ppm. Either the initiator fragment or the trialkoxysilyl group integration was compared to the repeat unit methylene and methine integration to determine functional molecular weights. Table 1 shows

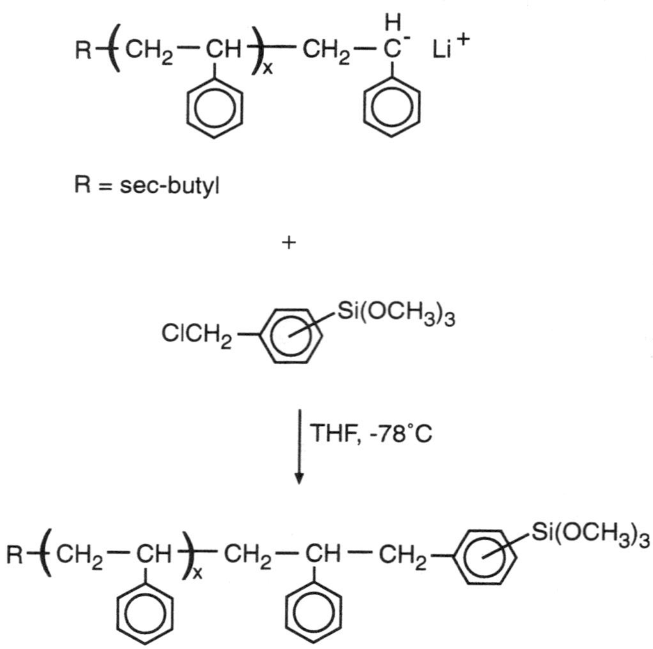

Scheme 1 End-capping reaction of polystyryllithium with CMPTMS.

Table 1 Molecular Weight Determinations for *p*-(Chloromethylphenyl) Trimethoxysilane-Terminated Polystyrene Oligomers[a]

M_n (GPC)[b]	M_n (NMR)[c]	M_n (NMR)[d]	M_w/M_n
3900	3000	3100	1.19
3100	2700	2700	1.13
3900	3500	3500	1.22

[a]Polymerization conditions: THF, –78°C, sec-butyllithium.
[b]Polystyrene standards, THF, 25°C, DRI detector.
[c]Ratio repeat unit resonance to initiator fragment (sec-butyl).
[d]Ratio repeat unit resonance to $Si(OR)_3$.

functional molecular weights for various end-capped polystyrene samples. Excellent agreement exists between functional molecular weights based on the initiator fragment and functional molecular weights based on the trialkoxysilyl end group. In addition, functional molecular weights compare favorably with number average molecular weights determined by size exclusion chromatography (SEC). These observations support an efficient and quantitative end-capping reaction. Molecular weight distributions are also narrow (1.10–1.20), which indicates a well-defined polymerization and efficient functionalization reaction. The preparation of low molecular weight oligomers with a narrow molecular weight distribution (~1.05) is difficult in polar solvents because of rapid rates of initiation and propagation.

Neutron activation analysis (NAA) also was used to verify the presence of both silicon and oxygen in the polymer and to calculate functional molecular weights. Functional molecular weights were obtained by comparing the percent silicon or oxygen to the percent carbon. These values agree within 5% of the M_f values determined by ^1H NMR spectroscopy.

III. SOL–GEL COUPLING AND ANALYSIS

^{29}Si NMR is the most discriminatory technique for characterization of these end-capped oligomers. A typical spectrum for a 3100 g/mol functionalized polystyrene oligomer is shown in Figure 2. In most cases, only a resonance that is associated with a phenyltrimethoxysilyl group is observed at approximately –54 ppm. However, often hydrolysis and condensation of the end groups during either precipitation or exposure to air lead to a small amount (2–10%) of dimer formation. The dimer SiOSi resonance appears at approximately –60

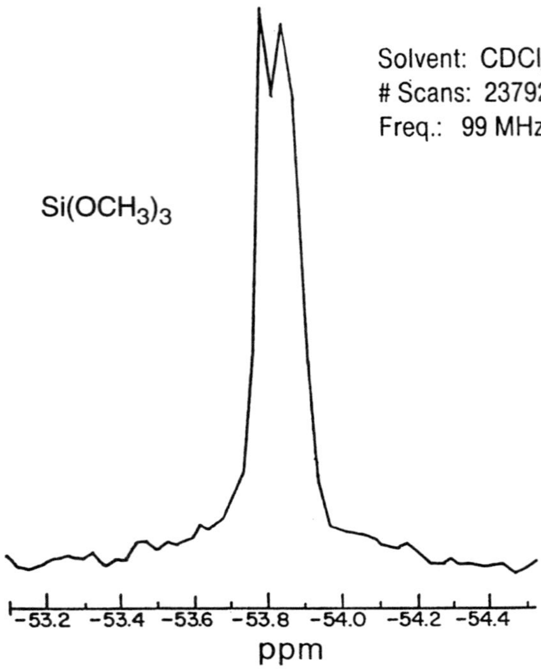

Figure 2 ^{29}Si NMR spectrum of a 3100 g/mol functionalized polystyrene oligomer.

ppm. ^{29} Si NMR also confirms that the displacement of the methoxy group with a second living carbanion during functionalization does not occur. A side reaction of this nature would lead to formation of a $CSi(OR)_2C$ resonance with a substantially different chemical shift. It should be noted that the splitting (8–10 Hz separation) in the trialkoxysilyl resonance is reproducible and is currently being investigated. Preliminary explanations attribute this observation to polymer tacticity at the functionalized chain end (i.e., heterotactic and syndiotactic adjacent units).

Thin-layer chromatography (TLC) is a useful technique for the qualitative characterization of functional oligomers. Polymers ascend conventional TLC plates in a fashion similar to low molecular weight organic compounds. The nature of the end group alters the interaction characteristics of the macromolecule with the silica gel plate. TLC demonstrates the absence of homopolystyrene in the functionalized oligomers. Quirk [24] has shown that this technique is capable of detecting 1% unfunctionalized oligomer. In fact, unfunctionalized oligomer was intentionally added to the functionalized oligomer in order to demonstrate the effectiveness of this technique.

The molecular weight of each macromonomer must be accurately known to calculate the number of arms in the polymer condensates. Practically, it is difficult to measure the true molecular weights of trimethoxysilyl-terminated polystyrenes directly because small amounts of material condense through the reactive end groups during or immediately after isolation of the macromonomer. To simplify characterization of the starting macromonomers, a small fraction of each living polystyrene was separated and terminated with a proton by addition to methanol. The remaining material was then terminated with (trimethoxysilyl)phenyl groups. The proton-terminated polystyrenes are not contaminated with coupled macromonomer.

SEC and ^{1}H NMR confirmed that residual [*p*-(chloromethyl) phenyl]trimethoxysilane-terminating reagent and related, small-molecule alkoxysilanes were not present in the macromonomers. NMR confirmed nearly complete functionalization of all but the 48,000 g/mole macromonomer, which contained on the order of 25% unfunctionalized material. The accuracy of NMR in the determination of the degree of functionalization in high molecular weight macromonomers can be questioned. It is only important to recognize in the following discussion that small amounts of unreactive, unfunctionalized material are present in *all* samples. No attempt was made to remove this unfunctionalized polymer, and we have assumed that it does not affect the chemistry of Equations (1) and (2) where R = oligomer.

$$R—Si(OCH_3)_3 + 3H_2O \qquad R—Si(OH)_3 + 3CH_3OH \tag{1}$$

$$R—Si(OH)_3 \qquad R—Si(OH)_2—O—Si(OH)_2—R + H_2O \tag{2}$$

The functionalized polymers were readily soluble in tetrahydrofuran with an excess of water (compared to silicon) added. Hydrolysis and condensation were catalyzed by addition of 1 *M* HCl. After allowing the solutions to dry to a film in air, the samples were heated in vacuo at various conditions. Scheme 2 depicts the hydrolysis and condensation to form dimer for a trialkoxysilyl-functionalized polystyrene macromonomer. Variables such as trialkoxysilyl group, molecular weight, functionality, and subsequent heat treatment have been addressed.

IV. CHARACTERIZATION OF STAR POLYMERS PREPARED VIA A SOL–GEL PROCESS

The solubility characteristics of hydrolyzed/condensed phenyltrimethoxysilyl-terminated polystyrene oligomers were particularly interesting. After hydroly-

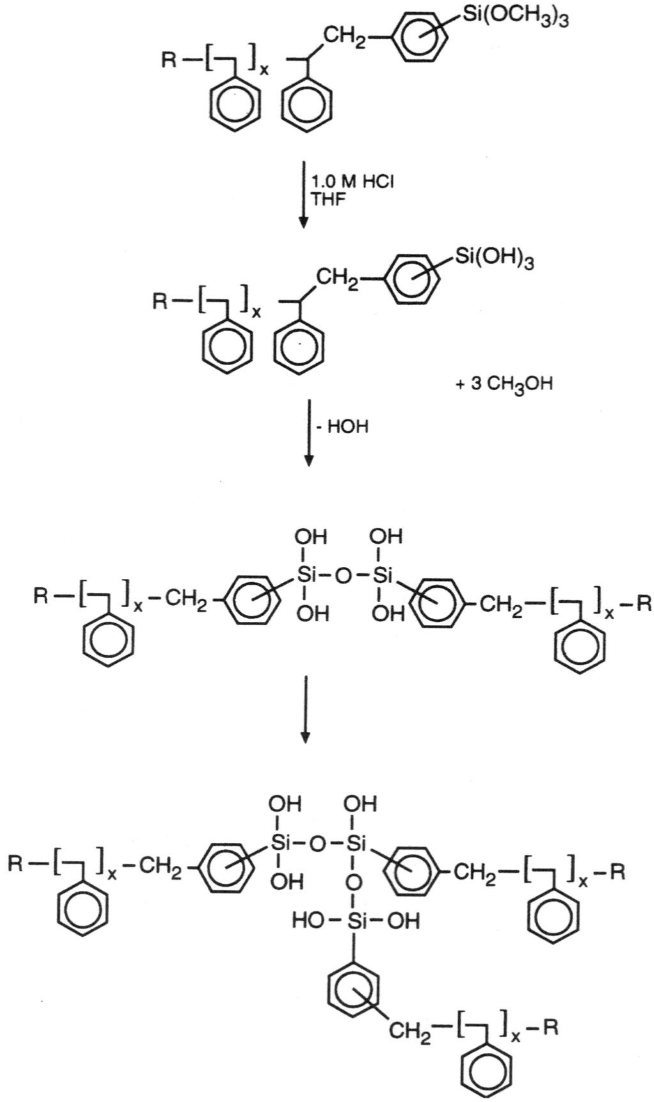

Scheme 2 Proposed star polymer growth mechanism based on ^{29}Si NMR data.

sis and condensation of the monofunctionally terminated oligomers, their films remained soluble in tetrahydrofuran. This characteristic facilitates the characterization of the reaction product by spectroscopic and gel permeation chromatographic techniques. Difunctional oligomers were rendered insoluble after hydrolysis and condensation and their films would only swell in tetrahydrofuran. Consequently, characterization of difunctional condensates was more difficult.

Solubility of the monofunctional condensates permits the facile characterization of hydrolysis and condensation reactions. A size exclusion chromatograph equipped with viscometric and light-scattering detectors was used to determine absolute molecular weight changes after hydrolysis and condensation. Figure 3 depicts the chromatograms of unreacted, phenyltrimethoxysilyl-terminated precursor (solid line) and the hydrolysis/condensation product after heating at 70°C for 3 h (dashed line). The most striking feature is the nar-

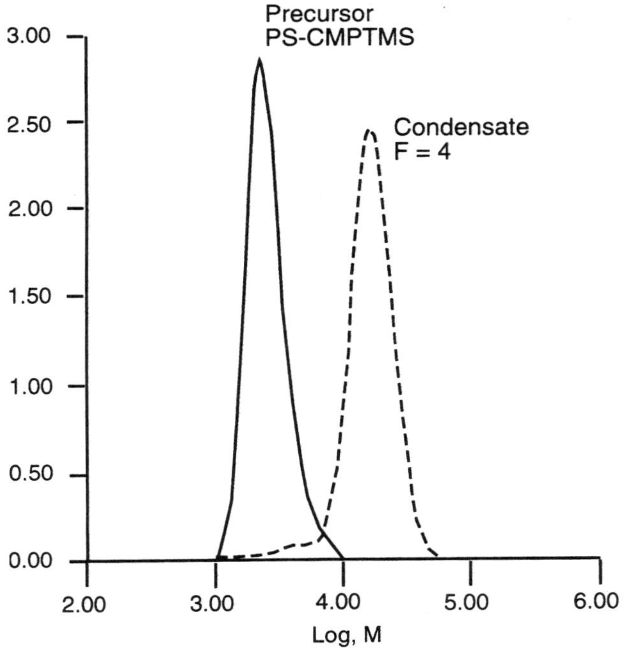

Figure 3 SEC chromatograms of precursor (*solid line*) and condensate (*dashed line*) after heating at 70°C for 3 hours.

row molecular weight distribution of the condensate. The weight average molecular weight of the condensate is approximately 4 times the molecular weight of the precursor (M_w = 5000 g/mol). This implies that a macromolecule with four branches was formed under these conditions. Absence of appreciable uncondensed precursor in the chromatogram of the condensate implies that the end-capping reaction was quite efficient. SEC evidence for quantitative functionalization was reproducibly observed for various phenyltrimethoxysilyl-terminated precursors with different molecular weights. The small shoulder on the low molecular weight side of the distribution may imply the presence of either a minor amount of unfunctionalized oligomer or functionalized oligomer (T°, uncondensed end groups) below the ^{29}Si NMR detection limit (<5 mol %).

The molecular weight of a functionalized macromonomer controlled the degree of condensation in the solid state. This observation was attributed to steric inhibition during acid-catalyzed cluster–cluster growth. In order to probe the effect of molecular weight on the hydrolysis and condensation kinetics in solution, different molecular weight macromonomers (3000 g/mol and 8,000 g/mol), which contained a single phenyltrimethoxysilane end group (Structure I in Figure 1), were synthesized. Figure 4 illustrates the effect of molecular weight on the condensation rate of the macromonomer in tetrahydrofuran at 30°C. The data implies that as the molecular weight of the oligomer increased,

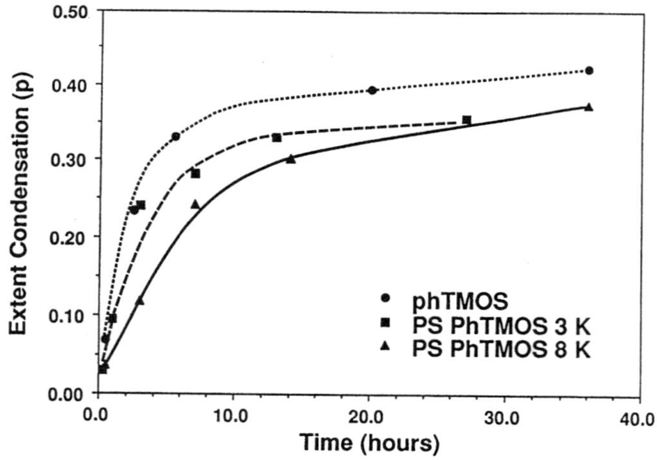

Figure 4 Effect of oligomer molecular weight on the condensation rate in THF at 30°C.

Table 2 Effect of Precursor Molecular Weight on Star Growth at 120°C/240 hours

Precursor M_n	*Condensate M_p*	*Number of arms*
2,100	35,100	17.0
3,700	56,600	15.5
8,400	84,200	10.0
49,800	137,000	2.7

Table 3 Effect of Processing on the Molecular Weight and Number of Arms

Sample conditions	M_w	M_w/M_n	*[n]*	*Number of arms*
Precursor	35,300	1.14	0.219	1
Cast film air-dried at 25°C	87,500	1.33	0.315	2.5
Cast film above and 3 h/75°C	95,300	1.33	0.311	2.7
Cast film above and 19 h/120°C	163,000	1.44	0.371	4.6
Cast film above and 72 h/120°C	214,000	1.44	0.375	6.1

the rate of condensation decreased. The extent of condensation (p) was determined using ^{29}Si NMR spectroscopy.

Table 2 further illustrates the effect of precursor molecular weight on the final estimated number of arms based on SEC peak molecular weight (M_p). All samples were condensed at identical conditions (i.e., 120°C for 240 hours).

Table 3 describes the effect of condensation temperature and time for a 35,300 g/mol CMPTMS-terminated polystyrene precursor. Condensation temperatures above the glass transition temperature (100°C) lead to higher degrees of condensation. The higher molecular weight precursor leads to a lower degree of branching, and 6.1 arms represents the maximum limit for this precursor. This observation is also illustrated in Table 2.

V. EFFECT OF ADJACENT BONDING ON THE SOL–GEL COUPLING REACTIVITY AND MECHANISM

Figure 5 illustrates the effect of replacing a hydrogen at the alpha position with a polystyrene oligomer. PS-BzTES (Figure 1II) hydrolysis was significantly slower than the hydrolysis of the model compound (benzyltriethoxysilane). In

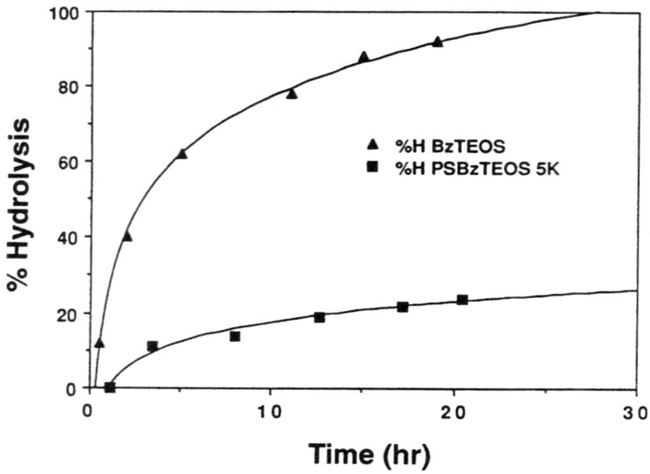

Figure 5 Hydrolysis rate for PS-BzTES (Figure 1-II)

addition, the partially hydrolyzed PS-BzTES oligomer did not undergo any condensation at 30°C in tetrahydrofuran during the time period shown in Figure 5. Consequently, it appears that when a trialkoxysilyl group is attached to the polymer backbone through a phenyl bridge, less steric inhibition is observed compared to the benzyl attachment.

When silicon trialkoxides hydrolyze (Equation (1)), the silicon nucleus becomes deshielded and ^{29}Si NMR resonances shift to a downfield, higher ppm. Figure 6 shows assignments for the three hydrolysis products of the phTMOS (−52–−55 ppm) model system. Condensation of hydrolysis products, as in Equation (2), results in an upfield, lower ppm, shift of ^{29}Si NMR resonances. Figure 7 shows assignments for PhTMOS with one condensation bond (T^1, −60 ppm—63 ppm; dimer formed during the polymerization process). T^2 silicons (two Si—O—Si bonds on the observed silicon) resonate approximately 8–9 ppm upfield. Cyclic species with three or four silicons in a ring (or cage) due to ring strain or small Si—O bond angles resonate further downfield (less negative) than other T^2 or T^3 silicons in larger rings, cages, or linear species. Dilute THF solutions of phenyltrialkoxysilyl monomers often completely hydrolyze before condensation begins (Figure 7).

Another interesting feature of these reactions is the appearance sequence of T^1 resonances. Fully hydrolyzed silicons form condensation bonds first. End groups with one methoxy appear after hydrolyzed end groups. This must be due

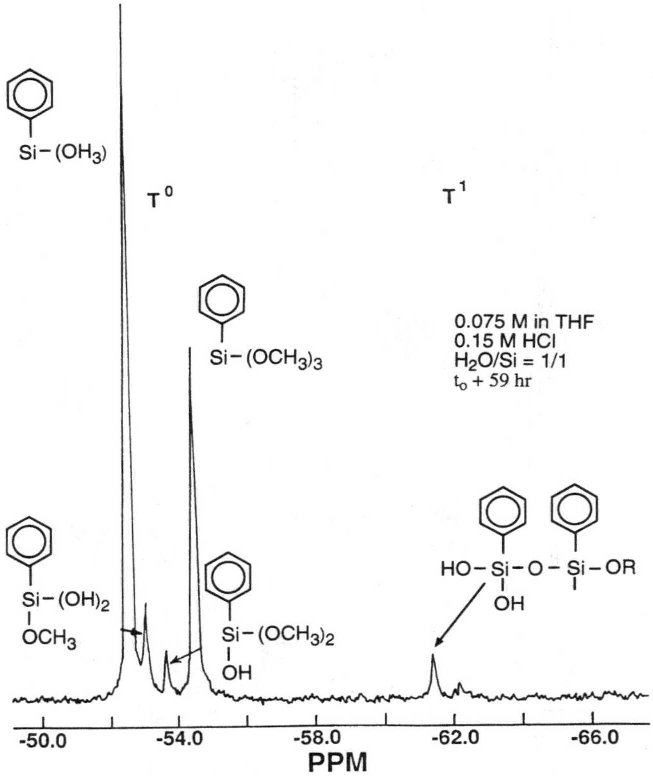

Figure 6 ^{29}Si NMR spectrum depicting various phTMOS hydrolysis products.

to the alcoholysis reaction between liberated methanol and silicon hydroxide. Steps to eliminate alcohol from the sols or steps to increase the condensation rate may be desirable to obtain maximum extent of reaction.

Monofunctional condensates were analyzed by ^{29}Si NMR in solution. The spectrum in Figure 8 implies that at least two different types of silicon are present in the branched molecule (precursor M_w = 5000 g/mol, condensate M_w = 20 200 g/mol). The chemical shifts are consistent with 60% T^2 and 40% T^1. T^2 represents a silicon atom that has undergone two condensation reactions (i.e., SiO*Si*OSi), and T^1 has one such linkage. This interpretation is consistent with assignments previously published for other sol–gel systems [25,27]. The presence of T^1 and T^2 implies that the condensate is not a single cyclic species

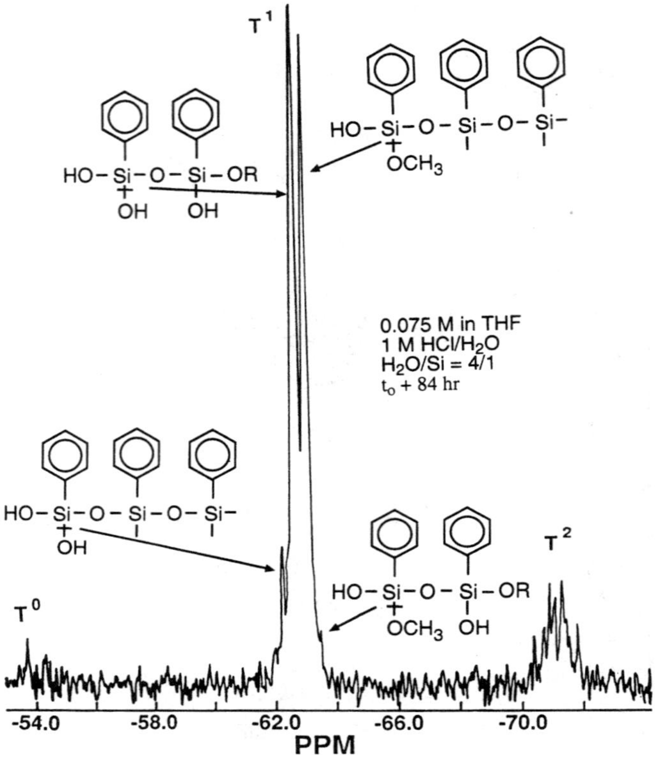

Figure 7 ^{29}Si NMR spectrum of phTMOS T^1 condensation products.

but may, for example, be a 20:80 mixture of cyclic/linear tetramers (cyclic T^2 and linear T^2 cannot be separated owing to the broad T^2 resonance, approximately 5 ppm). Although the star molecule has a narrow molecular weight distribution, it is probable that the distribution consists of different size species (e.g., trimer or pentamer) with different core structures. The condensate may simply consist of a linear silicon–oxygen backbone with pendant polystyrene branches as depicted in Scheme 2. In addition, since the core contains both T^1 and T^2 atoms, the exclusive formation of a silica-like tetrahedral structure (only T^3) or a cyclic (only T^2) structure in the core is impossible. In fact, the formation of T^3 is relatively difficult; it forms only at the expense of T^1 and T^2.

The final end group addressed was the attachment of the trialkoxysilyl group via a methylene bridge. PI-ATES (Figure 1III) contains such a methyl-

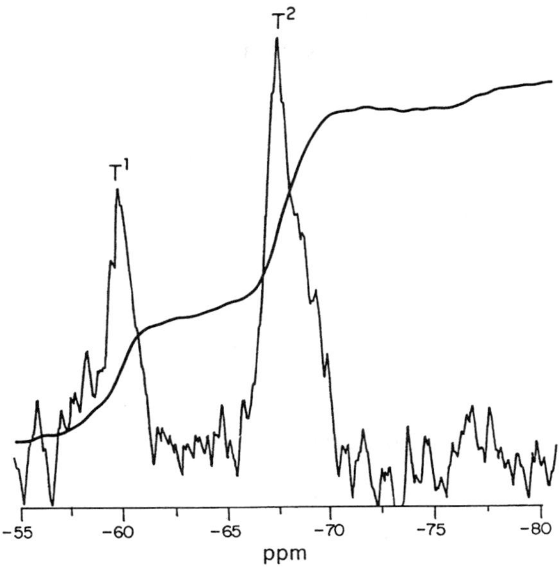

Figure 8 ^{29}Si NMR spectrum of a PS-phTMOS (M_w = 5000 g/mol precursor).

ene alpha to the silicon atom. Scheme 3 illustrates the functionalization of living polyisoprenyllithium with triethoxychlorosilane in hydrocarbon solvents. *Cis*-4,1-, *trans*-4,1-, and 4,3- addition adjacent to the terminal silicon atom was evident in the ^{29}Si NMR spectrum (Figure 9). Figure 9 also illustrates the loss of cis/trans evidence in the ^{29}Si NMR spectrum by addition of styrene units adjacent to the Si atom. However, it was found that PI-ATES was ideally suited for protonation of the terminal double bond to generate a positive tertiary carbenium ion beta to the silicon atom. ^{29}Si NMR confirmed the transformation of a T silicon atom ($RSiO_3$) to a Q silicon (SiO_4). Beta-silicon stabilization of intermediate carbenium ions during electrophilic attack of trialkylallylsilanes has been addressed earlier [28,29]. Scheme 4 depicts the loss of trialkoxysilyl functionality during acid-catalyzed hydrolysis.

The functionalized oligomers offer potential in the modification of inorganic surfaces, and subsequently aid in the adhesion promotion of organic coatings. Preliminary adhesion studies have shown that the precursors react quantitatively with silanol-containing surfaces. Extraction of the functionalized surfaces with THF did not remove the covalently bound polymers. Improved adhesion of a polystyrene homopolymer was evident when the molecular

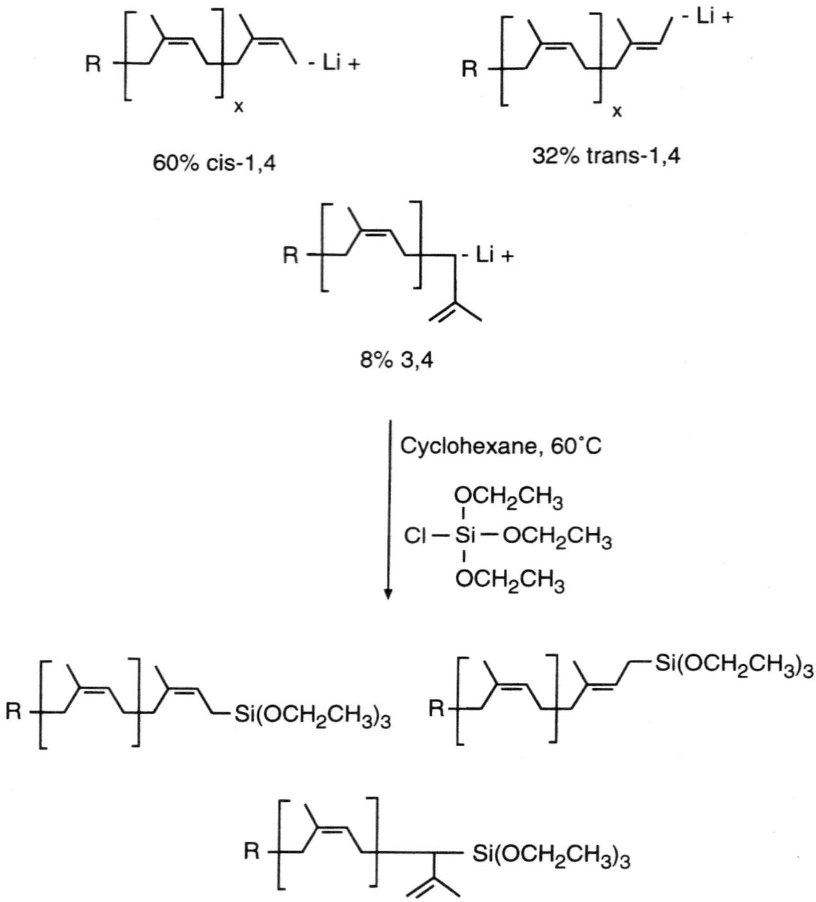

Scheme 3 Functionalization of polyisoprenyllithium with triethoxychlorosilane in hydrocarbon solvents.

weight of the covalently bound precursor was above the critical entanglement molecular weight (~20K g/mol for polystyrene). The star polymers also offer unique applications in the area of polymeric dispersants. Figure 10 depicts a model derivatization of the star core with a dye-containing chlorosilane. ^{29}Si NMR was used to confirm the covalent attachment of the desired molecule. Reaction generally occurred first at the T^1 and T^2 silicon centers because of the

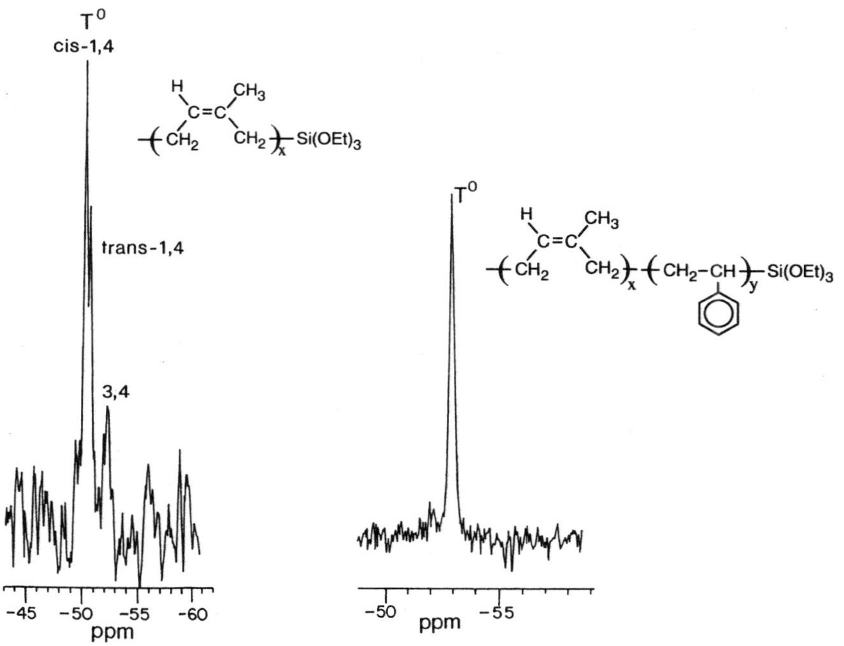

Figure 9 ^{29}Si NMR spectrum of triethoxysilyl-terminated polyisoprene.

steric accessability. It was demonstrated that the star polymers were capable of dispersing these organic molecules in various solvents at high solids concentration because of the low solution viscosity of the highly branched star molecule.

Table 4 depicts the effect of branching on absolute M_w and intrinsic viscosity $[n]$. Both condensates derived from 5000 and 2800 precursors have intrinsic viscosities significantly less than a linear analogue of equivalent molecular weight.

VI. CONCLUSION

Trialkoxysilyl-terminated oligomers were successfully prepared by conventional anionic techniques with predictable molecular weights, narrow molecular weight distributions, and high functionality. The monofunctionalized oligo-

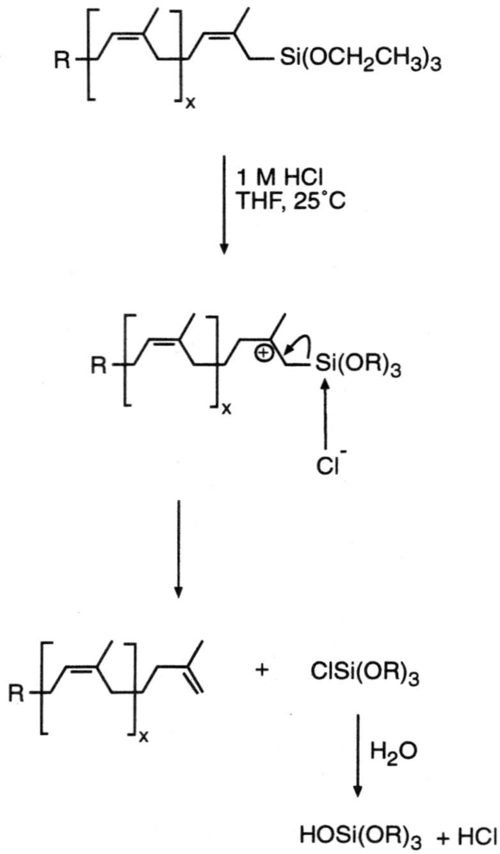

Scheme 4 Proposed loss of functionality mechanism for triethoxysilyl-terminated polyisoprene.

mers exhibited extraordinary condensation behavior producing polymers with narrow molecular weight distribution leading to star-shaped polymers.

Sol–gel chemistry and processing was a facile and efficient methodology for the preparation of branched, star-shaped homopolymers. The hydrolysis and condensation reactions are efficiently catalyzed with either simple acids or bases, and it was possible to control the nature of the silicate-like hub by judicious selection of the catalyst. The sol/gel condensation reactions to form the branched polymers occurred readily in the solid state above the glass transition temperature. The sol–gel process generates a silicate-like hub that contains

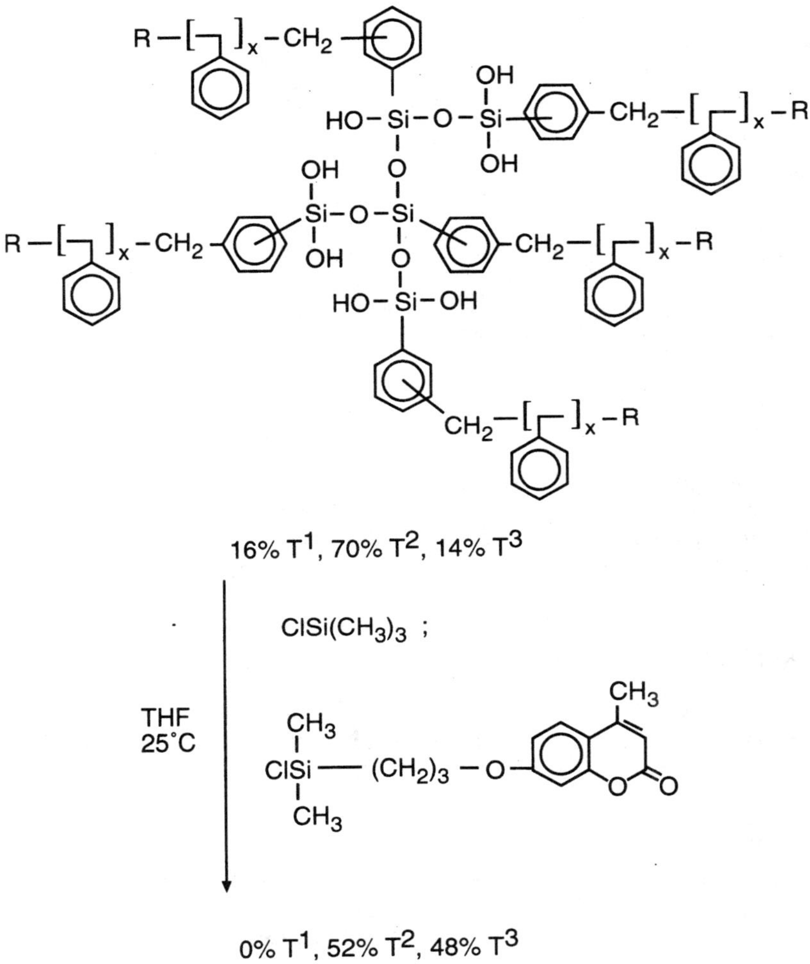

Figure 10 Model derivatization of a silanol-containing star core.

residual silanol groups, and these groups are readily reactive with suitable compounds to disperse inorganic and organic molecules and to facilitate adhesion to inorganic substrates.

Three different alpha environments adjacent to a trialkoxysilyl functionality were investigated. The hydrolysis and condensation rates in solution were compared to the corresponding model compounds in order to probe molecular

Table 4 Viscometric Analysis of the Star-Shaped Condensates[a]

Sample	M_w	$[n]$
Precursor	5,000	0.062
Condensate	20,200	0.092
Linear control	20,200	0.146
Precursor	2,800	0.047
Condensate	17,600	0.070
Linear control	17,600	0.133

[a]M_w determined by (SEC/LALLS (SEC intrinsic viscosities determined by THF).

weight effects. It was found that as the molecular weight was increased, the condensation also decreased. In addition, it was determined that the condensation rate was drastically decreased by attachment of the functionality via a benzyl configuration. Finally, loss of functionality occurred during acid-catalyzed hydrolysis of the allyltrialkoxysilyl-terminated polyisoprene oligomer.

ACKNOWLEDGMENT

We acknowledge Sally Miller and Michele Thomas (Analytical Technology Division) for SEC measurements and Richard Turner for technical discussions.

REFERENCES

1. Yamashita, Y., and Hane, T., *J. Polym. Sci., Polym. Chem. Ed.* **11,** 425 (1973).
2. Smith, S. D., York, G., Dwight, D. W., and McGrath, J. E., *Chemical Reactions on Polymers,* ACS Symposium Series 364, American Chemical Society, Washington, D.C., 1988, p. 85.
3. Richards, D. H., and Szwarc, M., *Trans. Faraday. Soc.* **55,** 1644 (1959).
4. Rempp, P., and Loucheux, M. H., *Bull. Soc. Chim. Fr.* 1457 (1958).
5. Morton, M., *Anionic Polymerization: Principles and Practice,* Academic, New York, 1983, p. 233.
6. Jurek, M. J., Geier, D. T., McGrath, B. E., and McGrath, J. E., *Polym. Prepr. (Am. Chem. Soc., Div. Polym. Chem.)* **27**(1), 315 (1986).
7. Hergonrother, P. M. J., *Polym. Sci., Polym. Chem. Ed.* **20,** 3131 (1982).
8. Hadjichristidis, N., and Fetters, L. J., *Macromolecules* **13,** 191 (1980).
9. Meunier, J. C., and Van Leemput, R., *Makromol. Chem.* **142,** 1 (1971).

10. Lutz, P., Masson, P., Beinert, G., and Rempp, P., *Polym. Bull.* **12,** 79 (1984).
11. Rempp, P., Lutz, P., Masson, P., and Franta, E., *Makromol. Chem., Suppl.* **8,** 3 (1984).
12. Ito, K., Tsuchida, H., and Kitano, T., *Polym. Bull.* **15,** 425 (1986).
13. Milkovich, R. Canadian Patent 712245, Aug. 24, 1965.
14. Martin, M. K., Ward, T. C., and McGrath, J. E., in *Anionic Polymerization: Kinetics, Mechanisms, and Synthesis,* McGrath, J. E., Ed., ACS Symposium Series 166; American Chemical Society, Washington, D.C., 1981.
15. Worsfold, D. J., Zillox, J. G., and Rempp, P., *Can. J. Chem.* **47,** 3379, (1969).
16. Quack, G., Fetters, L. J., Hadjichristidis, N., and Young, R. N., *Ind. Eng. Chem. Prod. Res. Dev.* **19,** 587 (1980).
17. Sheridan, M. M., Hoover, J. M., Ward, T. C., and McGrath, J. E., *Polym. Prepr.* **25**(2), 102 (1984).
18. Quack, G., and Fetters, L. J., *Polym. Prepr.* **18**(2), 558 (1977).
19. Hadjichristidis, N., Guyot, A., and Fetters, L. J., *Macromolecules* **11,** 668 (1978).
20. Hadjichristidis, N., and Fetters, L. J., *Macromolecules* **13,** 191 (1980).
21. H. Schmidt, *J. Non-Cryst. Solids,* **73,** 681 (1985).
22. Long, T. E., Kelts, L. W., Turner, S. R., Wesson, J. A., and Mourey, T. H., *Macromolecules* **24**(6), 1431–1434 (1991).
23. Mourey, T. H., Miller, S. M., and Balke, S. T., *J. Liq. Chromatogr.* **13,** 435 (1990).
24. Quirk, R. P., Chen, W., and Cheng, P., in *Reactive Oligomers,* ACS Symposium Series 282, American Chemical Society, Washington, D.C., 1985, p. 139.
25. Pouxviel, J. C., Boilot, J. P., Beloeil, J. C., and Lallemand, J. Y., *J. Non-Cryst. Solids* **89,** 345 (1987).
26. Kelts, L. W., and Armstrong, N. J., *J. Mater. Res.* **4**(2), 423 (1989).
27. Vega, A. J., and Scherer, G. W., *J. Non-Cryst. Solids* **111,** 153 (1989).
28. Lambert, J. B., Wang, G., Finzel, R. B., and Teramura, D. H., *J. Am. Chem. Soc.* **109,** 7838 (1987).
29. Klaver, W. J., Hiemstra, H., and Speckamp, W. N., *Tetrahedron* **44**(21), 6729 (1988).

8
Hyperbranched Polymers

Young H. Kim and Owen Webster
DuPont Central Research and Development Experimental Station, Wilmington, Delaware

I. INTRODUCTION

A new class of branched architectural polymers—dendritic or hyperbranched polymers—is attracting attention lately with the expectation that their unique structures may result in unusual properties, thereby leading to novel applications. Two general methodologies have been developed for preparing highly branched polymers. Employing the same strategy used for preparation of branched small molecules, such as “cascade compounds” [1] and “arborols” [2–9], one can prepare higher molecular weight versions in a stepwise manner, involving protection, coupling, and deprotection cycles. Both divergent and convergent synthetic approaches have been employed successfully. Substances obtained by these methods are structurally homogeneous and are called *dendrimers*. However, single-step direct polymerization of AB_x-type monomers, where x is 2 or greater, also gives highly branched polymers as shown in Scheme 1, as long as A reacts only with B in another molecule [10,11]. (Reaction between A and B in the same molecule will result in termination of polymerization by localized cyclization.) This approach produces highly branched polymers possessing one unreacted A functional group and an $(x-1)n+1$ number of unreacted B functional groups at the surface of the polymer, where n is the degree of the polymerization. Polymers obtained by this method, which are structurally inhomogeneous, are called *hyperbranched polymers*. This method

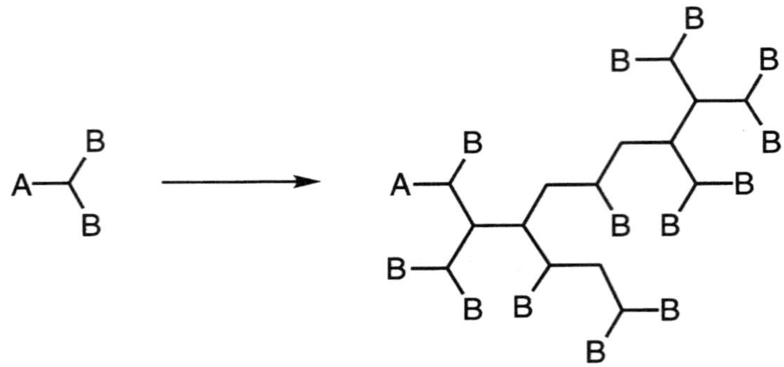

Scheme 1 The number of unreacted A's: 1. The number of unreacted B's: (functionality – 1) × dispersity + 1.

does not produce a polymer of well-defined structure but has the advantage of rapidly providing it in a large quantity. Similarly, copolymerization of A_2 and B_3 or other multifunctional monomers also can give hyperbranched polymers if the polymerization is kept below the gel point by limiting polymer conversion or the multifunctional monomer stoichiometry is manipulated. Obviously, a dendrimer is the perfectly symmetrical isomer of a group of hyperbranched polymers.

The properties of hyperbranched polymers include their intrinsic globular structure and the large number of surface functional groups, as in dendrimers. For most uses hyperbranched polymer can be used to replace dendrimers, and they can be prepared much more economically. For this reason, hyperbranched polymers will receive much attention in the future. The synthesis of hyperbranched polymers is much more diverse and challenging than the synthesis of dendrimers.

II. GENERAL STRUCTURAL CONSIDERATIONS

Unlike dendrimers prepared by stepwise synthesis, hyperbranched polymers have diverse molecular weights, isomerism, and geometrical shapes.

A. Degree of Branching

In an ideal case for a hyperbranched polymer, only cross reaction between groups A and B in the other molecule is allowed. By conventional terms in polymer branching [11], the degree of branching (α) of an AB_x type monomer with a functionality f equals the fraction of B groups reacted (p_B). Since the fraction of A group (p_A) reacted is the same as that of the B group divided by the functionality of the B group, x or $f - 1$,

$$\alpha = \frac{p_A}{f-1} \tag{1}$$

or

$$p_A = \alpha(f-1) \tag{2}$$

The average functionality (f_{avg}) for the stoichiometrically unbalanced case [12] is equal to twice the total number of functional groups that are not excess divided by the total monomer molecules (N_0).

$$f_{avg} = \frac{2N_0 f_A}{N_0} \tag{3}$$

According to Carother's equation, the critical extent of reaction p_C is

$$p_C = \frac{2}{f_{avg}} \tag{4}$$

Therefore, polymerization of AB_x-type monomers with $x \geq 2$ would never reach the gel point. This is due to the fact that there is always only one A functional group in each molecule throughout polymerization.

We could adopt α as an indication of branching, but such a notation could generate confusion when monomers of various number of functionalities are dealt with. For example, a fully branched polymer of AB_2 would have a branching factor of 0.5, but an AB_3 monomer 0.33. In order to describe the degree of perfection of the branching, a normalized branching factor was introduced [13].

The branching factor f_{br} is equal to the mole fraction of fully branched monomers relative to all possible branching sites. In other words, it can also be described as the ratio of the sum of all the monomers that are fully branched and terminal versus total monomer units [14]. If we call the mole fraction of monomers at the terminal position T, unbranched linear monomer units L, and fully branched monomer units B, then,

$$f_{br} = \frac{T+B}{N_0} \tag{5}$$

Thus, the branching factor for a fully branched polymer, such as a dendrimer, is 1, and that of a linear isomer of high molecular weight is 0. Since $N_0 = T + B + L$, if one can measure the mole fraction of the L units, f_{br} can be calculated from $1 - L/N_0$. Thus, the degree of branching can be determined if at least one type of branched monomer units could be quantified. For a given mol-

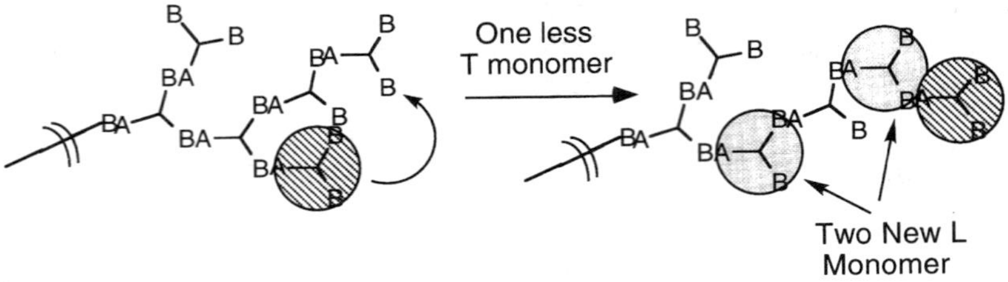

Scheme 2

ecular weight, each one less terminal monomer unit creates two unbranched monomer units, as shown in Scheme 2. Therefore, the number of terminal monomers for a polymer having L unbranched monomer units become $T - L/2$, while the number of terminal monomers of a fully branched polymer is $T = N_0/2 + 1$. The ratio of terminal monomers of these two cases $(T - L/2)/T$ equals $1 - L/(N_0 + 2)$. In a case where DP >> 2, the ratio $1 - L/(N_0 + 2)$ is close to $1 - L/N_0$, the f_{br} [15].

The mole fractions of each types of monomer units can be measured by a spectroscopic method, such as nuclear magnetic resonance (NMR). However, spectroscopic methods may not be applicable in many cases where the difference in the monomer is subtle or the spectra are too complex for interpretation. For certain degradable polymers, selective degradation after tagging the functional groups can give a precise account of branched, linear, or terminal monomers. For example, hydrolysis of a hyperbranched polymer obtained from 4,4-bis(4′-hydroxyphenyl) pentanoic acid, after methylation of the terminal hydroxyl group, gave a full accounting of the branching, as shown in Scheme 3 [16].

B. Isomerism

Isomerism is probably the most important distinction between dendrimers and hyperbranched polymers. Since the addition of each monomer takes place randomly, a large number of geometrical isomers can be formed, even for a given molecular weight and branching factor. Isomerism causes further dispersity in the aspect ratio of the polymer molecule. Scheme 4 shows two polyphenylenes that have the same molecular weight and f_{br} and differ in geometry. This variation of geometry has an importance influence on the solution as well as solid-state packing structure and related properties of the polymer. For example, the

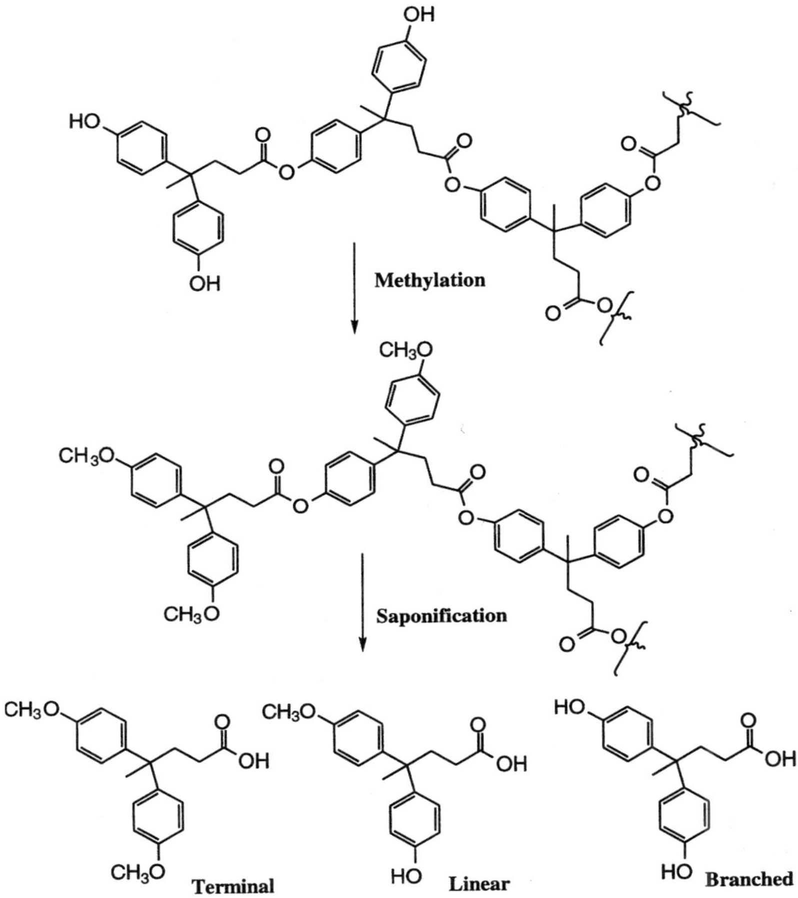

Scheme 3

packing order influences not only the relaxation process but also the solubility of the polymer.

Unlike the dispersity (DP) of molecular weight and branching, that of isomerism is difficult to denote with a number. As the DP increases, the number of isomers also increases. Flory has shown [11] that the number of configurations can be estimated by first treating two B groups differently. The n number of AB_x-type monomers contains nx B and n A groups, and can form $n - 1$

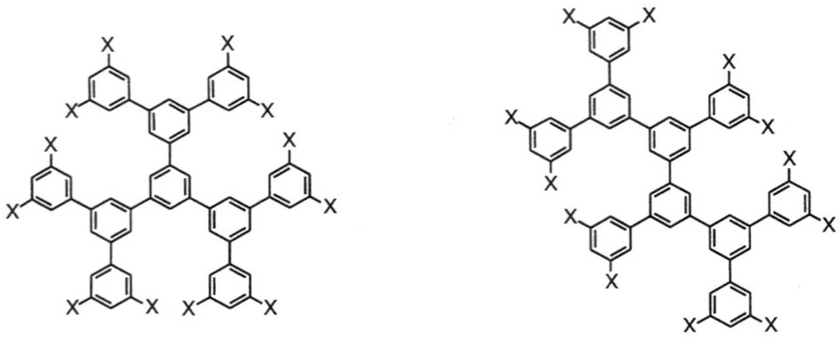

Scheme 4

number of linkages. The total number of sets created by these linkages is the number of nx choices taken n at a time.

$$_{nx}C_{n-1} = \frac{nx!}{(nx - n + 1)!(n - 1)!} \tag{6}$$

Multiplication of this expression with $(n - 1)!$, which is the number of independent ways to make $n - 1$ linkages, avoiding the reaction of A and B within the same molecules, gives the total number of arrangements. Dividing this number by the number of permutations of the monomer units for a given configuration $n!$ gives the total number of configurations.

$$\frac{nx!}{(nx - n + 1)!n!} \tag{7}$$

Table 1 shows calculated numbers of possible isomer configurations for various AB_x monomers. Obviously, the number of isomers increases with the complexity of the monomers and the molecular weight of the polymers.

Table 1 Number of Possible Isomers in Hyperbranched Polymers

		Number of configuration x in AB_x		
No.	*Generation number for* AB_2 *dendrimers*	*2*	*3*	*4*
4	1	1.40E + 01	5.50E + 01	1.40E + 02
10	2	1.68E + 04	1.43E + 06	2.73E + 07
22	3	9.15E + 10	4.05E + 15	4.52E + 18
46	4	8.74E + 24	1.09E + 35	

Some attempts have been made to describe the isomeric structural characteristic designation using graph theory. Graph theory, pioneered by A. Cayley in 1857, has been used for many years in enumeration of isomer of alkanes and other organic molecules [17]. Cayley introduced the idea of using rooted and unrooted treelike structures to describe branched alkanes and derivatives in order to enumerate all possible isomeric structures. For example, graph theory successfully estimates the total number of isomers of decane to be 75, for the case of tetravalent alkanes, and that of monofunctional decane to be 507. Among the various topological indices [18], the Wiener index [19] seems to be the most convenient for designating the geometry of hyperbranched polymers. The centric index, which has been introduced to define the shape of the molecule, could also be used. However, the centric index distinguishes poorly between geometrical isomers. Wiener's polarity number is defined as the number of pairs of the monomer unit that are separated by three units. The path number w is defined as the sum of the distances between any two repeating units in the polymer molecule and is commonly known as the Wiener index. The smaller the Weiner w path number is, the more compact the molecule is. Several research groups have published systems of mathematical logic, for calculating the Wiener index [20] or hyper-Wiener index [21] for dendritic polymers. An example of Weiner indices for isomerism is shown in Table 2. The example deals with oligomers $N_0 = 10$ of AB_2-type monomer, which is equivalent to second-generation dendrimer. It can be found from Tables 2 and 3 that in general the polar indices increases and the path number decreases with increasing complexity in branched structure.

For hyperbranched polymers, a method of subgraph enumeration of the orbital and wedgeal population and its relationship with some structural properties such as molecular weight and volume is given by Diudea [22]. General formulas derived from the Randic algorithm by using progressive vertex degrees and orbit numbers or from the Klein–Lukovits–Gutman formula were used to describe the fractal dimensions of surfaces of hyperbranched polymers [23].

C. Molecular Weight Distribution

Hyperbranched polymers are closer to conventional polymers in terms of molecular weight distribution than dendrimers. Because of variations in the degree of branching, the molecular weight distribution could be broader than that of the linear polymers.

The effect of branching on the molecular weight was also described by Flory. The total number of polymer molecules N is $N_0(1 - p)$, where N_0 is the

Table 2 Isomers and Wiener Indices Of DP = 10 Oligomers

	Branching factor	*Polarity number*	*Path number (W)*	*Number of isomers*
Linear				
	0.2	7	165	1
single branching				4
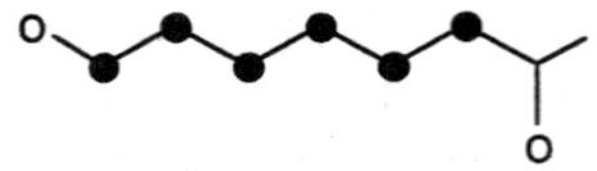	0.3	7	158	
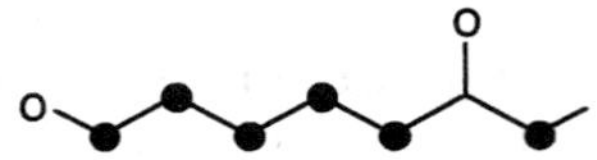	0.3	8	153	
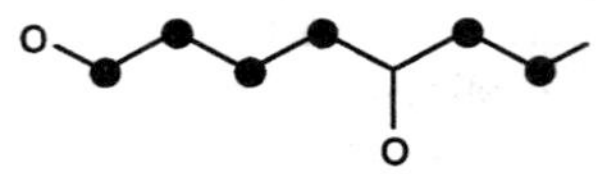	0.3	8	150	
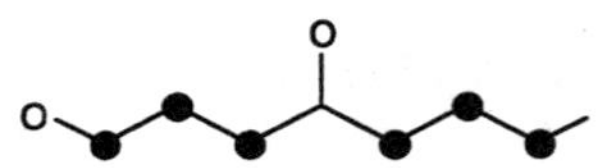	0.3	8	149	
Double branching				10
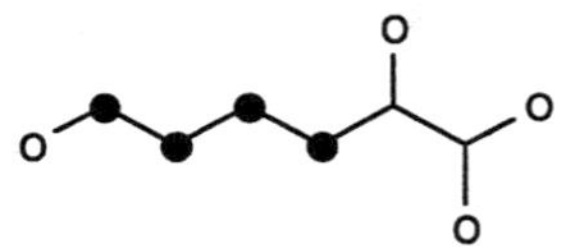	0.5	9	134	
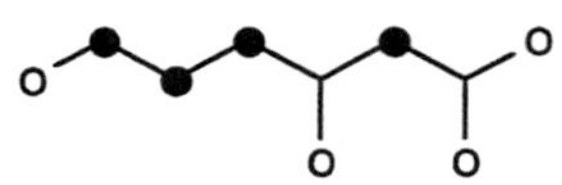	0.5	8	142	
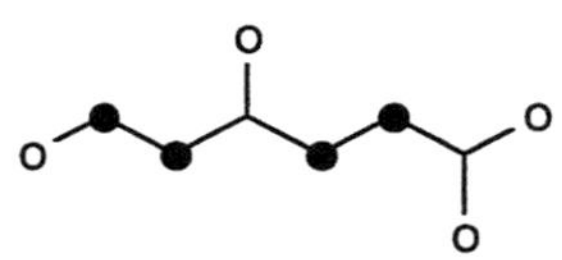	0.5	8	143	
	0.5	8	146	
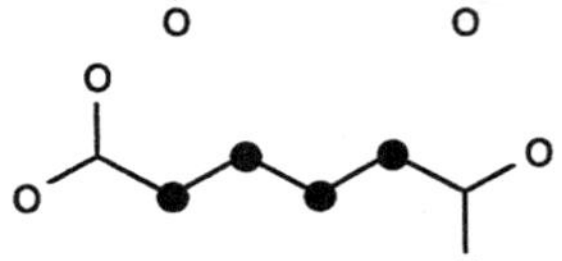	0.5	7	151	

Table 2 Continued

	Branching factor	*Polarity number*	*Path number (W)*	*Number of isomers*
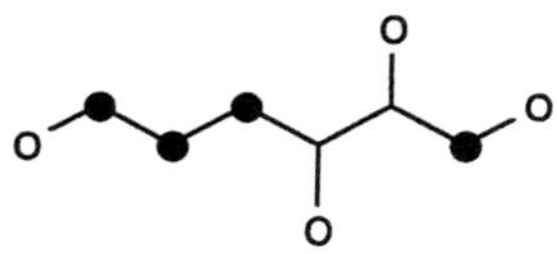	0.5	9	137	
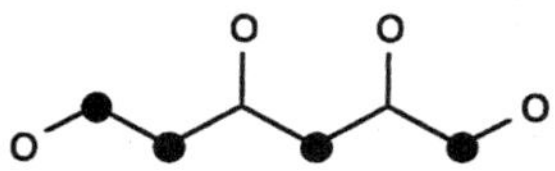	0.5	9	138	
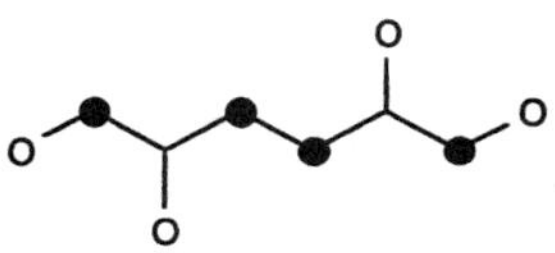	0.5	9	141	
	0.5	8	146	
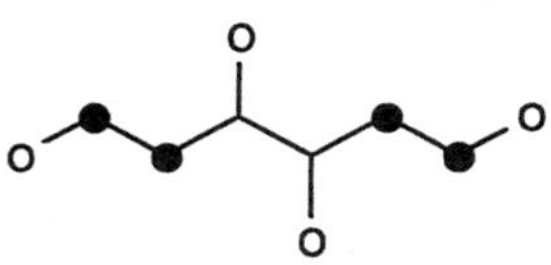	0.5	10	135	
Triple branching				6
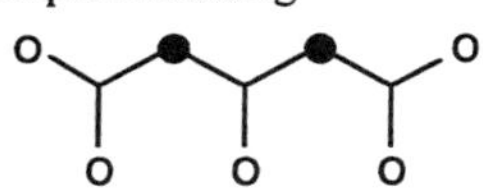	0.7	8	136	
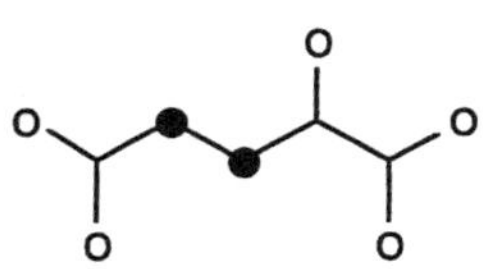	0.7	9	137	
dimeric branching				
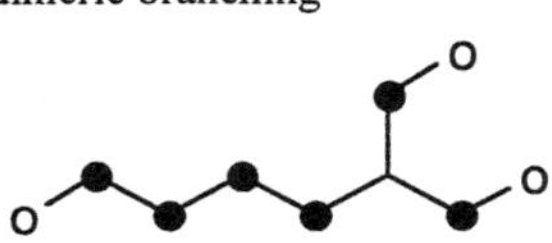	0.3	9	145	2
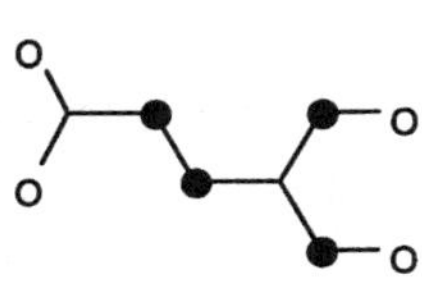	0.5	9	138	5

(continued)

Table 2 Continued

	Branching factor	*Polarity number*	*Path number (W)*	*Number of isomers*
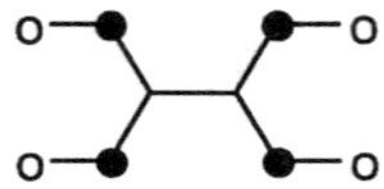	0.5	12	125	1
Trimeric branching				
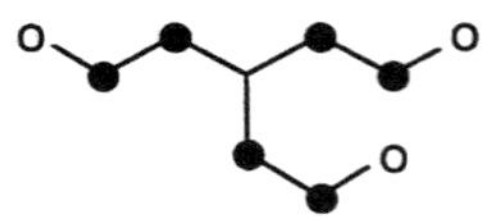	0.3	9	138	1
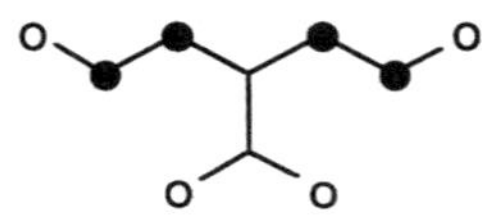	0.5	10	131	1
Fully branched				
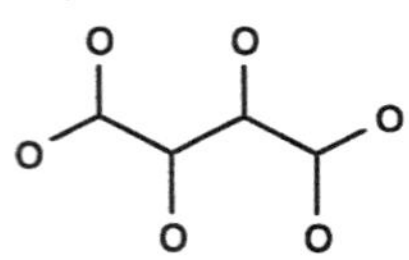	1.0	13	121	1
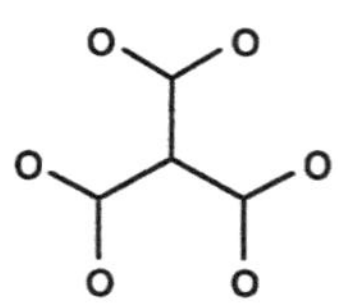	1.0	13	117	1
Total number of isomers				33

White circle, termininal unit (T); black circle, linear unit (L)

total number of the monomer unit. Therefore, the number average degree of polymerization is

$$x_n = \frac{N}{N_0} = \frac{1}{(1-p)} = \frac{1}{1-\alpha(f-1)} \tag{8}$$

Similarly, the weight average degree of polymerization and polydispersity (PD) are

$$x_w = \frac{1-\alpha^2(f-1)}{[1-\alpha(f-1)]^2} \tag{9}$$

Table 3 Wiener Indices of DP = 22 Fully Branched Oligomers

Structure	*Weiner index*
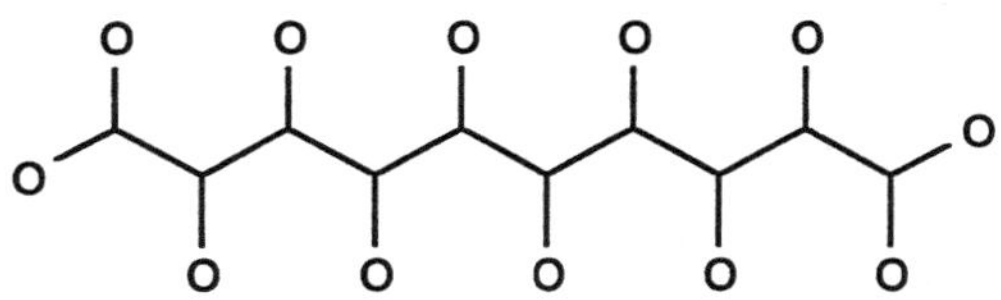	867
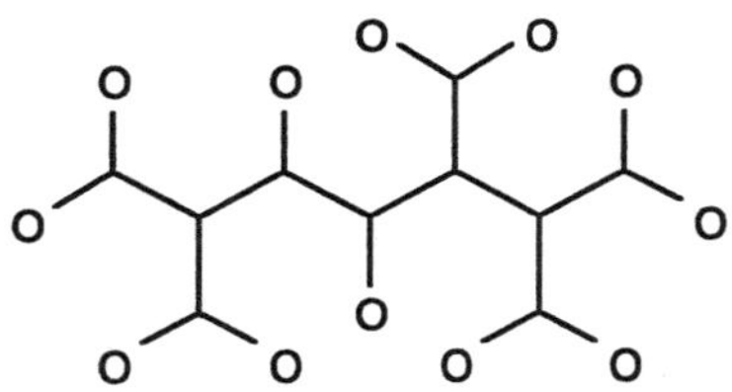	985
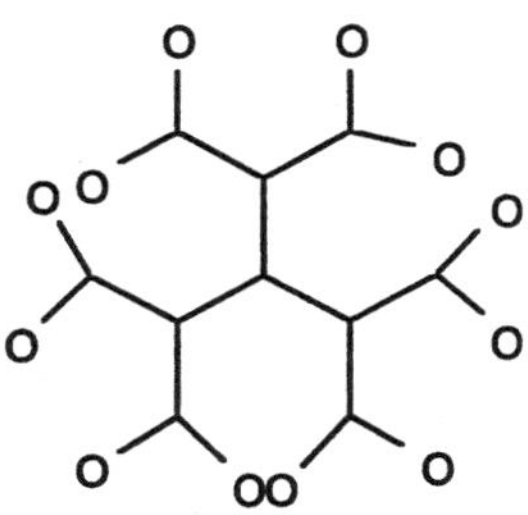	909

$$PD = \frac{x_w}{x_n} = \frac{1 - \alpha^2(f-1)}{1 - \alpha(f-1)} \tag{10}$$

The polydispersity of a hyperbranched polymer will increase eventually to infinity at the infinite molecular weight, as shown in Figure 1. Thus, the polydispersity of the hyperbranched polymer is expected to increase to infinity at the infinite polymer weight. In the case of AB_2-type three-functionality monomers, one would expect a significant increase in the polydispersity when the monomer conversion reaches about 90%. The molecular weight distribution of imperfect dendrimers, which have a structure similar to hyperbranched polymers but are obtained by multistep synthesis, was found to be narrow [24].

The gel permeation chromatography (GPC) molecular weight measurement reflects the radius of gyration of a polymer due to not only its molecular weight but also the size differences produced by geometrical isomerism. Even though the absolute molecular weight of a hyperbranched polymer can be

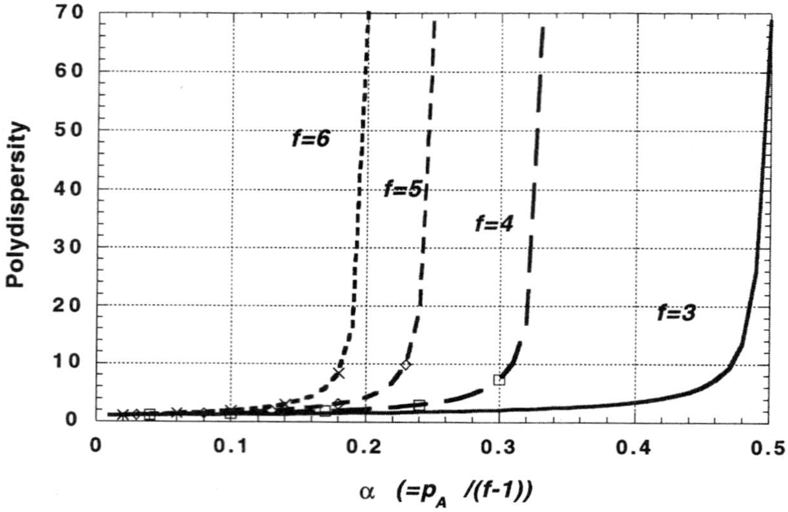

Figure 1 Polydispersity of hyperbranched polymers as a function of extent of reaction. (The *f* number gives the functionality of the monomer.)

obtained by using a universal calibration method, it may not be legitimate to define the polydispersity of a hyperbranched polymer based solely on a GPC measurement.

III. SYNTHESIS

Even though controlled polymerization of A_x ($x \geq 3$) or copolymerization below the gel point could also give a highly branched polymer, we will limit our discussion to AB_x-type monomers. We do not intend to survey all the literature, but rather to give highlights in this field, and the articles selected may represent only prevailing trends. In principle, practically all known polymerization methods—condensation, addition, ring opening, etc.—can be employed to polymerize AB_x-type monomers. Some examples of such monomers are shown in Scheme 5.

A. Condensation Polymerization

It is conceptually most obvious to identify condensation polymerization monomers that have an AB_x-type structure. For this reason these monomers are the ones studied most extensively. Probably most early works in this area have

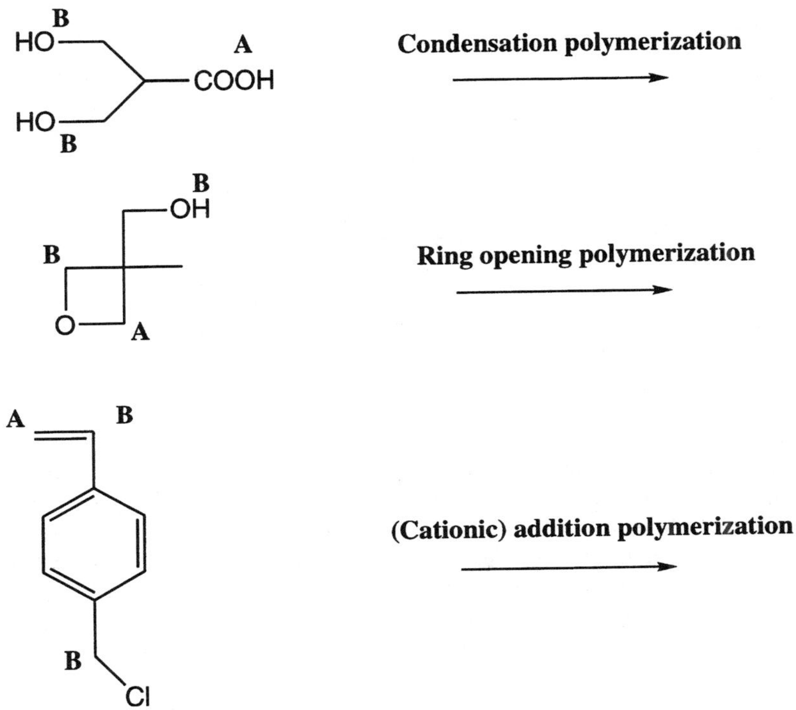

Scheme 5

not been designed to prepare hyperbranched polymers intentionally, since the unusual properties of these polymer have not always been appreciated. Many classes of monomers fit into this category.

1. *Polybenzyls*

Thermoset polymers made of benzylic units has been known for a long time, (e.g., Bakelite). Controlled cationic polymerizations of benzylic compounds are known to produce soluble polymers, which should have a highly branched structure. Benzylhalide or benzyl alcohol can generate a benzylic cation that will undergo condensation via aromatic nucleophilic substitution either in the ortho or para position, leading to hyperbranched polymers (Scheme 6). Unfortunately, most of these polymers have not been characterized extensively.

Friedel and Craft reported the formation of insoluble polymer from benzylchloride in 1885. Until recently, there has been much work on controlling

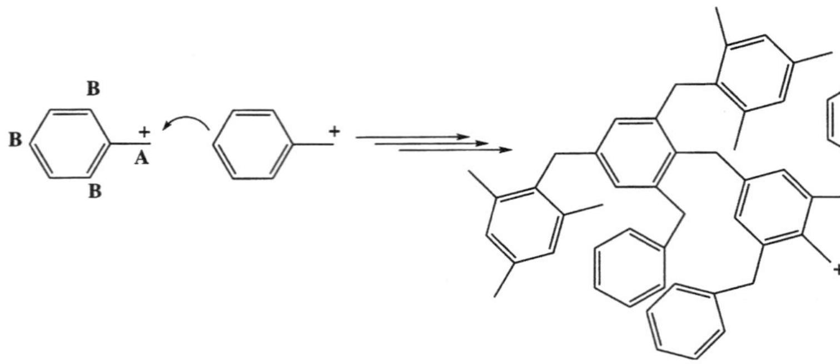

Scheme 6

this polymerization and characterization of the polymer. Ingold and Ingold reported a glassy polymer obtained by treating benzylfluoride with catalytic amounts of protonic acids [25]. Low molecular weight soluble polymers, which were speculated to be mixtures of various isomers, were reported by reaction of benzylchloride and its derivatives with about 0.5 mol % of $AlCl_3$ and other Lewis acid catalysts [26]. These polymers are amorphous and have softening temperatures in the range of 75–80°C. In contrast, linear polybenzyls were reported to be crystalline with melting temperatures higher than 300°C [27,28]. Haas and co-workers have concluded that the structure of polybenzyl can best be described as a complex mixture of repeating units, a small fraction having a highly branched structure and the rest being monosubstituted repeating units [29].

2. *Polyesters*

The possibility of using hydroxyl-terminated hyperbranched polymers as multifunctional crosslinking agents was been recognized as early as 1972. Polymers obtained by condensation of a polyhydroxy monocarboxylic acid, such as bis(hydroxymethyl)propionic acid, were used in coating compositions [30]. Similarly, aliphatic hyperbranched polymers derived from higher molecular weight AB_x-type monomers, prepared by Michael addition of ω-hydoxy amine to arylates, as well as fully aromatic polyester (polyarylate), were reported. Thus, heating $(HOCH_2CH_2N)_2CH_2CH_2COOCH_3$ in the presence of Fascat 4102 catalyst at 125°C for 1 hour gave a polymer with a glass transition temperature of −36°C and η_{inh} 0.11 in 83% yield. These polymers are reported to be useful in drug delivery and as rheology modifiers [31].

Numerous method and structure variations for hyperbranched polyarylates are reported. Initially, copolymerization of an AB_2-type monomer, bis(trimethylsiloxy)benzoyl chloride, with an AB monomer, 3-(trimethylsiloxy)benzoyl chloride, in an attempt to control the crystallization of polyarylates was reported by Kricheldorf et al. [32]. Cornell researchers utilized the same chemistry for a thermal self-condensation of 3,5-bis(trimethylsiloxy)benzoyl chloride to generate hyperbranched polymers in 80% yields with polystyrene-equivalent molecular weight of 30,000–20,000. The polydispersity and the molecular weight was found to vary extensively with the polymerization temperature. The degree of branching of the polyesters as determined from NMR experiments was 55–60%. Their glassy nature and high solubility were attributed to their shape [14,33]. With a catalytic amount of dimethylformamide (DMF) or $Me_3N{\cdot}HCl$, the polymerization was better controlled and gave polymers with higher reproducibility and controllable molecular weights. The molecular weight of the polymers increases with increasing temperature, reaction time, and catalyst concentration. Functionalization of these free phenolic hydroxyls was readily accomplished in a homogeneous solution, and the properties of the resulting polymers were highly dependent on the types of end groups [34]. In addition to 3,5-bis(trimethylsiloxy)benzoyl chloride, Turner et al. polymerized much more readily available 3,5-diacetoxybenzoic acid by thermal condensation, followed by hydrolysis to get a phenolic polymer similar to that obtained from 3,5-bis(trimethylsiloxy)benzoyl chloride. Mark–Houwink "α" values of less than 0.5 for the series of polymers are significantly lower than those for a typical rigid-chain linear aromatic polymer. The polydispersity was found to broaden significantly at high conversions, as predicted [35]. By switching the functionality, melt condensation of 5-acetoxyisophthalic acid and 5-(2-hydroxyethoxy)isophthalic acid yielded hyperbranched polyarylates with carboxylic acid terminal groups. The polymer from 5-acetoxyisophthalic acid showed a T_g at 239°C, with molecular weights in the 20,000–80,000 range. The polyester based on 5-(2-hydroxyethoxy)isophthalic acid showed a lower T_g. Comparisons of Mark–Houwink plots of these polyesters and NMR studies confirmed branched structures. These polymers were readily converted into water-soluble ammonium or sodium salts [36]. The polymerization conditions for this system have been fine-tuned with various activating living groups. Polycondensation of the trimethylsily ester of 5-acetoxyisophthalic acid, bis(trimethylsilyl) 5-acetoxyisophthalate, gave products with higher viscosity and glass temperature than those obtained with the parent acid [37]. In addition, whereas polycondensations of the free acid above 250°C resulted in partial crosslinking, the silylated monomer yielded perfectly soluble polyesters even at 280°C. A higher order of branching was introduced by polycondensations of silylated 3,5-diacetoxybenzoic acid with acetylated tetraphenols [38]. Using this method, various copolymers with AB-type monomers were also prepared. Copolymers of trimethylsilyl 3-acetoxybenzoate with either trimethylsilyl 3,5-diacetoxybenzoate or bis(trimethylsilyl) 5-acetoxyisophthalate showed nearly random incorporation of the trifunctional "branching units." However, analogous copolycondensations with silylated 2-(4-*tert*-butylphenoxy)terephthalic acid failed. The T_g's of the copolymers vary largely with the mole fraction of branching units and with the nature of the end groups. In the case of phenolic OH and acetate end groups, the relationship between the T_g and the number of branching units passes through a minimum T_g around 100–120°C [39] (Figure 2).

A macromonomer, the trimethylsilyl ester of *N*-(3′-acetoxybenzoyl)-3-aminobenzoic acid, was polycondensed in situ at 260 or 280°C to give hyperbranched poly(ester-amide)s, which are amorphous materials with glass transition temperatures in the 190–200°C range [40]. In some cases, homopolymerization lead to gelation, and NMR spectroscopy suggested that gel formation might have originated, to a significant extent, from ester–amide interchange reactions. Copolymerization of this monomer with a controlled feed ratio of acetylated tetraphenols as star centers was claimed to give molecular weight control. A systematic effect of the ratio of AB_2 and B_4 monomers on the viscosity and T_g was expected; however, in the range of a 30–90 mole ratio between B_4 and AB_2 monomers, no significant effect was realized [41]. On the

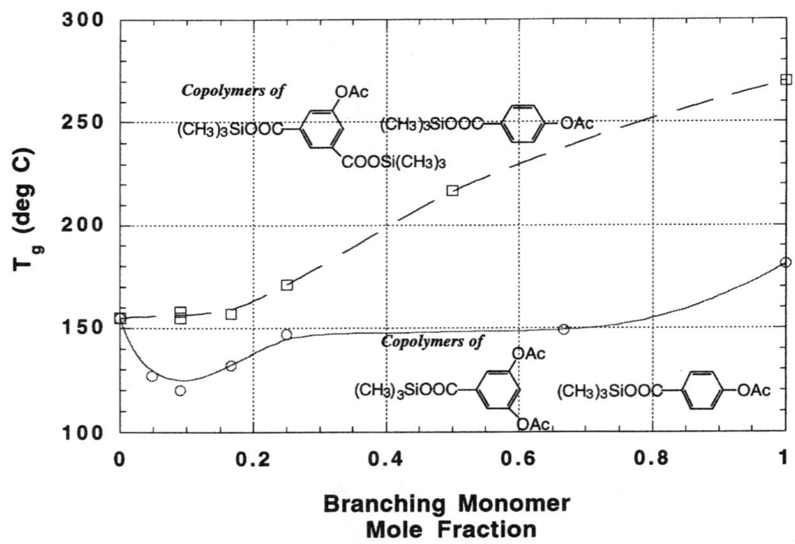

Figure 2 Effect of branching on the glass transition temperatures of aromatic polyesters. (From Ref. 39.)

other hand, it was suggested that copolymerization of AB_2 and B_3 monomers (e.g., dimethyl 5-(2-hydroxyethoxy)isophthalate and trimethyl 1,3,5-benzenetricarboxylate) could give better geometrical control of hyperbranched polymer synthesis [42]. See Scheme 7.

Another kind of hyperbranched macromonomer containing an oligomeric segment in the monomer unit has been reported. Thus, 3,5-dihydroxy-(oligo)ethyleneoxy ethyl benzoate was condensed in the molten state to give a hyperbranched structures with varying lengths of functional spacer. The polymer complexed with alkali metals picrate [43].

A systematic investigation of hyperbranched polyester as a curing agent has been developed in Sweden. Low molecular weight allyl ether-maleate

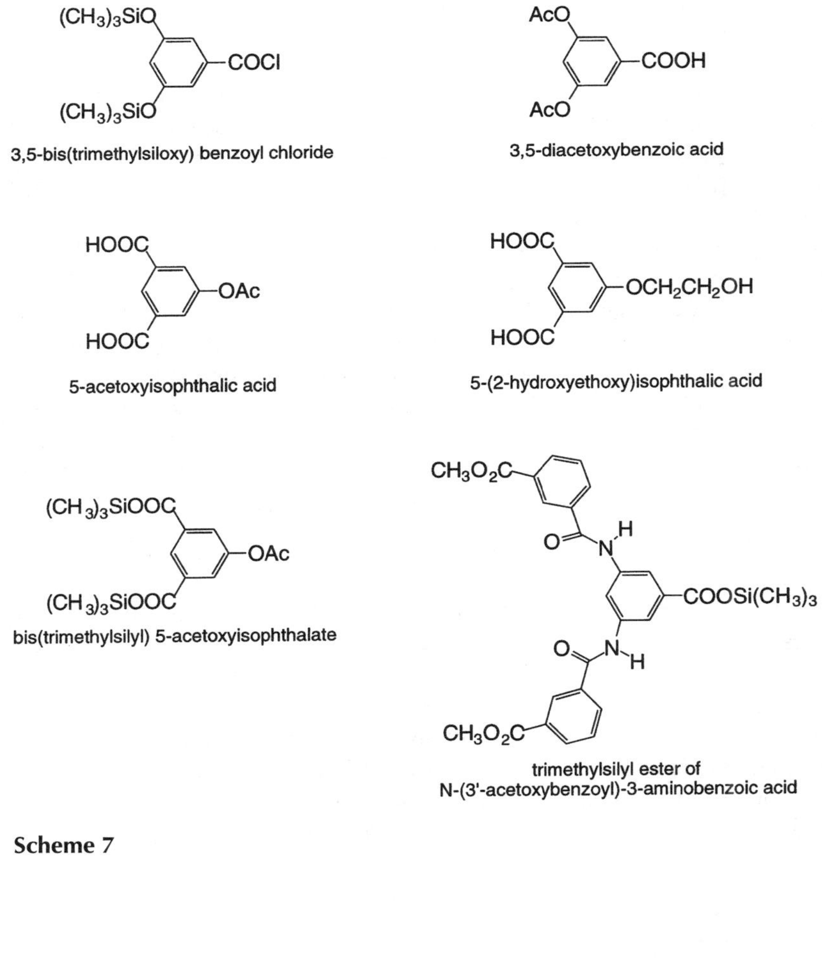

Scheme 7

functional hyperbranched ester prepared in a two-step procedure showed much promise in curing and film formation [44]. Hyperbranched aliphatic polyesters with molecular weights in the range of 1,200–44,300 were synthesized in the molten state from 2,2-bis(hydroxymethyl)propionic acid and 2-ethyl-2-(hydroxymethyl)-1,3-propanediol as core molecules using acid catalysis. A drastic effect on molecule weight control by use of the core molecules was claimed. Thus, there was an excellent correlation between the observed and calculated molecular weight based on the ratio of B_4 and AB_2 monomers. More sterically crowded alcohols were found to give poor molecular weight control. The resulting polymers with a branching factor in the range of 0.8 exhibited an almost Newtonian behavior in the molten state. They also showed a lower Mark–Houwink's "α" constant than linear polymers, but the viscosity of the solution was greatly affected by the surface functionality [45]. The polymerization is a pseudo-1-step reaction where stoichiometric amounts corresponding to each generation were added successively [46]. A kinetic study of polymerization of 2,2-bis(methylol)propionic acid indicated that the reaction rates were strongly dependent on the miscibility of the monomer and the polyol core material. Trimethylolpropane (TMP) core material exhibited the highest rate of reaction, because of the low melting temperature of TMP and good solubility and mass transfer of heterophase 2,2-bis(methylol)propionic acid in the TMP melt [47]. A series of hyperbranched aliphatic polyester resins with acrylate functionality and terminal groups such as benzoates, propionates, or hydroxyls showed a dependence of their T_g's and viscosity on the nature of the end groups. The resin could be UV-cured, and the T_g of the hard-cured films was correlated to the difference in the structure of terminal groups in the resins [48]. The viscosity of alkyds based on hyperbranched polyesters and unsaturated fatty acids exhibited substantially lower viscosity compared to similar mixtures with conventional curing agents. They also had excellent curing properties with amazingly short curing times, thanks to the large number of reactive groups in the surface layer of the polymers [45,49,50].

3. *Polyamides*

Even though some polyamide dendrimers, including polypeptide dendrimers, are reported [51–53], not many aliphatic polyamides have been investigated yet.

In the area of aromatic polyamides, condensation of 3,5-diaminobenzoyl chloride or 3-aminoisophthalyl chloride in an amide solvent gave hyperbranched polymer (Scheme 8). The GPC analysis of these polymers in a solvent containing a complexing salt showed that the molecular weight ranged between 24,000 and 46,000 with polydispersities of 2.0–3.2. GPC in the

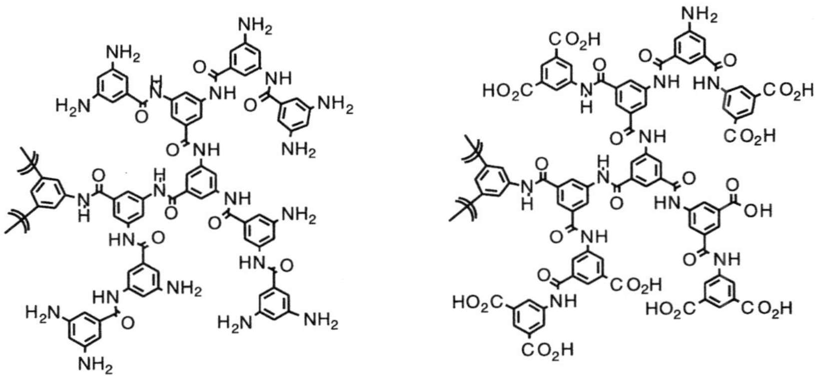

Scheme 8

absence of a complexing salt indicated a high degree of polymer aggregation. Amide solutions containing more than 40 wt % of polymer exhibited nematic phase texture under a polarizing microscope [54].

4. Polyethers

A few hyperbranched polymers are known in this category. Ring-opening polymerization of 2-hydroxy methyl oxetane under basic conditions gave only low molecular weight polymer [55]. Most of the work in this category of polymers has been concentrated in polybenzyl ether or polyphenylene ethers.

3,5-Dihydroxy benzylbromide was polymerized with K_2CO_3 and 18-crown-6 in various solvents [56]. The yield and molecular weight were quite solvent dependent. Acetone seemed to be the best solvent, but as much as 16% of C-alkylation, besides the desired O-alkylation was found to take place in acetone during polymerization. The phenolic end groups of the hyperbranched polyethers were modified in an attempt to modulate the physical properties. On capping the terminal groups, the molecular weight of the polymer increases as expected, but surprisingly, the glass transition of the polymer is little affected (see Table 4) [57].

Low molecular weight aromatic polyethers prepared from 2,4,6-tribromophenol (Scheme 9) have been known since the 1920s [58,59]. This phenol is an example of an AB_3 type of monomer and can be polymerized via an oxidative free radical intermediate. A low molecular weight polymer obtained with 1 equivalent of NaOH and a catalytic amount of $K_3Fe(CN)_6$ showed a T_g = 148.8°C by differential scanning calorimetry (DSC). High molecular weight polymer can be obtained in high yield by adding KOH to neutralize the phenol,

Table 4 Effect of the Nucleophilc Functional Group on the Molecular Weight and Glass Transition Temperature of Polybenzyl Ether

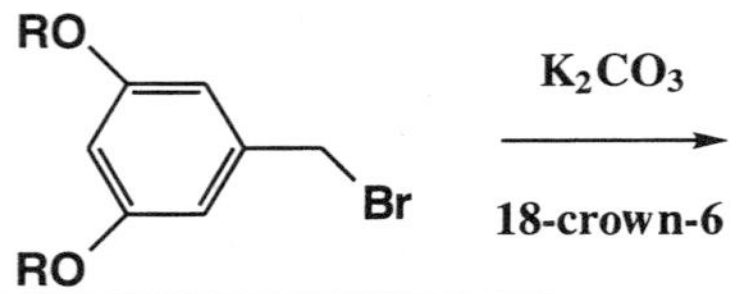

Functional group	*Observed* M_w	*Calculated* M_w	T_g
Hydroxy (R=H)	6500		311
Acetyl (R=Ac)	6200	7600	343
Benzyl (R=Bzl)	9800	8900	324
Silyl (R=TMS)	14000	9600	343

Source: Ref. 57.

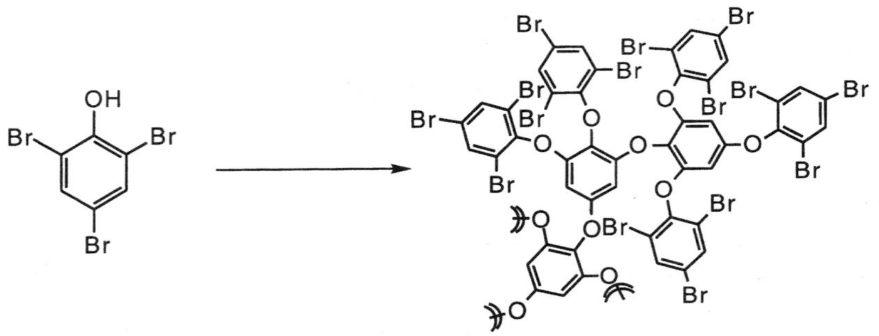

Scheme 9

thus assuring complete dissolution of the phenol in water prior to the addition of the initiator. Under standard phase transfer conditions with tetrabutylammonium fluoride as the phase transfer catalyst, only low molecular weight polymer (M_p = 1100) was obtained [60].

5. Polyphenylenes

Polyphenylene is one of the most intractable polymers known. Hyperbranched polyphenylenes, on the other hand, exhibit excellent solubility properties. Compared to aliphatic hyperbranched polymers, rigid monomer segments offer more open and accessible cavities where solvation can be accommodated for dissolution. The term *hyperbranched* polymer was first used to describe these

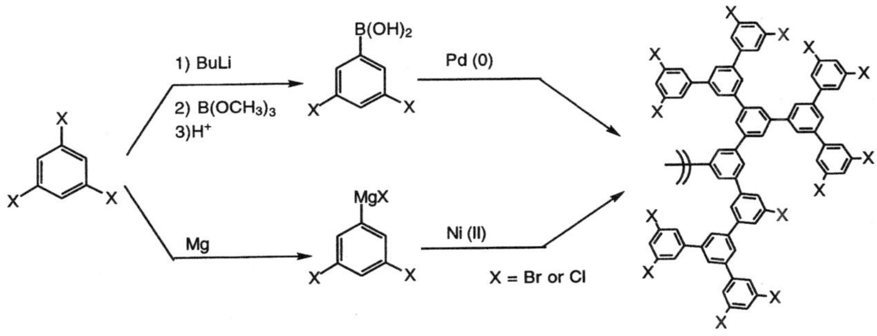

Scheme 10

unique polymers [61]. Highly branched polyphenylenes were synthesized from AB_2-type monomers, e.g., (3,5-dibromophenyl)boronic acid and 3,5-dihalophenyl Grignard reagents, as shown in Scheme 10. These monomers were polymerized by Pd(0) and Ni(II)-catalyzed aryl–aryl coupling reactions, respectively. Polymers with molecular weights 5000–35,000 and polydispersities smaller than 1.5 were obtained, with a branching factor of about 0.8. Unlike the other condensation polymerizations, the polydispersity of the polymer got narrower as the polymer molecular weight increased, implying that the "sticky factor" in the fractal growth for this type of polymerization might be different from reversible polymerization. These polymers were thermally stable to 550°C and soluble in many organic solvents. A T_g at 236°C was found. The bromine functional group was converted to various groups, through lithiation or coupling reactions, to give other functional polymers with vastly different relaxation and solubility characteristics. Some of these derivatives were used as multifunctional initiators to prepare star polymers, for example, via ring-opening polymerization of propiolactone and anionic polymerization of methyl methacrylate [62]. Reaction of the polymer with BuLi, followed by CO_2, resulted in a water-soluble carboxylated polymeric lithium salt. This polymer showed properties resembling those of micelles, and was proposed as a unimolecular micelle [13]. Surfactant properties were also observed in a Langmuir monolayer experiment and from its effect on $CaCO_3$ crystal formation [63].

6. *Polyurethanes*

An in situ AB_2 monomer having an isocyanate group, and two hydroxyl groups was prepared by thermal decomposition of 3,5-dihydroxy benzoyl azide

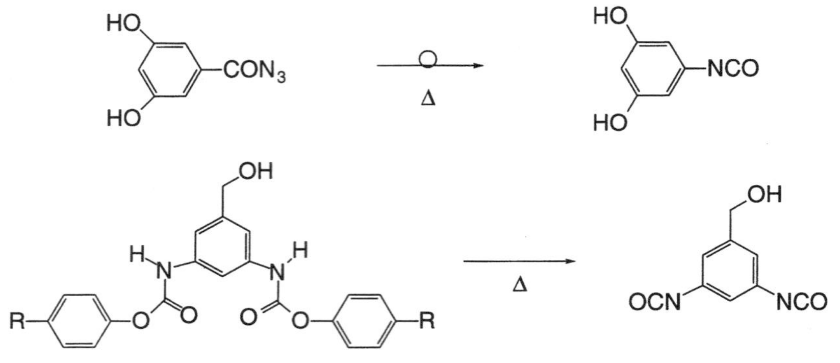

Scheme 11

(Scheme 11). It was polymerized to form a hyperbranched polyurethane [64]. Similarly, another type of AB_2-blocked isocyanate monomer 3,5-bis((benzoxy carbonyl)imino)benzyl alcohol gave a hyperbranched polyurethane, M_w = 34,000, when refluxed in tetrahydrofuran (THF) with a tin catalyst at 1 *M* concentration. At higher concentrations, however, a crosslinked polymer formed due to a small amount of unidentified isocyanate side reactions. Soluble, albeit much lower molecular weight, polymers were prepared, independent of the monomer concentration, when an excess amount of end-capping alcohol was added at the start of the polymerization to reduce the free concentration of isocyanate. The end-capping groups change the physical properties of the polymers such as glass transition and solubility; thus, the polymer made with hexyl alcohol and decyl alcohol showed T_g's at 111 and 83°C respectively [65].

7. Polyether Ketones

Several papers on polyether ketone dendrimers are in the literature. Both for dendrimer and hyperbranched polymers, most polymerizations were carried out using the well-known nucleophilic substitution reaction with phenolate and an activated aromatic fluoride [66]. Chu and Hawker has prepared an iso-structure of AB_2-type monomers, one with two hydroxyl groups and one fluoride group, and the other with two fluoride groups and one hydroxyl group, which in principle should give polyether ketone with an identical backbone structure, but with different terminal groups. (See Table 5.) These were polymerized in a mixture of toluene and *N*-methylpyrrolidone with potassium carbonate. This is the first example in which clever molecular design was used to enhance the

Table 5

	M_n (D)	End group	T_g	Solubility
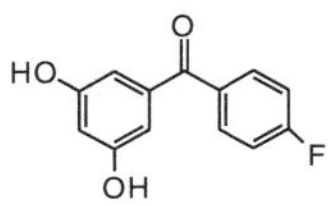	95,000 (2.2)	OH	127	Soluble in DMF, KOH solution
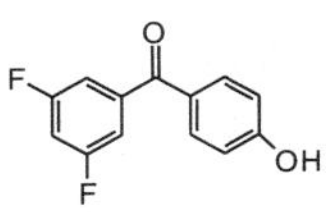	20,000 (1.8)	F	162	Insoluble in DMF, KOH solution

DMF, dimethylformamide.

understanding of branching density by examining dependence on the bulkiness of reactants [67].

Similarly, several AB_2 monomers were polymerized in hot dimethylacetamide. The products had molecular weights of 7000–36,000 and polydispersities in the range of 1.50–4.50. They were highly soluble in organic solvents. The polymers were thermally stable to 500°C under N_2 and had glass temperatures ranging from 135 to 231°C depending on the backbone and the terminal groups (Scheme 12) [68,69].

8. Heteroatom Polymers

Incorporation of metal or heteroatoms in dendrimers has been a popular pursuit, especially in attempts to prepare soluble polymeric catalysts. This type of effort has not been seen much in the hyperbranched area as yet. Silicon-con-

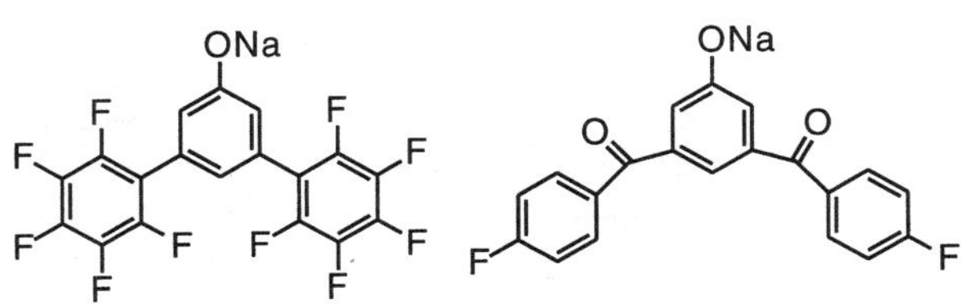

T_g 152 143

Scheme 12

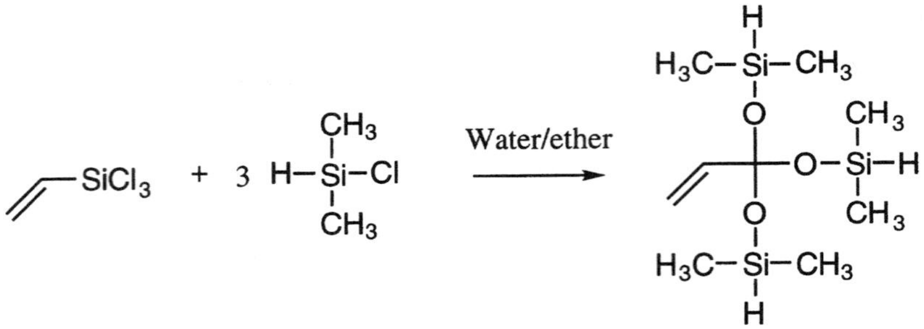

Scheme 13

taining polymers are the only reported hyperbranched heteroatom polymers so far.

An AB_3 monomer (Scheme 13), prepared from the reaction of 1 equivalent of allyltrichlorosilane with 3 equivalents of dimethylchlorosilane contains 1 alkene group and 3 SiH moieties. It was polymerized with a catalytic amount of $H_2PtCl_6 \cdot 6H2O$ to give a polysiloxane of moderate molecular weight. The hydride-terminated polymer was unstable and crosslinked on standing. Subsequent reaction with allylic derivatives gave end-capped materials that were stable to further end-group reaction [70]. To address the stability issue by using hydride terminal groups, an alternative monomer, containing three vinyl groups and one hydride, was also polymerized. Unfortunately, these monomers were found to readily undergo chain-terminating cyclization into a stable six-membered siloxane ring, which interfered with formation of high molecular weight polymers (Scheme 14). Incorporation of a vinyl silane with a long alkyl

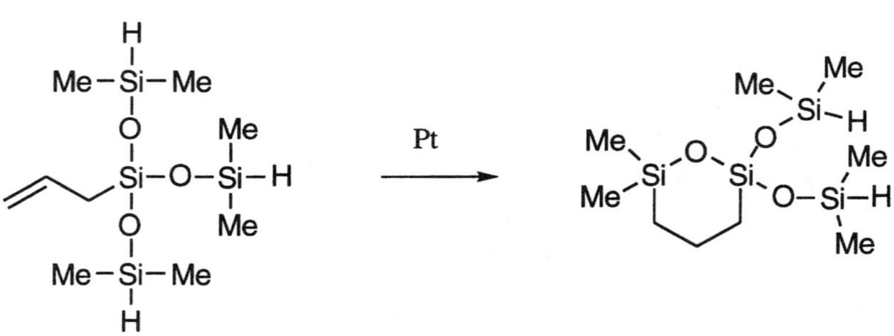

Scheme 14

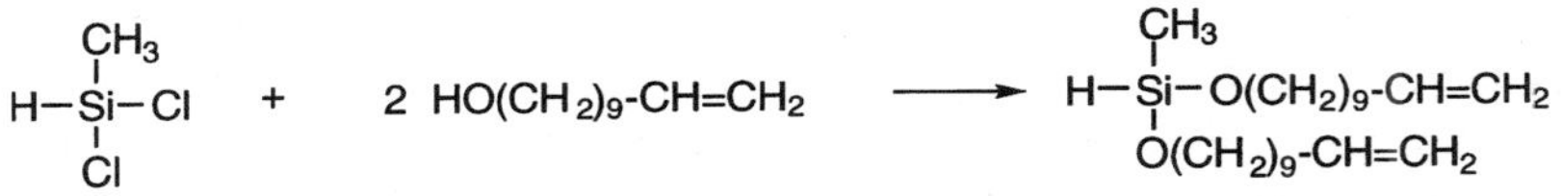

Scheme 15

spacer made it possible to form high molecular weight polymers [71]. On the other hand, homopolymeriztion of an AB_2-type allylic silane $H(CH_3)Si(CH_2CH{=}CH_2)_2$ (Scheme 15) was reported to give polymer with no gel formation [72].

Since allylic groups facilitate chain-terminating six-membered cyclization, polymerization of vinyl silane was attempted. Unusual hydrolytic cross-coupling of two chlorosilanes has provided an efficient route to two monomers, vinyltris(dimethylsiloxy)silane and tris(dimethylvinylsiloxy)silane. The hydrosilation reaction under Pt catalysts of these AB_3 monomers afforded hyperbranched poly(siloxysilane) polymers, with hydride or vinyl functional groups on the outer sphere. The functional groups on the surface of both polymers was modified further with a wide variety of reagents [73].

Polymerization of divinyl silane in which the vinyl groups were separated from the silicon by a long alkyl chain gave a slightly different result [71,72]. Methyl-di-10-undecenyloxysilane under dicobaltoctacarbonyl catalysis gave potentially degradable hyperbranched polymers. Unlike $H(CH_3)Si(CH_2CH{=}CH_2)_2$, this monomer was found to form gels during polymerization under Pt catalysis, presumably because of monomer rearrangement. Degradation of the polymers in the presence of dry alcohol or in the presence of water gave mono(dialkoxymethylalkylsilane)s or a polysiloxane [74].

Polysilane hyperbranched polymers containing Si atoms connected to three and four other Si atoms were prepared. Copolymers made from $RSiCl_3$ (R = *n*-hexyl, Ph and Me) and $MePhSiCl_2$ resemble hybrid materials of polysilyne and polysilanes [75].

B. Ring-Opening Polymerization

In spite of the straightforward concept of ring-opening polymerization and the ready availability of different types of monomer, this mode of polymerization has not been utilized much for hyperbranched polymer synthesis. The ring-opening polymerization of AB_2-type monomer, such as hydroxyl methyl oxetane [55] or other cyclic monomers provides two new reactive groups after each ring-opening reaction, thus offering continuing branching sites. Symmetric and unsymmetric structures are conceivable. A cyclic monomer, 5,5-dimethyl-6-ethenylperhydro-1,3-oxazin-2-one, was polymerized at room temperature in THF in the presence of $Pd_2(dba)_3$ catalyst and primary or secondary amine initiator. The polymerization readily proceeds, most likely via a II-allyl palladium complex that multiplies the propagating ends of N–H groups by exclusion of CO_2 [76]. Even though the monomer is cyclic, and the polymerization proceeds by ring opening, this is not a ring-opening polymerization, but rather is similar to a comb-graft-type polymerization. (See Scheme 16.)

C. Addition Polymerization

Addition polymerization of monomers that contain an initiating function and a propagating function in a same molecules has been shown to give hyperbranched polymers. Unlike the AB_x-type structure required by condensation polymers, addition polymerization renders hyperbranched polymer via multiple reactive sites that are generated as a result of vinyl addition reaction in addition to the already existing initiating function. Some vinyl monomers containing a pendant group that was transformed into an initiating moiety by the action of an external stimulus were self-polymerized.

Hyperbranched polymers were prepared from 3-(1-chloroethyl)-ethyl benzene by cationic polymerization [77], and from 4-[2-(phenyl)-2-(1-2,2,6,6-tetramethylpiperidinyloxy)ethyloxy]methyl styrene [78] or *p*-(chloromethyl)-styrene [79] by free radical polymerization. In the cationic living polymerization, the benzylic chloride of 3-(1-chloroethyl)-ethyl benzene is activated with $SnCl_4$ in the presence of Bu_4NBr [80,81], followed by low-temperature living

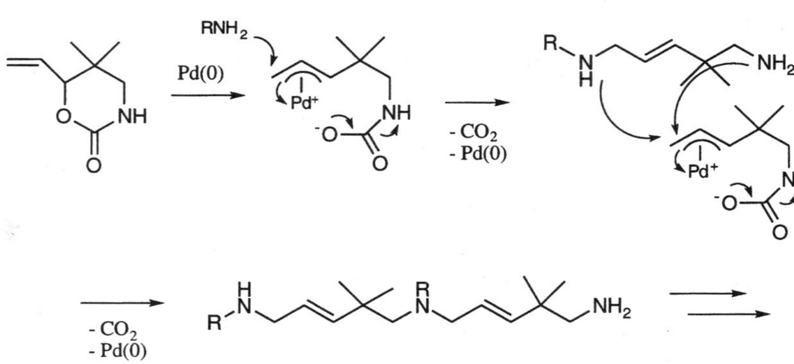

Scheme 16

cationic addition polymerization of styrenic double bond. The chloride functional group in the monomer as well as those in the polymer chain were activated to provide a situation similar to polymerization of an AB_x-type monomer (Scheme 17). The living nature of the polymerization allowed architectural control as well as molecular weight control. The Mark–Houwink α constant for this series of polymer was 0.43 in contrast to 0.7 for a linear polystyrene.

Recent advances in free radical living polymerization have allowed more facile preparation of hyperbranched polymers (Scheme 18). Thus, a styrene containing an initiating groups for a living free radical polymerization,

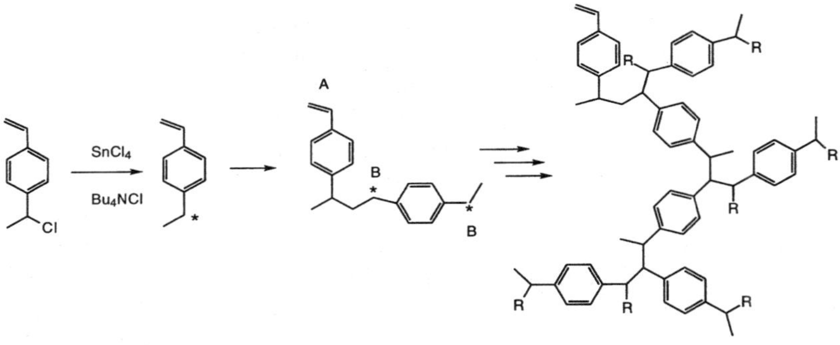

Scheme 17

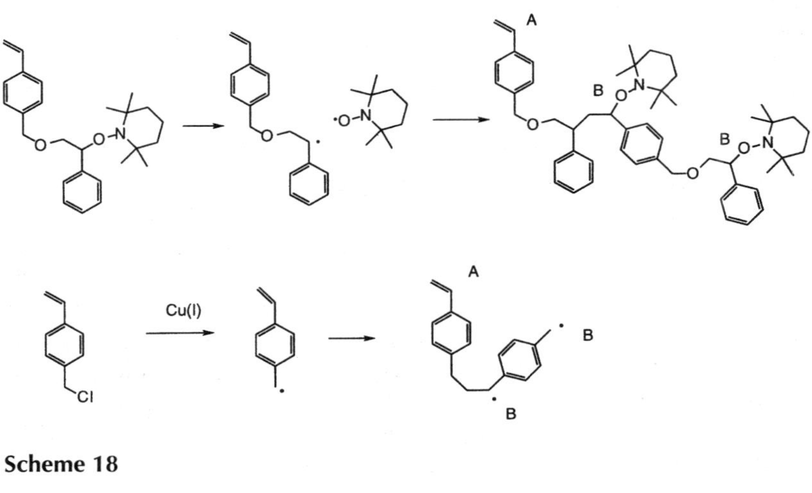

Scheme 18

TEMPO (2,2,6,6-tetramethylpiperidinyloxy) [82], has been shown to afford hyperbranched structures. Copolymerization with styrene was also shown to afford branched polymers with a controlled branch and chain length. Atom transfer radical polymerization (ATRP) [83] of *p*-(chloromethyl)styrene similarly provided hyperbranched polymers [79].

IV. COMB-BURST DENDRIMER OR ARBORESCENT GRAFT POLYMERS

Another intriguing concept in building high molecular weight highly branched polymers has been approached by two independent groups. These groups have demonstrated that highly branched polymers can be prepared by grafting a polymer chain that is already grafted to other polymers. Thus the polymer would have branches of branches, so that there would be no chain entanglement even at extremely high molecular weights. Tomalia has prepared such polymers using a convergent approach, and called them comb-burst dendrimers [84], while Gauther et al. have obtained structurally similar polymers and named them arborescent graft polymers [85,86]. These technologies also rely on controlling the chain end group by living polymerization or the equivalent for quantitative functionalization and effective grafting. This approaches could utilize a large number of addition or ring-opening polymerization monomers, and the possibilities for architectural design are nearly unlimited.

Comb-burst polyoxazoline dendrimers were prepared by repeated sequential grafting and hydrolysis of performed living poly(2-ethyl-2-oxazoline) oligomers on a polyethylenimine core. Acidic hydrolysis of the oxazoline amide group provided new graft sites for the living oxazoline [84]. This approach encountered some difficulties with respect to the efficiency of grafting as generation numbers increased and in the separation of grafted polymer from the occluded oligomers (Scheme 19).

Arborescent graft polystyrenes were prepared by successive chloromethyl-anionic grafting sequences on polystyrene (Scheme 20). This approach led to well-defined hyperbranched macromolecules with branching functionalities of f larger than 5000 and mol wt $> 10^7$ g/mol, while maintaining a molecular weight distribution below 1.3. The stiffness of the polymer chain was modulated by varying the degree of grafting sites along the chain, resulting in controllable stiffening of the molecule structure. Static and dynamic light-scattering experiments showed that the polymer behaved like a hard sphere in dilute solution. Measurements in the semidilute range, however, showed a progressive structural stiffening effect as the degree of branching increased [87].

These shapes of polymer can be prepared if the chain termini or the chain

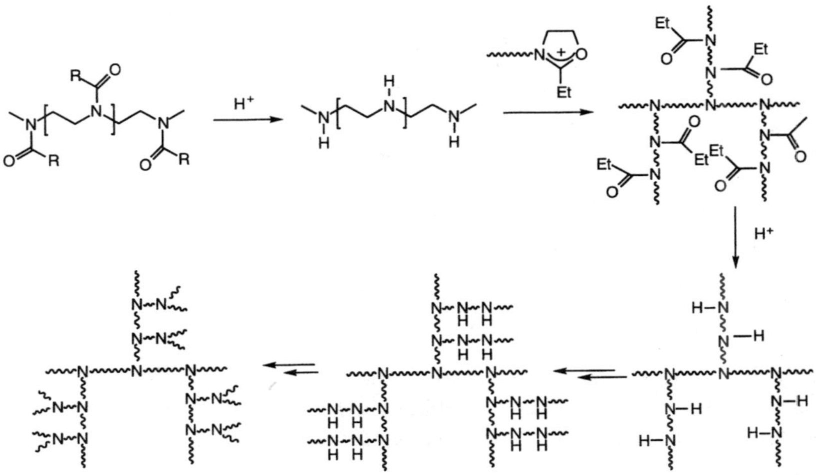

Scheme 19

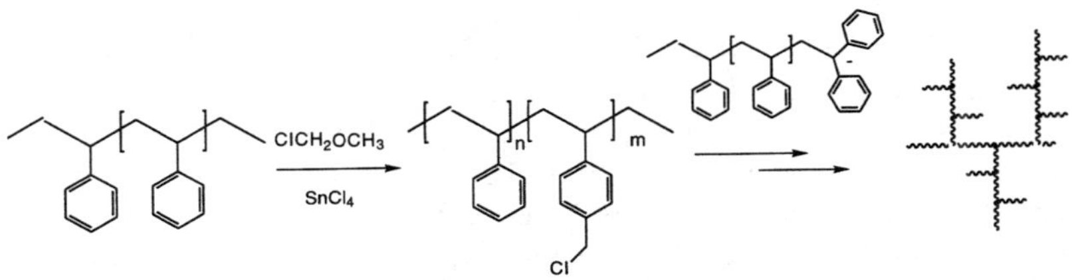

Scheme 20

backbone contains well-defined initiating functional groups. For example highly branched polyethylene oxide was prepared by this method. A multifunctional initiator was used to form first-generation branched polyethylene oxide. This living polymer was end-capped with a latent multifunctional group. Upon reactivation, the chain ends had provided sites for further branching (Scheme 21) [88].

The most intriguing aspect of all these polymers is that their properties depend more on the chain length between the branching points than on the total molecular weight. Thus, if the chain length between the branching point is less than the critical molecular weight for the chain entanglement, the polymer behaves like a low molecular weight polymer (with the slope of a plot between

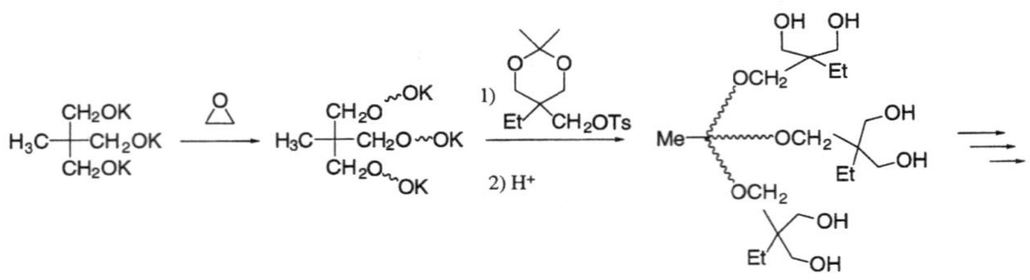

Scheme 21

log η and log MW being 1), regardless to the total molecular weight. When the molecular weight between the branching point increases beyond the entanglement molecular weight, the slope of a plot between log η and log MW becomes 3.4.

V. PHYSICAL PROPERTIES

Since hyperbranched polymers cannot engage in chain entanglement to the same degree as linear polymers, their usefulness in conventional structural applications is limited. In order to find uses other than in crosslinking chemistry, a better understanding of the physical properties of these polymers is necessary. A few intriguing properties have surfaced recently. Some polymers show colloid-like properties. The large number of terminal functional groups greatly influences the molecular relaxation process of the polymer, as exemplified by a wide range of T_g's depending on the nature of the end group. Polymer miscibility and solubility is another issue that needs to be addressed. The Flory–Huygens theory is based on the cohesive interaction energy parameter of the polymer chain. Since most of the interaction in the hyperbranched polymer resides at the chain ends, rather than the chain itself, conventional Flory–Huygens theory may not be able to predict the miscibility of the hyperbranched polymer. Even though physical understanding of these types of polymers is highly desired, only a limited amount of information is available.

A. Aggregation Properties

Some of these polymers exist as unusual colloid-like aggregates. Some organometallic hyperbranched polymers were found to aggregate even at very low concentrations [89]. A carboxylated hyperbranched polyphenylene had an oriented structure owing to partial segregation of the polymer at the air–water interface, resulting in an usual nucleation of the $CaCO_3$ crystal structure [90].

Several hyperbranched polymers show mesophases. Thermotropic hyperbranched polymers, comprised of mesogenic branched monomer units, were described in a series of publications. Phase transfer–catalyzed polymerization of 6-bromo-1-(4-hydroxy-4′-biphenylyl)-2-(4-hydroxyphenyl)hexane, 13-bromo-1-(4-hydroxyphenyl)-2-[4-(6-hydroxy-2-naphthalenylyl)phenyl]tridecane, and 13-bromo-1-(4-hydroxyphenyl)-2-(4-hydroxy-4″-*p*-terphenylyl)tridecane followed by in situ alkylation of their phenolate chain ends provided hyperbranched polymers. The minimum energy conformations of the flexible branching points of these polymers are anti and gauche. The gauche conformer leads to a hyperbranched polymer with a conventional treelike architecture. Above their glass transition temperature, these hyperbranched polymers are considered to minimize their free energy by lowering their free volume via a conventional nematic mesophase that is generated by the conformational change of their structural units from gauche to anti (Scheme 22) [91–94].

On the other hand, aromatic polyamides exhibit polymer aggregation in the absence of a complexing salt, resulting in mesomorphism [54]. Similarly, a protonated hyperbranched amide with an amine center showed mesomorphism [95].

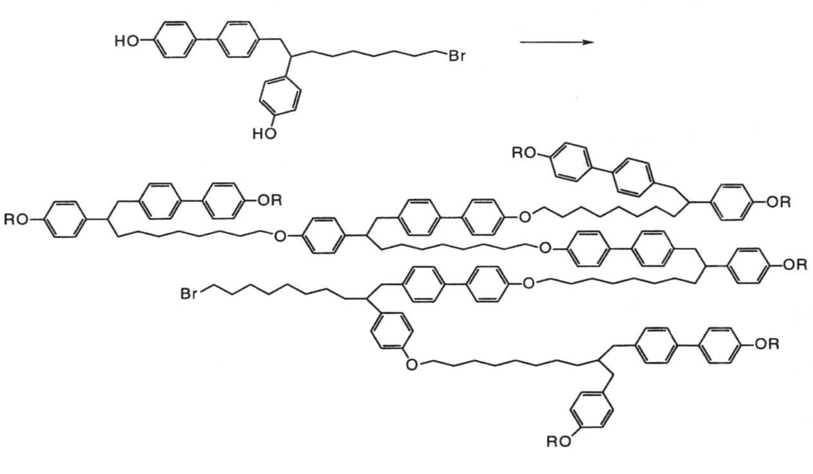

Scheme 22

B. Relaxation Properties

The strong correlation between the degree of branching and the nature of the terminal groups at the glass transition temperature has intrigued investigators for some time. Even though there is little understanding on the relaxation mode of the polymer molecules near the glass transition temperature, several empirical correlations were found to exist. The variation in the glass transition temperature as a function of molecular weight and types of chain end for dendritic polyethers and polyesters was explained by using a modified chain end free-volume theory. Although the modified Flory equation for a T_g considering the volume of end groups could explain the molecular dependence of polybenzylether dendrimer, the effect of the functional group could not be explained [96]. In comparing the T_g's of hyperbranched polyphenylenes (Table 6) and *p,p′,p″*-trisubstituted triphenylbenzenes, T_g's were found to increase with end-group polarity for both the polymers and the models, with almost linear correlation between these two sets of materials [97]. More detailed dielectric permittivity and loss measurement for hyperbranched polyesters of 2,2-bis(hydroxymethyl)propionic acid showed three relaxation transitions, the glass–rubber transition and two subglass processes with Arrhenius temperature dependence having an activation energy of 96 ± 2 kJ/mol. The low-temperature process could be assigned to motions of the terminal OH groups or reorientation of the ester groups [98].

In order to compare general properties of hyperbranched polymers and dendrimers, Wooley et al. examined a model hyperbranched polyester and corresponding dendrimer. Polymers prepared from 3-hydroxy-5-(*tert*-butyldimethylsiloxy)benzoic acid, as branching point, showed that thermal properties, such as T_g and those shown by thermogravimetic analysis (TGA), were independent of polymer architecture. However, the dendritic and hyperbranched materials demonstrated comparative solubilities that were much greater than that found for the linear polymer [99]. Their conclusions on the thermal properties may contradict some other findings. For examples, the T_g of hyper-

Table 6 The Effect of the Functional Group on T_g's of Hyperbranched Polyphenylene

Functional groups	*Br*	*H*	*CH_3*	*$(CH_3)_3Si$*	*p-Anisol*	*CH_2Cl*	*α-Vinyl phenyl*
T_g (°C)	221	121	177	141	223	182	96

branched polyphenylenes is much lower than that of dendrimers of comparable molecular weights (Table 7). This discrepancy may have originated from the fact that the hyperbranched model compound Wooley et al. used is a pure substance, whereas a real hyperbranched polymer is a mixture of millions of isomers of molecular weight and isomeric structures. This large number of isomers increases the state of entropy for the system and is expected to lower the transition temperature, as indicated in Eq. (11).

$$T = \frac{\Delta H}{\Delta S} \tag{11}$$

C. Polymer Solubility and Miscibility

Molecular modeling of the polymer reveals a large number of empty cavities between branches. We speculated that a complex would form between these polymers and other aromatic groups through intercalation of the aromatic rings. Existence of such complexation was confirmed by a NMR study of a water-soluble version of polymer **3** and *p*-toluidine [13]. If this interaction takes place with a polymer containing aromatic groups, it will cause a physically reversible crosslinking of the polymer chains. On the other hand, in a molten state where such static interaction would not be sustained, a spherically shaped hyperbranched polymer could affect the rheology of the other polymer. Polystyrene (PS) and polyvinyl chloride (PVC) were chosen to test this hypothesis. When PS was blended with 5% of brominated hyperbranched polyphenylene, its melt viscosity and thermal stability under shear were greatly improved. The activation energy of flow of PS was substantially lower in the presence of the hyperbranched polymer. In contrast, however, the hyperbranched polymer has no noticeable influence on the melt viscosity of PVC [62]. There are also no changes in T_g's of PVC or polystyrene blends containing up to 30 wt % of the hyperbranched polymer The difference between PS and PVC in rheological effect suggests that compatibility of a polymer with the hyperbranched polymer may be an important parameter for such an effect.

Some of these questions have been addressed using hydroxyphenyl- and acetoxyphenyl-terminated hyperbranched polyesters. The phase behavior of blends of hydroxyphenyl- and acetoxyphenyl-terminated hyperbranched polyesters with linear polymers such as polycarbonate, polyesters, and polyamides indicated that the hydroxy-terminated hyperbranched polyester blend miscibility was identical to that of poly(vinylphenol), suggesting that strong interactions due to hydrogen bonding, more than chain architecture, dominate the blend miscibility. The acetoxy-terminated hyperbranched polyester showed less miscibility than the hydroxy-terminated polymers owing to the absence of

Table 7 The Glass Thermal Transition Temperatures of Polyphenylene Dendrimers and Hyperbranched Polymers

Branched phenylenes	T_m (°C)	DH_f (kJ/mol)	T_g (°C)	DC_p (J/°C·g)
4Ph 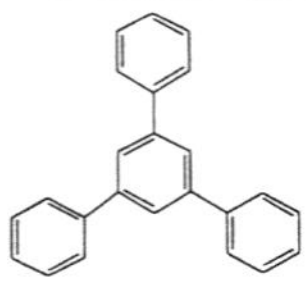	174.0	32.3	no T_g or 100 (?)	
10Ph 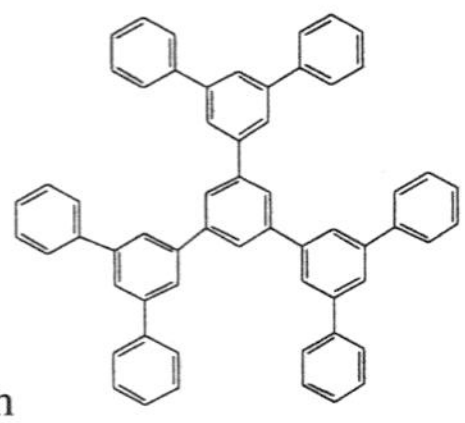	271.2	67.6	126	0.28
22Ph 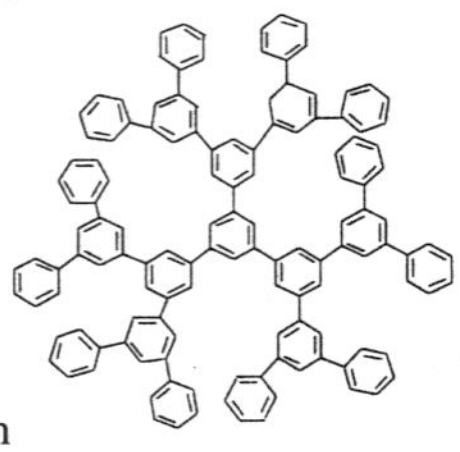	338.9	74.2	190	0.22
46Ph 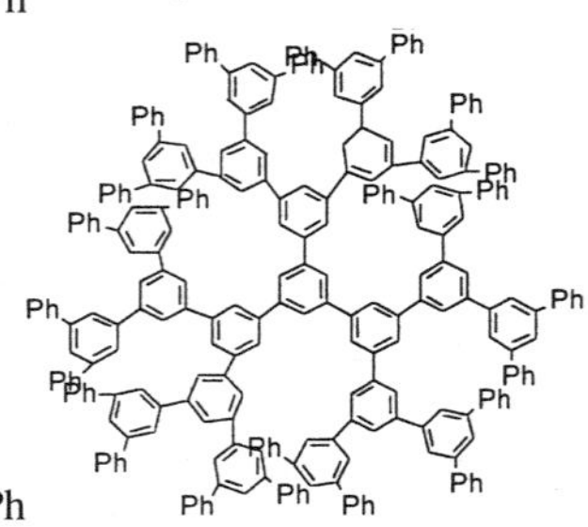	511.5	—	220	0.11
1,3,5-Polyphenyelene-H	None	—	121	0.26

Source: Refs. 101 and 102.

hydrogen-bonding interactions, but showed still more miscibility than the linear analogs. In addition, a blend of linear bisphenol A polycarbonate with the hyperbranched polyester resulted in increased tensile and compressive moduli and decreased strain-to-break and toughness [100].

Note added in proofs:

Since the manuscript was completed there has been much progress in hyperbranched polymers. Readers may investigate the vibrant activities through a few review articles [103–107] and articles on the polymerization mechanism, especially the cyclization reaction [108–109], new characterizations, classification of branching [110–113], and some new polymerization methods [114–116].

REFERENCES

1. Buhleier, E.; Wehner, W.; Voegtle, F. *Synthesis* **155** (1978).
2. Newkome, G. R.; Yao, Z.; Baker, G. R.; Gupta, V. K. *J. Org. Chem.* **50,** 2003 (1985).
3. Newkome, G. R.; Baker, G. R.; Saunders, M. J.; Russo, P. S.; Gupta, V. K.; Yao, Z. Q.; Miller, J. E.; Bouillion, K. *J. Chem. Soc., Chem. Commun.* **752** (1986).
4. Newkome, G. R.; Yao, Z.; Baker, G. R.; Gupta, V. K.; Russo, P. S.; Saunders, M. J. *J. Am. Chem. Soc.* **108,** 849 (1986).
5. Newkome, G. R.; Baker, G. R.; Arai, S.; Saunders, M. J.; Russo, P. S.; Theriot, K. J.; Moorefield, C. N.; Rogers, L. E.; Miller, J. E. *J. Am. Chem. Soc.* **112,** 8458 (1990).
6. Newkome, G. R.; Behera, R. K.; Moorefield, C. N.; Baker, G. R. *J. Org. Chem.* **56,** 7162 (1991).
7. Newkome, G. R.; Moorefield, C. N.; Baker, G. R.; Behera, R. K.; Escamillia, G. H.; Saunders, M. J. *Angew. Chem., Int. Ed. Engl.* **31,** 917 (1992).
8. Newkome, G. R.; Lin, X.; Yaxiong, C.; Escamilla, G. H. *J. Org. Chem.* **58,** 3123 (1993).
9. Newkome, G. R.; Baker, G. R. *NATO ASI Ser., Ser. C* **456,** 59 (1995).
10. Flory, P. J. *J. Am. Chem. Soc.* **74,** 2718 (1952).
11. Flory, P. J. *Principles of Polymer Chemistry;* Cornell University Press: Ithaca, 1953; Chapter 9.
12. Pinner, S. H. *J. Polym. Sci.* **21,** 153 (1956).
13. Kim, Y. H.; Webster, O. W. *J. Am. Chem. Soc.* **112,** 4592 (1990).
14. Hawker, C. J.; Lee, R.; Frechet, J. M. J. *J. Am. Chem. Soc.* **113,** 4583 (1991).
15. Kim, Y. H. *Macromol. Symp.* **77,** 21 (1994).
16. Kambouris, P.; Hawker, C. J. *J. Chem. Soc., Perkin Trans.* **1,** 2717 (1993).
17. Biggs, N. L.; Lloyd, E. K.; Wilson, R. L. *Graph Theory 1736–1936;* Clarendon: Oxford, 1976.
18. Trinajstic, N. *Chemical Graph Theory;* CRC Press, Inc.: Boca Raton, FL, 1983.
19. Wiener, H. *J. Am. Chem. Soc.* **17** (1947).
20. Gutman, I.; Yeh, Y.-N.; Lee, S.-L.; Chen, J.-C. *Match* **30,** 103 (1994).

21. Diudea, M. V.; Parv, B. *J. Chem. Inf. Comput. Sci.* **35,** 1015 (1995).
22. Diudea, M. V. *Match* **30,** 79 (1994).
23. Farin, D.; Avnir, D. *Angew. Chem.* **103,** 1409 (1991).
24. Mansfield, M. L. *Macromolecules* **26,** 3811 (1993).
25. Ingold, C. K.; Ingold, E. H. *J. Chem. Soc.,* 2249 (1928).
26. Jacobson, R. A. *J. Am. Chem. Soc.* **54,** 1513 (1932).
27. Vansheidt, A. A.; Mellnikova, E. P.; Gladkovskii, G. A. *Vysokomolekul Soedin,* **4,** 1303 (1962).
28. Overhult, W. C.; Ketley, A. D. *Makromol. Chem.* **95,** 143 (1966).
29. Haas, H. C.; Livingston, D.; Saunders, M. *J. Polym. Sci.* **15,** 503 (1955).
30. Baker, A. S.; Walbridge, D. J. U.S. Patent 3669939 (1972).
31. Figuly, G. D., U.S. Patent 5136014 (1992).
32. Kricheldorf, H. R.; Zang, Q.-Z.; Schwarz, G. *Polymer* **23,** 1821 (1982).
33. Frechet, J. M. J.; Hawker, C. J.; Uhrich, K., WO Patent 9208749.
34. Wooley, K. L.; Hawker, C. J.; Lee, R.; Frechet, J. M. J. *Polym. J.* **26**(2), 187 (1994).
35. Turner, S. R.; Voit, B. I.; Mourey, T. H. *Macromolecules* **26,** 4617 (1993).
36. Turner, S. R.; Walter, F.; Voit, B. I.; Mourey, T. H. *Macromolecules* **27,** 1611 (1994).
37. Kricheldorf, H. R.; Stoeber, O. *Macromol. Rapid Commun.* **15,** 87 (1994).
38. Kricheldorf, H. R.; Stoeber, O.; Luebbers, D. *Macromolecules* **28,** 2118 (1995).
39. Kricheldorf, H. R.; Stoeber, O.; Luebbers, D. *Macromol. Chem. Phys.* **196,** 3549 (1995).
40. Kricheldorf, H. R.; Loehden, G. *J. Macromol. Sci., Pure Appl. Chem.* **A32,** 1915 (1995).
41. Kricheldorf, H. R.; Loehden, G. *Macromol. Chem. Phys.* **196,** 1839 (1995).
42. Feast, W. J.; Stainton, N. M. *J. Mater. Chem.* **5,** 405 (1995).
43. Kumar, A.; Ramakrishnan, S. *Synthesis and Characterization of Hyperbranched Poly(ether-ester)s with Functional Spacers;* Bhardwaj, I. S., Ed.; Allied Publ.: New Delhi, India, 1994; Vol. 1, pp. 266.
44. Johansson, M.; Malmstroem, E.; Hult, A. *J. Polym. Sci., Part A: Polym. Chem.* **31,** 619 (1993).
45. Hult, A.; Johansson, M.; Malmstroem, E. *Macromol. Symp.* **98,** 1159 (1995).
46. Malmstroem, E.; Johansson, M.; Hult, A. *Macromolecules* **28,** 1698 (1995).
47. Malmstroem, E.; Hult, A. *Macromolecules* **29,** 1222 (1996).
48. Johansson, M.; Hult, A. *J. Coat. Technol.* **67,** 35 (1995).
49. Pettersson, B. *Proc. Water-Borne, Higher-Solids, Powder Coat.* 21st *Symp.,* 753 (1994).
50. Hult, A.; Malmstroem, E.; Johansson, M. *Polym. Mater. Sci. Eng.* **72,** 528 (1995).
51. Rao, C.; Tam, J. P. *J. Am. Chem. Soc.* **116,** 6975 (1994).
52. Tam, J. P.; Zavala, F. P. WO Patent 9011778.
53. Denkewalter, R. G.; Kolc, J.; Lukasavage, W. J., U.S. Patent 4289872.
54. Kim, Y. H. *J. Am. Chem. Soc.* **114,** 4947 (1992).
55. Kim, Y. H., unpublished results.

56. Uhrich, K. E.; Hawker, C. J.; Frechet, J. M. J.; Turner, S. R. *Polym. Mater. Sci. Eng.* **64,** 237 (1991).
57. Uhrich, K. E.; Hawker, C. J.; Frechet, J. M. J.; Turner, S. R. *Macromolecules* **25,** 4583 (1992).
58. Hunter, W. H.; Woollett, G. H. *J. Am. Chem. Soc.* **43,** 135 (1921).
59. Staffine, G. D.; Price, C. C. *J. Am. Chem. Soc.* **82,** 3632 (1960).
60. Kim, Y. H. *Highly Branched Aromatic Polymers;* Newkome; G., Ed.; JAI Press Inc.: Hampton Hill, Middlesex, 1995; Vol. 2, p. 123.
61. Kim, Y. H., U.S. Patent 4857630 (1987).
62. Kim, Y. H.; Webster, O. W. *Macromolecules* **25,** 5561 (1992).
63. M., D. J.; Oliver, P.; Mann, S.; Devries, A.; Hauschka, P.; Westbroek, P. *J. Chem. Soc. Faraday Trans.* **89,** 2891 (1993).
64. Kumar, A.; Ramakrishnan, S. *J. Chem. Soc., Chem. Commun.* 1453 (1993).
65. Spindler, R.; Frechet, J. M. J. *Macromolecules* **26,** 4809 (1993).
66. Morikawa, A.; Kakimoto, M.; Imai, Y. *Macromolecules* **26,** 6324 (1993).
67. Chu, F.; Hawker, C. J. *Polym. Bull.* **30**(3), 265 (1993).
68. Miller, T. M.; Neenan, T. X.; Kwock, E. W.; Stein, S. M. *J. Am. Chem. Soc.* **115,** 356 (1993).
69. Miller, T. M.; Neenan, T. X.; Kwock, E. W. *Macromol. Symp.* **77,** 35 (1994).
70. Mathias, L. J.; Carothers, T. W. *J. Am. Chem. Soc.* **113,** 4043 (1991).
71. Mathias, L. J.; Carothers, T. W. *Silicon-Based Stars, Dendrimers, and Hyperbranched Polymers;* Newkome; G., Ed.; JAI Press: Hampton Hill, Middlesex, 1995; Vol. 2, p. 101.
72. Muzafarov, A. M.; Gorbatsevich, O. B.; Rebrov, E. A.; Ignat'eva, G. M.; Chenskaya, T. B.; Myakushev, V. D.; Bulkin, A. F.; Papkov, V. S. *Vysokomol. Soedin., Ser. A, Ser. B* **35,** 1867 (1993).
73. Rubinsztajn, S. *J. Inorg. Organomet. Polym.* **4,** 61 (1994).
74. Muzafarov, A. M.; Golly, M.; Moeller, M. *Macromolecules* **28,** 8444 (1995).
75. Maxka, J.; Chrusciel, J.; Sasaki, M.; Krzysztof, M. *Macromol. Symp.* **77,** 79 (1994).
76. Suzuki, M.; Ii, A.; Saegusa, T. *Macromolecules* **25,** 7071 (1992).
77. Frechet, J. M. J.; Henmi, M.; Gitsov, I.; Aoshima, S.; Leduc, M. R.; Rrubbs, B. *Science* **269,** 1080 (1995).
78. Hawker, C. J.; Frechet, J. M. J.; Grubbs, R. B.; Dao, J. *J. Am. Chem. Soc.* **117,** 10763 (1995).
79. Gaynor, S. G.; Edelman, S.; Matyjaszewski, K. *Macromolecules* **29,** 1979 (1996).
80. Ishihama, Y.; Sawamoto, M.; Higashimura, T. *Macromolecules* **26,** 744 (1993).
81. Ishihama, Y.; Sawamoto, M.; Higashimura, T. *Polym. Bull.,* 361 (1990).
82. Georges, M. K.; Veregin, R. P. N.; Kazmaier, P. M.; Hamer, G. K. *Macromolecules* **26,** 2987 (1993).
83. Wang, J. S.; Matyjaszewski, K. *J. Am. Chem. Soc.* **117,** 5614 (1995).
84. Tomalia, D. A.; Hedstrand, D. M.; Ferritto, M. S. *Macromolecules* **24,** 1435 (1991).

85. Gauthier, M.; Moeller, M.; Burchard, W. *Polym. Prepr.* **34**(1), 60 (1993).
86. Gauthier, M.; Möller, M. *Macromolecules* **24,** 4548 (1991).
87. Gauthier, M.; Moeller, M.; Burchard, W. *Macromol. Symp.* **77,** 43 (1994).
88. Jean-Luc, S.; Gnanou, Y. *Macromol. Symp.* **95,** 137 (1995).
89. Campagna, S.; Giannetto, A.; Serroni, S.; Denti, G.; Trusso, S.; Mallamace, F.; Micali, N. *J. Am. Chem. Soc.* **117,** 1754 (1995).
90. Didymus, J. M.; Oliver, P.; Mann, S.; DeVries, A. L.; Hauschka, P. V.; Westbroek, P. *J. Chem. Soc., Faraday Trans.* **89,** 2891 (1993).
91. Percec, V.; Kawasumi, M. *Macromolecules* **25,** 3843 (1992).
92. Percec, V.; Chu, P.; Kawasumi, M. *Macromolecules* **27,** 4441 (1994).
93. Chen, F.-L.; Jamieson, A. M.; Kawasumi, M.; Percec, V. *J. Polym. Sci., Part B: Polym. Phys.* **33,** 1213 (1995).
94. Percec, V.; Chu, P.; Ungar, G.; Zhou, J. *J. Am. Chem. Soc.* **117,** 11441 (1995).
95. Stebani, U.; Lattermann, G. *Adv. Mater.* **7**(6), 578 (1995).
96. Wooley, K. L.; Hawker, C. J.; Pochan, J. M.; Frechet, J. M. J. *Macromolecules* **26,** 1514 (1993).
97. Kim, Y. H.; Beckerbauer, R. *Macromolecules* **27,** 1968 (1994).
98. Malmstroem, E.; Liu, F.; Boyd, R. H.; Hult, A.; Gedde, U. W. *Polym. Bull.* **32**(5–6), 679 (1994).
99. Wooley, K. L.; Frechet, J. M. J.; Hawker, C. J. *Polymer* **35,** 4489 (1994).
100. Massa, D. J.; Shriner, K. A.; Turner, S. R.; Voit, B. I. *Macromolecules* **28,** 3214 (1995).
101. Miller, T. M.; Neenan, T. X.; Zayas, R.; Bair, H. E. *J. Am. Chem. Soc.* **114,** 1018 (1992).
102. Miller, T. M.; Neenan, T. X. *Chem. Mater.* **2,** 346 (1990).
103. Kim, Y. H. *J. Polym. Sci., Part A: Polym. Chem.* **36,** 1685 (1998).
104. Kricheldorf, H. R. *Pure Appl. Chem.* **70,** 1235 (1998).
105. Kricheldorf, H. R. *Macromol. Symp.* **122,** 15 (1997).
106. Malmstroem, E.; Hult, A. *J. Macromol. Sci., Rev. Macromol. Chem. Phys.* **C37,** 555 (1997), C37(3).
107. Frechet, J. M. J.; Hawker, C. J. *Compr. Polym. Sci., 2nd Suppl.* 71 (1996).
108. Gooden, J. K.; Gross, M. L.; Mueller, A.; Stefanescu, A. D.; Wooley, K. L. *J. Am. Chem. Soc.* **120**, 10180 (1998).
109. Feast, W. J.; Keeney, A. J.; Kenwright, A. M.; Parker, D. *Chem. Commun.* 1749 (1997).
110. Yan, D.; Zhou, Z. *Macromolecules* **32,** 819 (1999).
111. Yan, D.; Zhou, Z.; Mueller, A. H. E. *Macromolecules* **32,** 245 (1999).
112. Yan, D.; Mueller, A. H. E.; Matyjaszewski, K. *Macromolecules* **30,** 7024 (1997).
113. Mueller, A. H. E.; Yan, D.; Wulkow, M. *Macromolecules* **30,** 7015 (1997).
114. Huber, T.; Boehme, F.; Komber, H. T.; Kronek, J.; Luston, J.; Voigt, D.; Voit, B. *Macromol. Chem. Phys.* **200,** 126 (1999).
115. Tian, D.; Dubois, P.; Jerome, R. *Macromol. Symp.* **130,** 217 (1998).
116. Suzuki, M.; Yobbshida, S.; Shiraga, K.; Saegusa, T. *Macromolecules* **31,** 1716 (1998)

9

Dendritic Architectures by the Convergent Method

Jean-Luc Six and Yves Gnanou
ENSCPB–CNRS–Université Bordeaux I, Talence, France

I. INTRODUCTION

The Staudinger concept of covalently bonded macromolecules introduced in 1922 and that pertaining to the random nature of all synthetic polymerization reactions—whether they proceed via chain or step growth—are the two cornerstones of polymer chemistry. Polymer chemists are thus used to dealing with averages to describe the samples they synthesize, because the samples are actually mixture of chains of different size. Similarly, the shape adopted by the macromolecule in solution is also subject to averages because the macromolecule continuously twists and turns around the chemical bonds of its backbone. One of the most popular areas of research since the early days of polymer chemistry was actually directed at minimizing the fluctuation of parameters such as the size and composition or shape of the macromolecules in order to prepare tailor-made polymers.

Despite significant progress, mainly due to the emergence of living polymerization techniques, no polymerization process can yield perfectly monodisperse products nor afford total control over the shape of the macromolecules obtained.

In sharp contrast with the classical approach and the practice of polymer chemists, researchers working at the interface between supramolecular chemistry and biochemistry, on the one hand, and polymer synthesis, on the other, have striven during the last 10 years to make truly homostructural and uniform macromolecules. This has led to the emergence of a novel domain of investigation concerned with the preparation of macromolecules having exactly defined molecular structure for which the size, the shape, the topology and the surface functionalization are unique. The dendrimers [1] and the polymers made by genetic engineering [2] are only two examples of synthetic macromolecules exhibiting the preceding features.

Nature provides numerous examples of molecular assemblies whose size, shape, disposition of chemical functions, etc., are precisely controlled, features that are mirrored in their unique properties and functions. In the quest to discover unexpected properties, several highly innovative research groups have conceived original methods of synthesis to achieve precision and control over the targeted topologies.

The case of dendrimers illustrates the efforts made to assemble supramolecular structures of nanoscopic scale that are free of any dispersity. The term *dendrimers* was coined by Tomalia [1] to designate three-dimensional layered arrangements of chemical bonds that arise from the introduction of a branching point at each monomer unit. More precisely, a dendrimer consists of four main components: a *central core, arms* of identical length, linking *branching points* that are therefore symmetrically distributed in the dendrimer, and *end-standing reactive functions*.

Two different approaches, based for both on stepwise reactions and functional group protection strategies—in a similar manner to the Merrifield method—have been chalked out to match the highly symmetrical and flaw-free nature of dendrimers.

The first approach [1], known as *divergent synthesis,* implies that the growth of the dendritic structure starts from a plurifunctional molecule and proceeds outward through successive coupling and activation reactions (Scheme 1). This activation step, which could be a deprotection reaction as well, is designed to generate two or more branching sites at the free terminus of the last introduced monomer units. Iteration of the very same coupling and activation/deprotection reactions progressively gives rise to the dendritic structure.

The second approach is based on the prior synthesis of treelike structures—also called *dendrons*—containing a specific function at their focal point, and on their subsequent coupling to a central core fitted with antagonist functions (Scheme 2). These treelike structures are themselves obtained by coupling a molecule containing one particular reactive group with the branch-

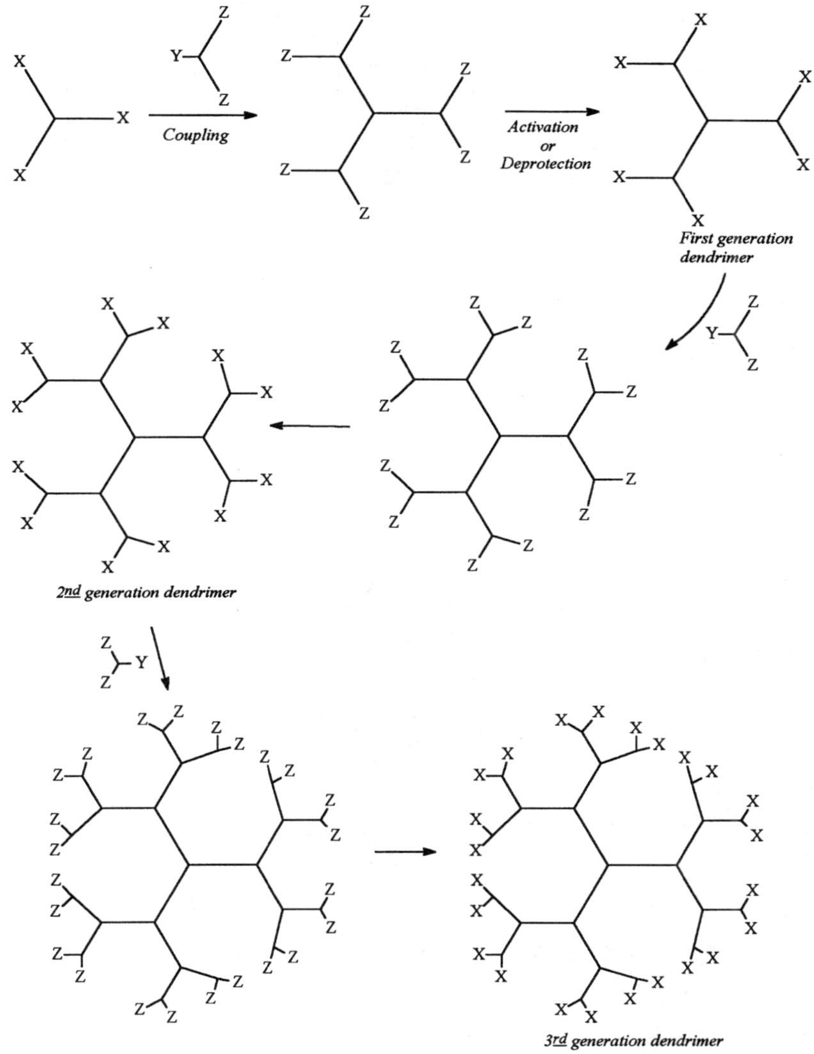

Scheme 1 Synthesis of dendrimers by the divergent method.

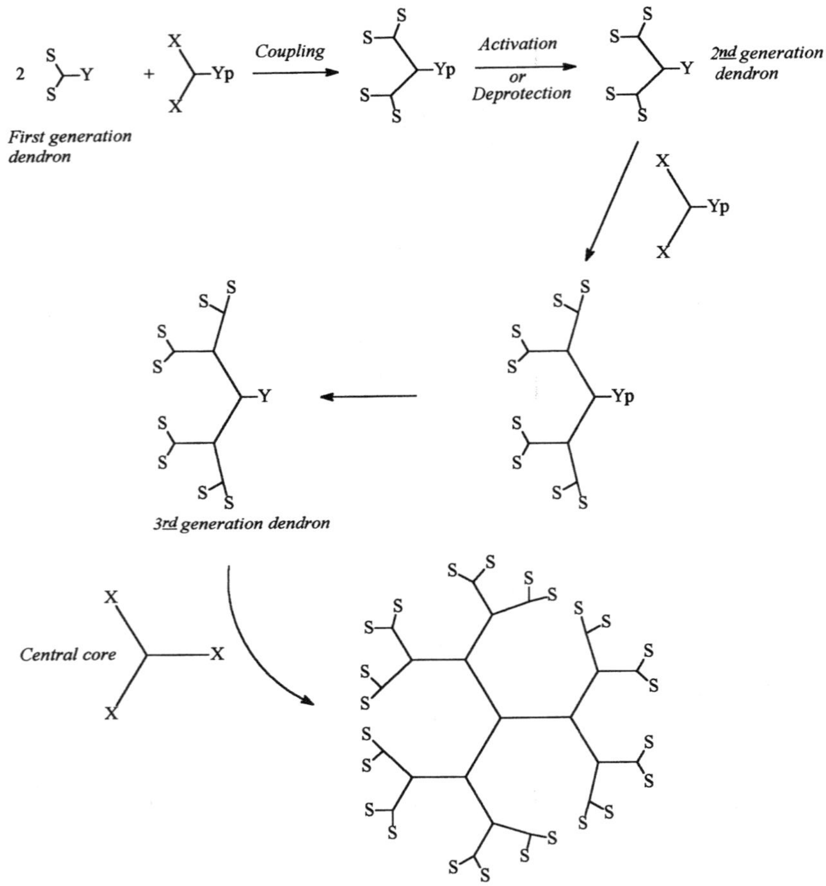

Scheme 2 Synthesis of dendrimers by the convergent method.

ing sites of a molecule containing another nonreactive function so that the resulting dendritic structure will carry this remaining reactive moiety. Upon activation of the latter, the preceding coupling reaction using the same branching molecule can be repeated, yielding dendritic fragments of a higher generation. The way the latter are built entails an inward growth that actually begins at the chain ends.

With the convergent method [3], the formation of each generation involves the same number of coupling reactions at the branching sites, whereas in divergent synthesis there is an ever-increasing number of reactions as the dendrimer grows.

Convergent synthesis therefore offers a better chance to control the purity and integrity of the dendrimer prepared. Considering the wealth of dendritic structures that have been generated using the convergent approach, it is obvious that it provides better prospects in terms of macromolecular engineering than divergent synthesis.

As a matter of fact, the two approaches used to build dendritic structures can be viewed as a means to generate species with $n \times f$ terminal functions from a precursor exhibiting f similar groups by condensation techniques. This feature has thus been used to construct branched polymers with perfectly defined structure. After a brief historical account, this paper will focus on the convergent approach to dendrimer synthesis and review the different dendritic structures that have been obtained via this method.

II. HISTORICAL BACKGROUND

The initial impetus to and the principles of the divergent mode of synthesis of arborescent molecules can be historically traced to a 1978 article by Vögtle [4], who first defined the concept of cascade process. He showed how to proliferate amine functions upon starting from benzylamine and repeating a sequence of two reactions: the Michael addition and the reduction of nitriles to amines. Vögtle had to stop at rather an early stage while attempting to synthesize his polyamine dendrimers because of the lack of selectivity of the Michael reaction.

Soon after Vögtle disclosed his cascade process, Denkewalter [5] ventured on this area and described the synthesis of branched polylysine, using standard peptide chemistry and a coupling/deprotection strategy. He started from benzhydrylamine as initiator, to which he coupled two molecules of bis(t-Boc)-L-lysine fitted with an activated ester (nitrophenyl).

Upon removal of the two *t*-BOC protecting groups, the two amine groups that arose were available for a subsequent amidation with two bis *t*-BOC-L-lysine building blocks. Repetition of these coupling/deprotection processes allowed Denkewalter to isolate polylysine dendrimers of high generation, although it seems that the products obtained were somewhat contaminated with unwanted species.

Several years after these first attempts to make arborescent structures, Tomalia revisited the chemistry used by Vögtle, giving particular attention to both the purification of the products obtained and the selectivity of each reaction carried out. He benefited from the efforts at Dow Laboratories to improve the Michael addition chemistry, which he directed toward the preparation of polyamidoamine dendrimers up to the ninth generation. The synthesis of these polyamidoamines, although reminiscent in its principle to that followed by Vögtle for his polyamines, differs in one main point: After adding two methyl acrylate units to primary amine via an exhaustive Michael's reaction, Tomalia used a second condensation reaction to generate the amine functions necessary to the growth of the subsequent generation, whereas Vögtle resorted to a deprotection strategy to obtain the same amine functions.

Tomalia [6a] actually started with ammonia as the central core and used its three sites to attach three ester-terminated branches. He then carried on with the amidation of these esters using a large excess of ethylene diamine that he carefully removed to isolate the first-generation polyamidoamine. He routinely repeated the same two reactions and the purification step to obtain polyamidoamines of a higher generation.

A similar strategy was followed by Newkome [7] to derive arborols, a term he coined to feature the treelike structure and the presence of alcohol functions. Following these early attempts, which all resorted to divergent synthesis, a wealth of similar structures emerged in recent years [1,6b].

It is worth pointing out that the dendrimers obtained in this way exhibit ratios of branching point to repeating units essentially equal to unity. This means that the interior volume of such dendritic species as well as the concentration of the reactive functions on their peripheral surface are fixed and depend on the size of the monomer unit linking two branching points. In an attempt to overcome the preceding limitations, dendritic macromolecules whose successive generations are true macromolecular chains have been recently synthesized using the divergent method [8]. Although iterative, as for any regular dendrimer, the principle of synthesis of this family of dendritic structures differs from that for classical divergent synthesis in one essential point: The arms in dendrimers made of macromolecular generations result from the chain polymerization of a monomer and not from stepwise reactions. The formation of each generation is preceded by the arborization of the arms ends of the previous generations. Dendrimers of poly(ethylene oxide) (PEO) have been prepared along this line upon successively repeating the polymerization of ethylene oxide and the branching of PEO arm ends [8]. Likewise, novel amphiphilic dendrimers made of an inner polystyrene layer and an external PEO corona have been synthesized by the same team [9].

Only in the early 1990s did the first account of the synthesis of dendrimers by convergent growth appear in the literature. Fréchet [10] and Neenan and Miller [11], who are the actual pioneers of this method of synthesis, disclosed almost concomitantly that starburst dendrimers could be generated via the convergent route. It rapidly became clear that convergent synthesis has some indisputable advantages over the divergent method. Most of these advantages concern the ease of purification and the isolation of the targeted dendrimer. Because each dendron is the result of a coupling reaction between two dendritic fragments of smaller generation, the separation of the expected material is straightforward owing to the fact that the latter is bound to exhibit much larger molar mass than that of the precursor. In divergent synthesis, the separation step is much more delicate, because incompletely reacted and exhaustively reacted materials involves only small molar mass differences. The formation of a given generation is indeed the result of the reaction between the dendrimer of the previous generation and a large number of small molecules.

But the convergent method also has its weaknesses. Steric hindrance affects the convergent synthesis more than the divergent method for the same reasons as those invoked for the ease of purification. The highest generation ever obtained using the convergent strategy rose to the sixth generation, whereas with the divergent method Tomalia went up to the ninth generation of polyamidoamine. In the following sections, the versatility of the convergent pathway to synthesize dendrimers as well as the difficulties encountered will be discussed, as well as the properties exhibited by the various dendritic structures.

III. DENDRIMERS BY THE CONVERGENT METHOD

A. Aryl Ether–Based Dendrimers

The first dendrimers ever prepared via convergent methodology were the aryl ether dendrimers of Fréchet [10]. This first study triggered a number of other investigations on similar aryl ether dendrimers and resulted in the creation of several original architectures, all including a dendritic part (see Section IV.C)

Fréchet and his co-workers first prepared aryl ether wedges fitted with a functional group at their focal point and then shaped the latter into dendrimers by reaction with a plurifunctional antagonist reactant. The molecule designed to reside at the periphery of the dendron is benzyl bromide, which was reacted with 3,5-dihydroxybenzyl alcohol to give the first-generation dendron (Scheme 3).

This dendritic fragment, after bromination of its benzyl alcohol function, was coupled to the same diphenol building block as above. This gave rise to the second-generation dendron. Repetition of the two steps of bromination and linking to 3,5-dihydroxybenzyl alcohol resulted in the formation of dendrons of higher generations. The final dendrimer was derived by coupling fourth-generation dendrons to 1,1,1-tris(4′-hydroxyphenyl)ethane. Dendrimers up to the sixth generation have been obtained in this way. These polyether dendrimers were then subjected to detailed characterization in dilute media. They were

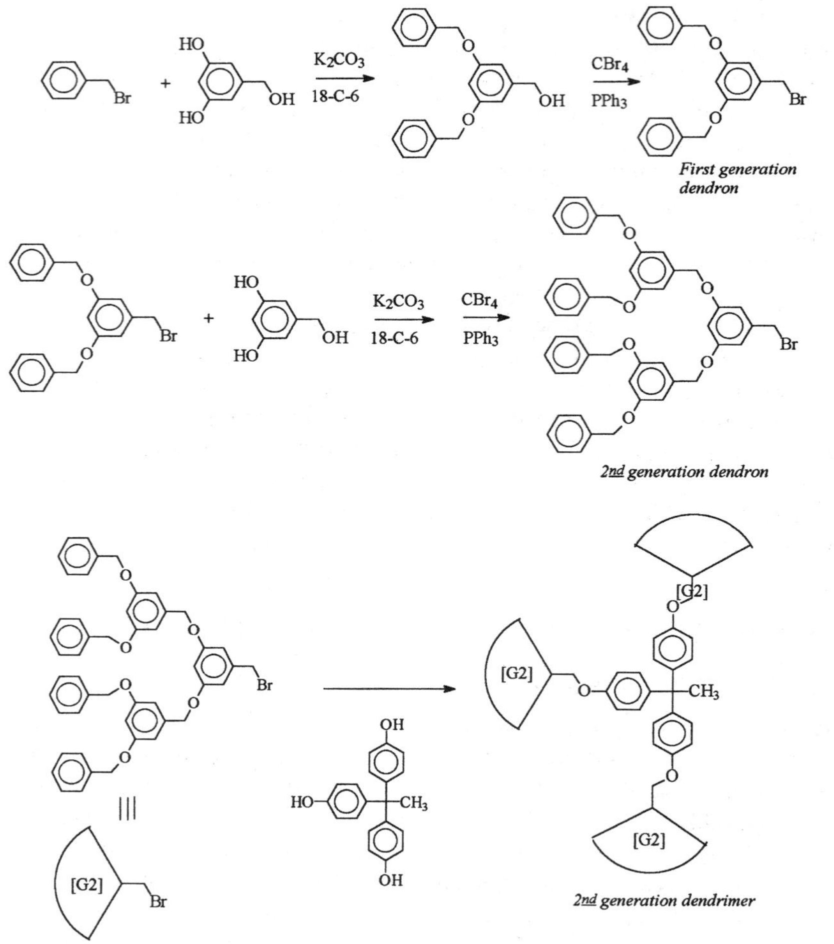

Scheme 3 Synthesis of second-generation aryl ether dendrimers.

shown to exhibit a very peculiar viscosity pattern, and depending upon the generation considered they behave quite differently. A bell-shaped curve was found for the variation of the intrinsic viscosity as a function of their molar mass. The increase in intrinsic viscosity with molar mass and the sudden decrease after generation 4 is reached were interpreted as the result of the transition from an extended structure to a more globular shape [12].

Further evidence for this progressive shape transition was given from the study of the variation of λ_{max} [13] for a series of dendritic polyether dendrons fitted with a solvatochromic probe at their focal point. This study was purposely conducted in solvents with low polarity, the change in λ_{max} serving to evaluate the ability of the solvent to penetrate into the interior of the dendron. As λ_{max} was found to increase with the generation, this feature was accounted for by the lack of penetration of the solvent and the greater influence of surrounding repeating units.

The polyether dendrons and dendrimers were also characterized in the solid state [14]. The variation of their T_g was found to depend on the ratio n_e/M where n_e represents the number of end groups and M the molar mass. As to the relationship between T_g, $T_{g\infty}$, and M, the traditional equation for linear samples has been modified to account for the dendritic structure.

Following the reports of Fréchet on aryl ether dendrimers, several groups embarked on studies in this area, adopting Fréchet's synthetic route to prepare dendrons and then coupling them to central cores of their choice. Inoue [15] attached third-generation aryl ether dendrons to a porphyrin molecule carrying not less than eight phenol functionalities. Likewise, L'Abbé [16] linked second-generation aryl ether dendrons to a trithiolate serving as the central core, whereas Meijer [17] attached the preceding dendritic wedges to multisubstituted pentaerythritol.

B. Polyesters with Dendritic Structure

For the preparation of polyesters with dendritic structures Fréchet and his coworkers [18] chose to start from benzoic acid and used 2,2,2-trichloroethyl 3,5-dihydroxybenzoate as a building block (Scheme 4). To obtain the first-generation dendritic wedge in relatively good yields, they had to overcome the lack of reactivity of the two compounds and discover the proper reaction conditions. With the use of dichloromethane as solvent and dicyclohexylcarbodiimide (DCC) along with 4-(dimethylamino)pyridinium *p*-toluenesulfonate (DPTS), the first-generation dendron was obtained in 89% yield (after purification) [18b]. The trichloroethyl benzoate function of the building block actually served to mask the benzoic acid functionality, which could be easily released upon treating the first-generation dendron with zinc in glacial acetic acid. This gave rise to the precursor of the second-generation dendron, with a benzoic acid function at its focal point. For the synthesis of dendrons of the second and third generation, the same synthetic scheme was employed as above. This scheme consists of coupling the dendritic precursor of the previous generation to the benzoate building block in a 2/1 ratio and subsequently activating the trichloroester function. The dendrimer of the third generation was then obtained in 72%

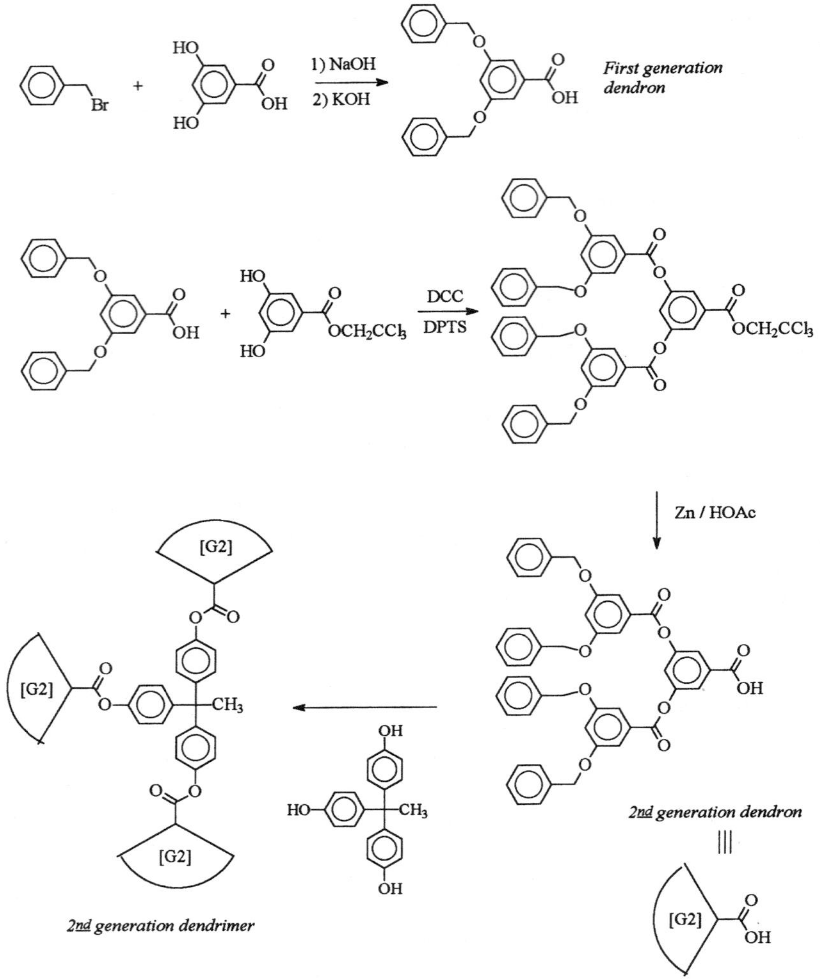

Scheme 4 Synthesis of second-generation aryl ester dendrimers.

yield, upon reacting 3 equivalents of third-generation dendrons with 1,1,1-tris(4′-hydroxyphenyl)ethane [18a].

Fréchet and his co-workers then characterized their dendritic polyesters both in the solid state [14,19] and in solution [19] and compared their behavior with that exhibited by linear samples of the same chemical structure. The physical properties of their aromatic polyesters seemed not to be affected by the overall architecture of the macromolecule, as the T_g's and the thermogravimetric behaviors did not vary when going from linear to dendritic polyesters.

The pathway followed by Miller and Neenan [20] to generate their aryl ester dendrimers was in many aspects quite similar to that described by Fréchet. The main difference concerns the type of reactants used. Miller and Neenan chose a derivative of 3-hydroxyisophthalic acid, namely 5-(*tert*-butyldimethylsiloxy)isophthaloyl chloride, as the building block—instead of the dihydroxybenzoic acid used by Fréchet—and started from phenol. The first-generation dendron was the result of the reaction of the diacid chloride with phenol; removal of the silyl-protecting group present at the focal point by HCl treatment gave rise to a phenol functionality. The latter group was subsequently used to grow the second-generation dendrimers upon repeating the coupling to diacid chloride. This series of reactions—coupling and deprotection—was pursued up to the third generation. Aryl ester dendrimers of the third generation were ultimately obtained in 25% yield upon coupling in a 3/1 ratio the third-generation dendrons to 1,3,5-benzenetricarbonyl trichloride.

The attempts of Bryce [21] to make aryl ester dendrimers decorated with external tetrathiofulvalene (TTF) revolved around the same approach as that of Miller and Neenan. Bryce et al. [21] started from a phenol containing two ester TTF moieties and used 5-*tert*-butyldimethylsilylisophthalic chloride as the building block. 1,3,5-Benzene tricarbonyl chloride was taken as the central core to couple second-generation dendrons. The second-generation dendrimer aryl ester dendrimers fitted with 12 terminal TTF units were then oxidized. Investigation of their redox behavior showed that simultaneous electron transfer did occur.

Feast and Stainton [22] also prepared aryl ester dendrimers, following a synthetic scheme slightly different from that successfully experimented with by Miller and Neenan. They separated two successive branching junctions with an aromatic spacer and not just an ester linkage, as in Miller and Neenan's dendrimers. The authors introduced these extension units in their aryl ester dendrimers to confer additional flexibility. The characterization of their third generation—the highest generation obtained—showed the presence of some dendritic wedge that could not be removed.

C. Aramid-Type Dendrimers

Interest in aromatic amide dendrimers surfaced after earlier studies by Tomalia and others showed that the dendritic architecture may induce unusual solid-state properties. It was therefore worth attempting to arrange in a dendritic structure those polymers known for their tendency to crystallize and to check whether their behavior would be modified.

Miller and Neenan [23] and later Feast's team [24] addressed this topic with a view of using such aramid dendrimers as additives for depressing the melting temperature of linear aromatic polyamides. The latter are indeed known to give semicrystalline materials with rather high melting temperature. Both groups [23,24] adopted the same synthetic scheme to get access to the targeted dendrimers. As in any convergent synthesis, they first prepared the dendrons that must then be coupled to a trifunctional central core in a second step. To form their first-generation wedges, they started from aniline and reacted the latter molecule with a 2-molar amount of 5-nitroisophthaloyl chloride, which was the pivotal building block. The second-generation dendron was grown after the nitro group fitting the focal point was reduced into amine, and was thus ready to get coupled with a 2-molar excess of the same building block as above. The elaboration of dendrons of a higher generation was not carried on further (perhaps the authors preferred to stop at early stages of the assembly of their aramide dendrimers). The last step consisted of reacting dendrons of first- and second-generation wedges with 1,3,5-benzenetricarbonyl chloride (Scheme 5). Whereas the coupling reactions giving rise to the central core functioned as expected in the first-generation dendrimer, it did not quantitatively occur with the dendron of a higher generation. Feast and co-authors invoked steric hindrance and the stiffness of the dendritic precursor to account for the failure in the attempt to shape third-generation dendrons into dendrimers [24a]. The second-generation aramid dendrimers were found to depress the melting temperature of the linear polyamides [24b], though blended in small amounts with nylon-6,6.

D. Polyphenylene-Based Dendrimers

The motivation behind the synthesis of polyphenylene dendrimers resembles in many aspects that invoked for making aramid dendrimers. Firstly, linear polyphenylene—even low molar mass samples—are insoluble and, secondly, hyperbranched polyphenylenes have been shown by Webster and Kim [25] to be, in contrast, readily soluble in most organic solvents. The strategy designed by Miller and Neenan [26] for linking aromatic rings was based on the Suzuki coupling reaction, whose antagonist aryl derivatives are aryl boronic acids and aryl halides, respectively. This reaction has to be catalyzed by $Pd(PPh_3)_4$. To make their first-generation dendrons, Miller and Neenan reacted phenyl boronic acid and 3,5-dibromotrimethylsilylbenzene, used as pivotal building block and branching agent (Scheme 6). Before growing the second generation, the trimethylsilyl function of the first-generation intermediate was converted into boronic acid and coupled with the same aryl dihalide as above. The obtained sec-

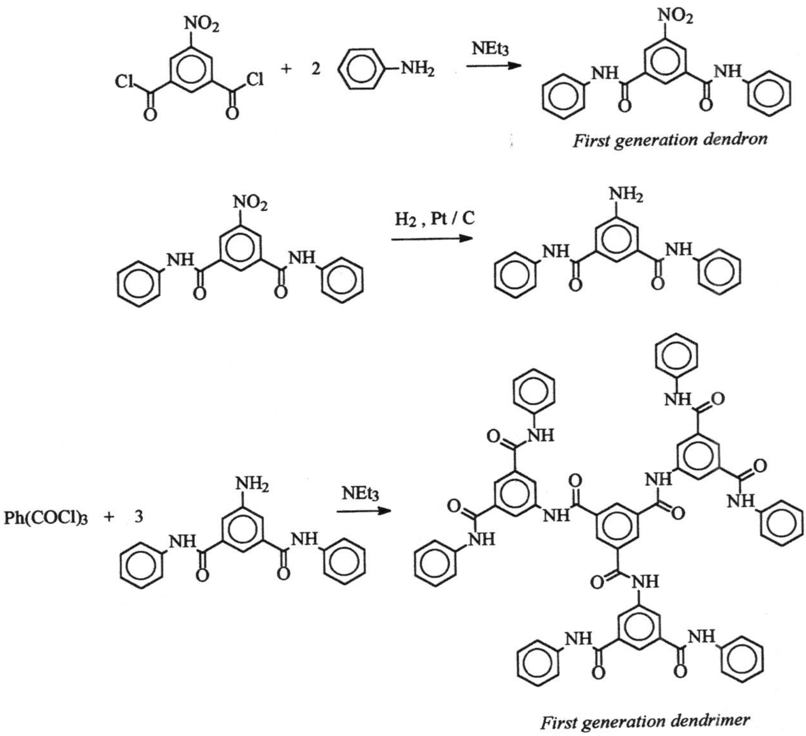

Scheme 5 Synthesis of aramid-type dendrimers.

ond-generation wedges were then linked to a central core fitted with an aryl bromide derivative, namely, 1,3,5-tris(3,5-dibromophenyl)benzene, affording the expected second-generation dendrimers.

Dendrimers made up of fluorinated phenyl rings were synthesized by a similar sequence of reactions. The solubility and thermal properties of these dendritic polyphenylenes have been evaluated. The largest hydrocarbon dendrimer was, for instance, soluble in toluene and stable up to 500°C [26].

Although directed toward completely different applications, the synthetic scheme followed by Rajca et al. [27] to prepare their polyarylmethane dendrimers is in the same vein as that conceived by Miller and Neenan for their polyphenylenes. As Rajca and his co-workers were interested in very high spin polyradicals, they considered that dendritic ordering of macromolecules should

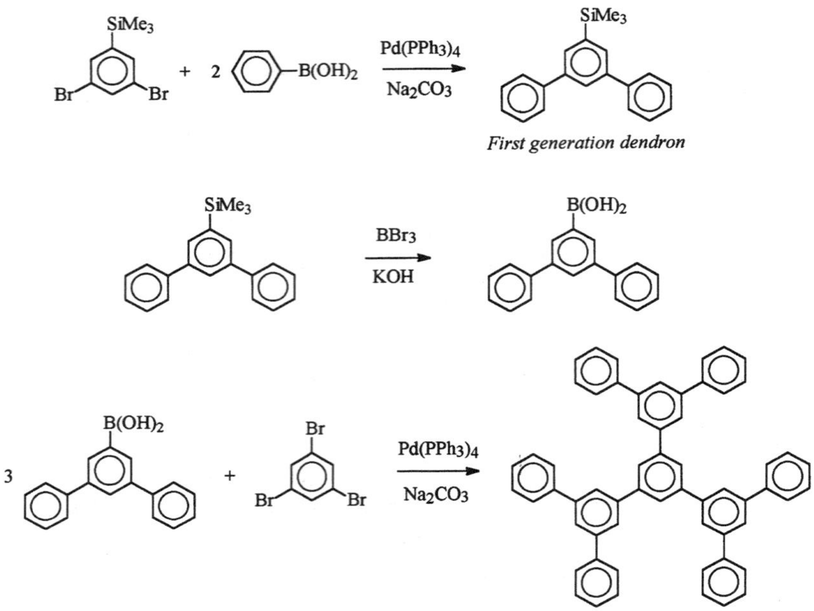

Scheme 6 Synthesis of first-generation polyphenylene-based dendrimers.

be superior to conventional linear arrangements, in regard to the objective of synthesizing organic polyradicals. They first targeted dendritic carbopolyanions, which they successfully oxidized into the corresponding polyradicals with iodine. Convergent synthesis again appeared as the best synthetic approach. These polyarylmethane dendrimers were obtained upon double addition of aryllithium to methyl 3-bromo-2,4-dimethylbenzoate. The triarylmethyl alcohol obtained as the first-generation dendrimer was then converted into the corresponding methyl ether. Exchange of bromide by lithium on the first-generation dendrons was followed by the reaction of the latter with methyl 3-bromo-2,4-dimethylbenzoate, yielding the second-generation dendrons. Repetition of the same reaction of exchange and coupling afforded dendrons of a higher generation with a controlled number of methyl ether functions. Cleavage of the latter by lithium and subsequent oxidation of polycarbanions by iodine gave rise to polyradicals systems, exhibiting very high spins. It seemed, however, that the chemistry of coupling must be further developed before perfectly defined systems become accessible.

E. Dendrimers with Miscellaneous Building Blocks

Two teams have endeavored to synthesize organometallic dendrimers with a metal atom present in each layer or at the periphery of the dendrimer by convergent synthesis. Liao and Moss [28] opted for these dendritic structures; they started from bromoalkyl complexes of $CpM(CO)_2$—$(CH_2)_3Br$ (where M = Fe or Ru and Cp = 5-C_5H_5) and chose 3,5-dihydroxybenzyl alcohol as the building block to couple with the preceding compound. After purification of the resulting first-generation dendron, which is nothing but a benzyl alcohol derivative fitted with two metal carbon σ-bonds, they converted the primary benzyl alcohol into the corresponding benzyl bromide. Second-generation dendrimers containing metal–carbon bonds at their periphery were assembled upon coupling the above ω-bromide dendrons with 1,1,1-tris(4-hydroxyphenyl)ethane.

Achar and Puddephat [29], on the other hand, investigated the synthesis of dendrimers containing a metal atom at each layer (Scheme 7). Their synthetic scheme was based on the oxidative addition of the C–Br bonds of aromatic bromomethyl to the platinum complexes. The first-generation dendrons consisting of two platinum centers were the result of the reaction between

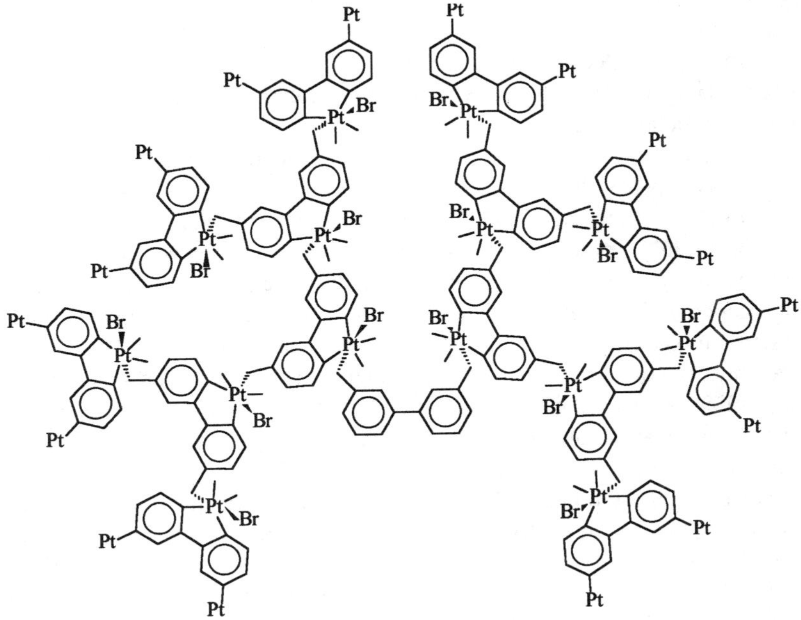

Scheme 7 Example of a dendritic structure containing a metal atom at each layer.

$PtMe_2(bu_2bpy)$ (with bu_2bpy = 4,4′-di *tert*-butyl-2,2′-bipyridine) and 4,4′-bis(bromomethyl)-2,2′-bipyridine. Second-generation materials were obtained by coupling the first-generation precursors to the same aromatic bromomethyl building block as above. To get access to the second-generation dendrimers that contained not less than 28 Pt centers, Puddephatt and his co-worker linked four second-generation dendrons to a tetrafunctional core, which turned out to be 1,2,4,5-tetrakis(bromomethyl)benzene [29].

In a different area Kakimoto's team [30,31] also resorted to the convergent method to prepare two families of dendrons and dendrimers. They first reported [30] the synthesis of siloxane-based dendrons and chose the hydrosilylation reaction as a means of assembling their molecules in a dendritic way. Diethylamino-(3-cyanopropyl) dimethylsilane and allylbis(4-hydroxydimethylsilylphenyl)methylsilane were respectively chosen as the starting material and building block as shown in Scheme 8. The cyanide-containing aminosilane that is bound to form the periphery of the molecule was obtained by hydrosilylation of allyl cyanide using chlorodimethylsilane and amination of the chlorosilane intermediate with diethylamine. The first-generation dendrons were assembled upon coupling ω-cyanide aminosilane to the silanol building block. Prior to the growth of the second-generation dendrons, the allyl function of the first generation was converted into aminosilane as by the preceding method, and the coupling to disilanols was repeated. The authors claimed to have isolated spherical starburst dendrimers by reaction of aminosilane dendrons with tris[4-(hydroxydimethylsilyl)phenyl]methyl silane, but no experimental details were given.

Kakimoto and co-workers [31] also prepared poly(ether ketone) dendrons and have based their synthetic strategy on nucleophilic substitution reactions. Phenol was taken as the material fitting the external part of the dendrons; the building block, 3,5-bis(4-fluorobenzoyl)anisole, was synthesized from 5-methoxy isophthaloyl chloride and fluorobenzene through Friedel–Craft reactions. First-generation dendrons were derived simply by coupling two phenol molecules to the building block. Release of the phenol functionality via demethylation of the first-generation anisole group was carried out prior to assembling the second generation, a step brought about upon coupling to the same building block as above. Dendrons up to fourth generation were built up by iteration of demethylation and coupling steps.

Dendrimers including a chiral part, carried by either the central core or the branching points distributed in the dendritic structure, have attracted recent interest. Three research groups ventured into this area, but the main contribution was due to Seebach's team [32]. These researchers relied on Fréchet's synthetic approach to first make aryl ether dendrons and then attach them to various chiral cores having either a true chiral center (Scheme 9A) or unsym-

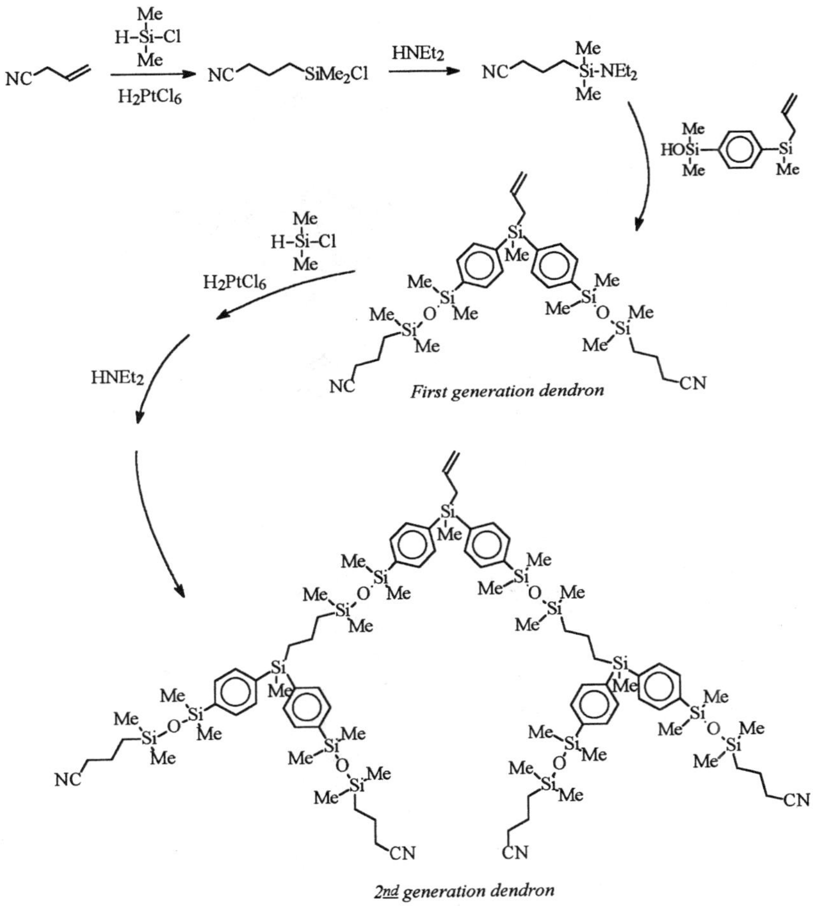

Scheme 8 Synthesis of siloxane-based dendrimers.

metrical spacers. The same team also prepared fully chiral dendrimers and used unsymmetrical building blocks [33]. They carried out the synthesis with a view to answering the following questions: "Will a chiral core cause the dendrimer to be optically active? Will the dendrimer containing unsymmetrical branching units exhibit chiral recognition?" It turned out that the optical activity decreases as dendrimers containing only one chiral center increase in size. In contrast, fully chiral second-generation dendrimers retain the same activity as that exhibited by the building blocks.

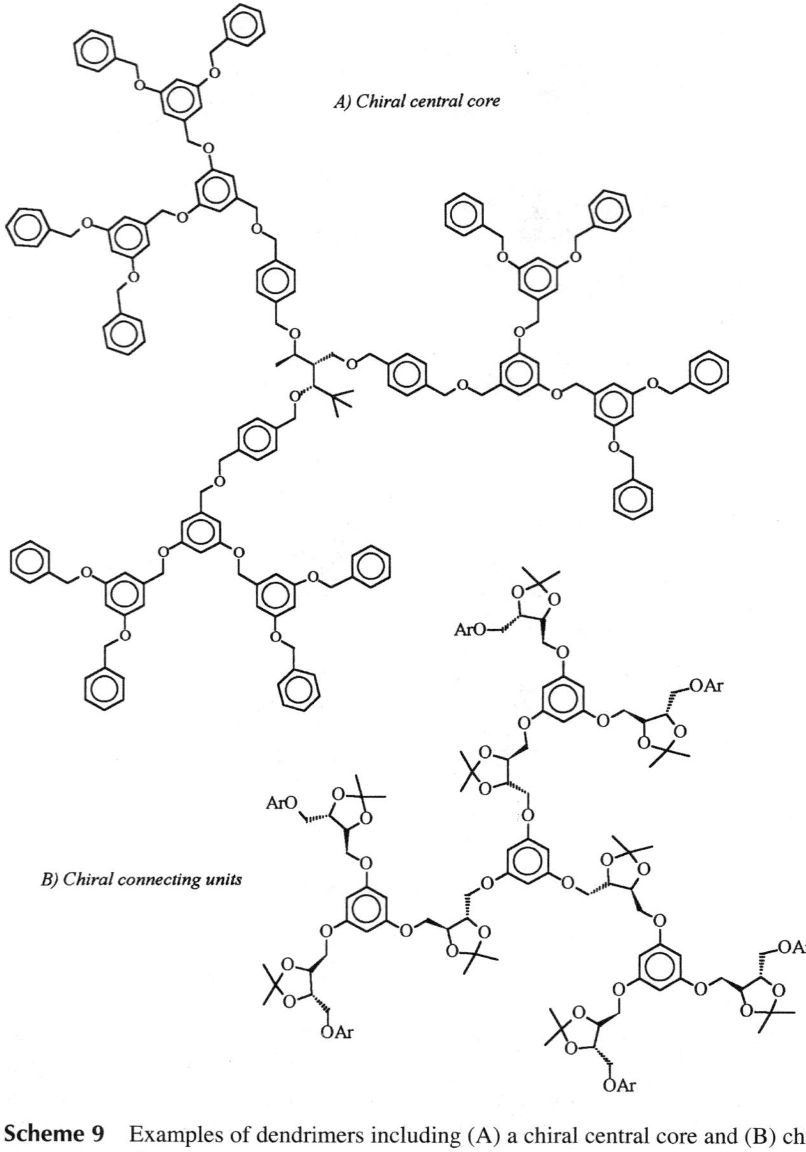

Scheme 9 Examples of dendrimers including (A) a chiral central core and (B) chiral connecting units.

Two other research groups showed interest in chiral dendritic molecules. Mitchell's group [34] reported the synthesis of glutamate-based dendritic molecules containing 15 chiral centers, all with identical configurations. Chow et al. [35] also contributed by tartaric acid–based dendrimers. The chiral element, which is a derivative of L-tartaric acid, was designed to serve as a connecting unit between two successive branching points, themselves arising from phloroglucinol (Scheme 9B). Investigation of the optical properties of these dendritic materials showed that they are proportional to the number of chiral tartaric units in the molecule.

Interest in the biological role of branched ribonucleic acids led Damha and his co-workers [36,37] to consider the synthesis of nucleotide-based dendrons by the convergent approach. The experimental scheme involved the synthesis of oligonucleotides in solid phase with an automated DNA synthesizer, the coupling of two adjacent polymer-bound nucleotide chains through the use of fully protected adenosine 2′,3′-*o*-bis(phosphoramidite) [36], and the repetition of the preceding two steps. Similar dendrons were obtained with thymidine [37] as the building block responsible for branching.

IV. DENDRIMERS WITH SPECIFIC FEATURES BY CONVERGENT SYNTHESIS

A. Dendritic Segment Block and Layer Block Copolymers [18b,38]

The possibility of associating layers of different chemical natures in a dendritic structure emerged soon after Fréchet and his co-workers showed the versatility of their convergent approach (Scheme 10). They chose to assemble aryl ether and aryl ester generations in the same dendritic structure. Using the same synthetic scheme as that experimented with for aryl ether dendrimers, they first isolated second-generation aryl ether dendrons and then coupled them with the building block that has served to make aryl ester dendrimers (Scheme 10A). As these dendritic intermediates were fitted with trichloroethyl ester functions at their focal point, Fréchet then switched to the chemistry he developed for aryl ester dendrimers. In more recent contributions, Fréchet and his co-workers [39,40] described the preparation of dendritic poly(ether-ester) layer block copolymers by a more rapid method associating both convergent and divergent approaches. The T_g's of these dendritic copolymers were measured and it turned out that they fall between the values characteristic to the two pure dendrimers [14].

Besides the possibility of assembling generations of different chemical nature in layers [18b,38], the convergent method also grants opportunities for

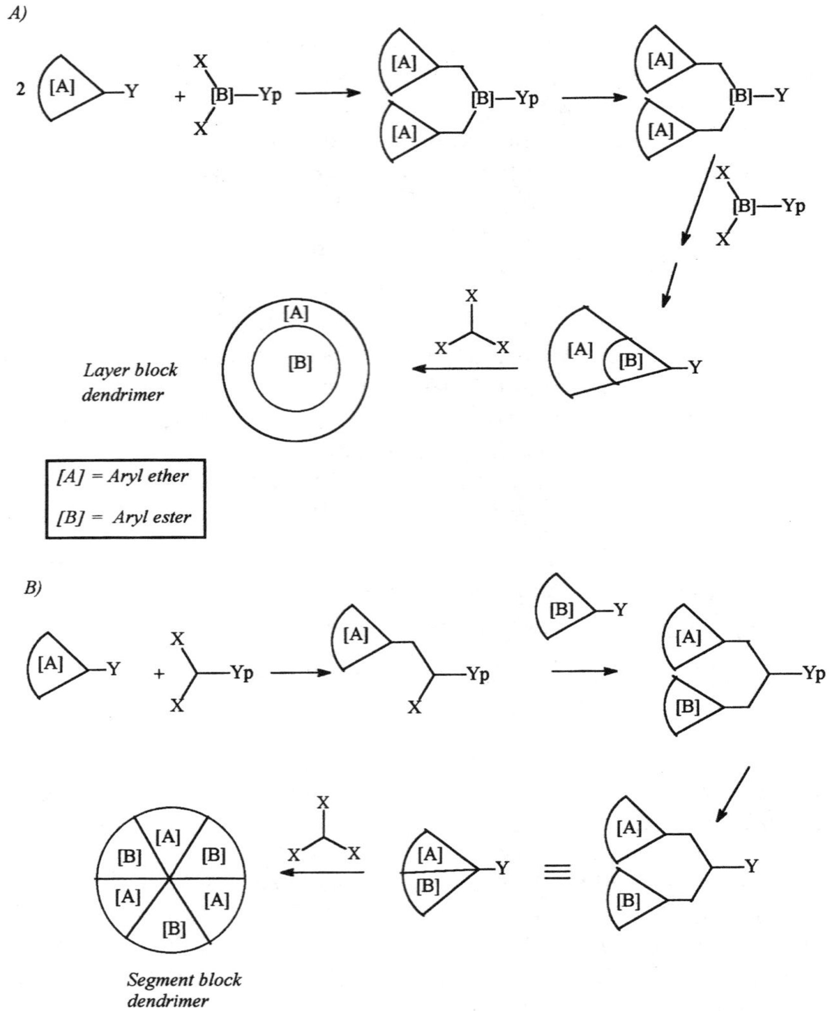

Scheme 10 Strategy utilized to prepare (A) layer block and (B) segment block dendrimers.

arranging dendrons in a radial alternating manner. From two dendrons made of ether and ester units respectively, attaching the latter hybrid dendrons to a plurifunctional central core was shown to give rise to the expected segment block dendrimers (Scheme 10B).

B. Dendrimers with Controlled Surface Chemistry [37,41–44]

One of the beauties of convergent synthesis is the possibility of assembling at the ultimate step dendrons with different surface functionality. Fréchet and his team have taken advantage of this feature to assemble dendrimers with various end-group functionalities (polar/nonpolar, hydrophilic/hydrophobic, etc.). They first showed that dendrimers obtained via convergent synthesis could be fitted with specific functions at their periphery, like those produced by the divergent method. Unimolecular micelles have been, for instance, created upon generating carboxylic acid functions on the outer surface of hydrophobic aryl ether dendrimers [41] (Scheme 11).

As for dendritic structures involving two unsymmetrical surface functionalities, several contributions from the same team described the method for making dendrimers covered with specific groups on the two opposite faces of the globular structure. Aryl ether dendrimers with high dipole moments [42] have been, for instance, obtained upon linking dendrons carrying polar cyano groups on their surface to wedges fitted with end-standing benzyloxy groups, so as to place two types of functions on opposite faces. Likewise, the synthesis of an amphiphilic dendritic block copolymer obtained upon connecting through a difunctional coupling agent two kinds of already made wedges—the first fitted with carboxylic acid functions and the second carrying hydrophobic dangling C-12 chains—have been reported by the same team [43]. In the same line, they prepared aryl ether dendrimers containing one, two, or three cyano groups from dendritic wedges fitted with one single cyano group at their periphery [44].

C. Hybrid Structures Made of Linear Chains and Dendritic Fragments by the Convergent Method

The dendritic intermediates in convergent synthesis do carry a reactive function at their focal point. These reactive functions have been utilized to derive several hybrid structures.

Macromonomers consisting of dangling dendritic wedges have been, for instance, obtained by introduction of an unsaturation at the focal point of the

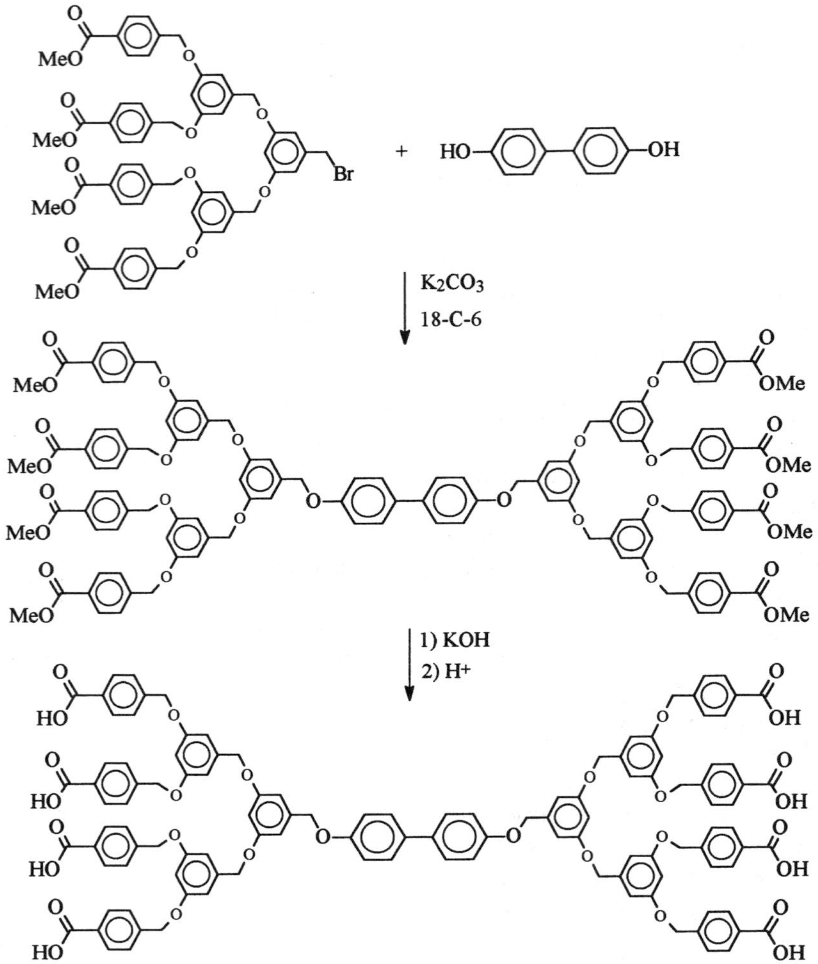

Scheme 11 Hydrophobic aryl ether dendrimer fitted with a hydrophilic surface.

dendrons and have been subsequently copolymerized by a free-radical process with styrene [45] (Scheme 12).

PS and PEO chains—obtained by anionic polymerization—have been coupled to aryl ether dendrons that carried a benzyl bromide function at their focal point [46–48]. ω-Hydroxy PEO chains have also been linked to aryl ether dendrons carrying a methyl ester function at their periphery [49]. With mono-

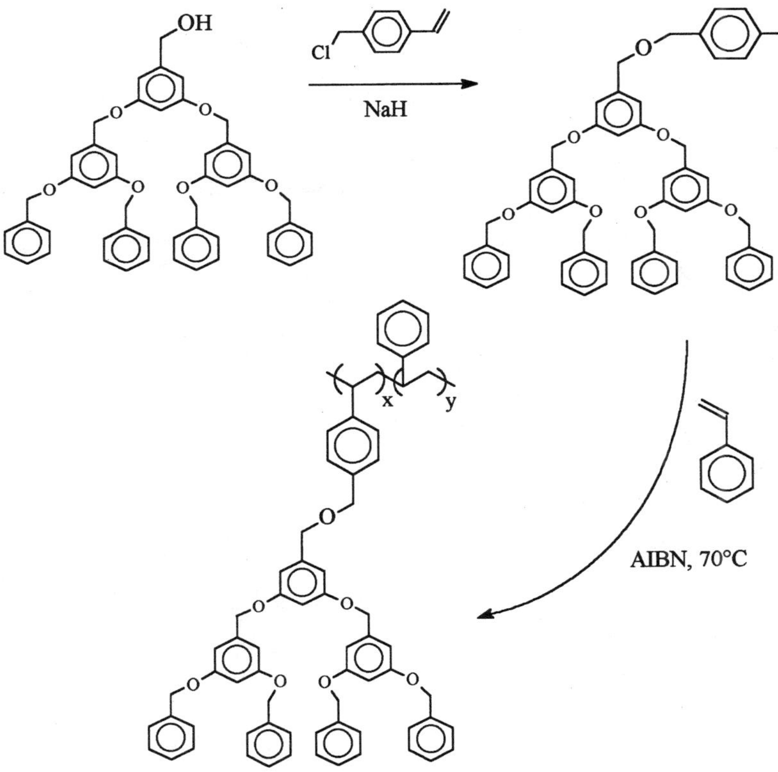

Scheme 12 ω-Styrenic aryl ether dendron and its copolymerization with styrene.

functional chains a sperm-shaped structure was formed (Scheme 13), whereas difunctional chains gave rise to barbell-shaped polymers (Scheme 14). Along the same lines, Moore and his co-workers [50] deactivated "living" poly(methyl methacrylate)s onto phenylacetylene dendrons carrying an aldehyde function. Also, a hybrid structure made of aryl ether dendrons and a poly(ε-caprolactone) chain has been described by Fréchet's group [51], who took advantage of the presence of an alcoholate function to initiate the anionic polymerization of ε-caprolactone.

Instead of anchoring dendrons at the end of the linear chains, Schlüter and his co-workers [52] attached Fréchet-type dendrons onto the rodlike, rigid backbone of poly([1,1,1]-propellane)s and polyphenylenes on which hydroxy functionalities had previously been generated. Schlüter's group made use of

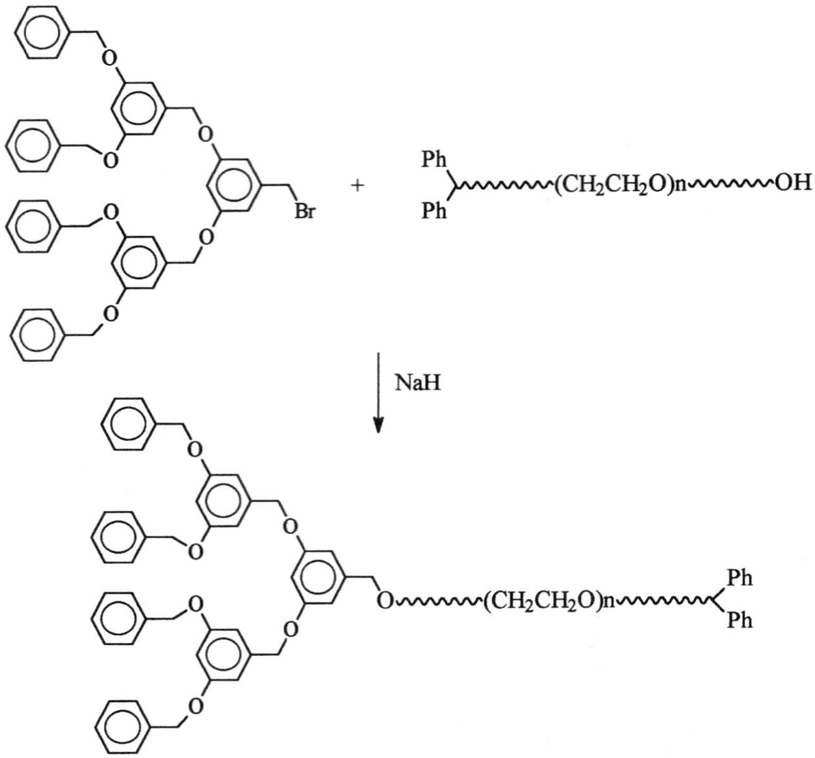

Scheme 13 Sperm-shaped polymer formed by coupling ω-hydroxy PEO chain to benzyl bromide–containing dendron.

the Williamson etherification reaction for this purpose and claimed that macrocylinders consisting of a rigid backbone wrapped by aryl ether wedges were obtained. They also attempted to produce poly(*p*-phenylene) decorated with dendritic fragments via step-growth polymerization of monomers previously modified with dendrons [53]. The polycondensation reaction was somehow hampered by the steric effect, so that only limited molar masses have been obtained. Of interest also is the hybrid structure that was derived by Fréchet and his co-workers by reaction of azide-containing dendrons with fullerenes [54].

Some of the hybrid linear-dendritic block copolymers have been subjected to a thorough characterization in solution [48,55] and in the solid state

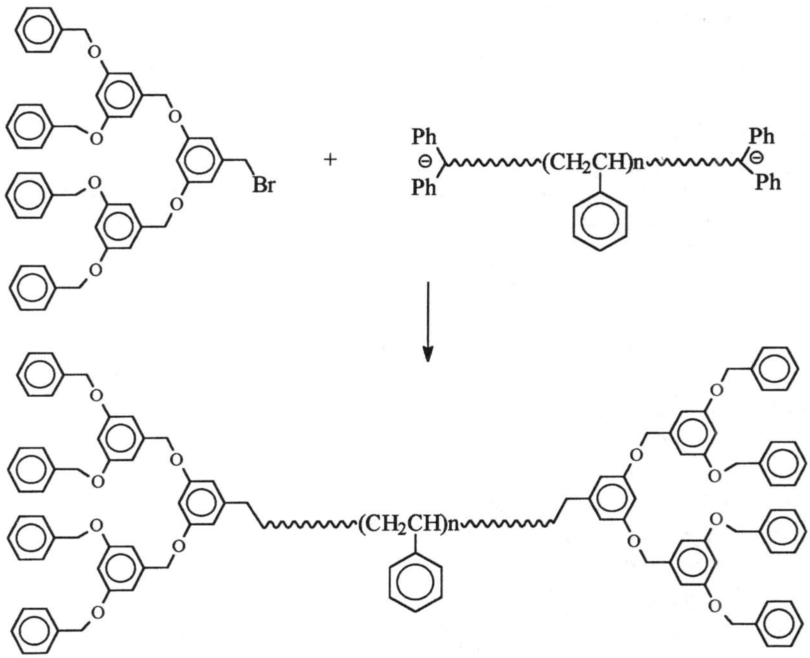

Scheme 14 Barbell-shaped polymer formed by deactivation of living polystyrene onto benzyl bromide containing dendrons.

[47,48]. Architectures consisting of linear PEO chains and one or two aryl ether dendrons have been investigated for their solution behavior in a methanol–water mixture. Depending upon the dendrons generated and their concentration, multimolecular micelles have been shown to form because of the presence of water, which is a nonsolvent for aryl ether dendrons [55]. In contrast, monomolecular micelles were observed in tetrahydrofuran (THF), which is a good solvent for the two blocks. The second type of hybrid structure, which has been comprehensively studied from the point of view of physicochemical properties, associated PS linear chains and aryl ether dendrons [48]. Radii of gyration ranging from 2.5 to 15 nm were measured for these barbell structures in THF. A shape transition from an extended globular form to that adopted by a statistical coil has been shown to occur as the size of the central PS chain increases.

V. OUTLOOK

The purpose of this review is to show the versatility of the convergent method of dendrimer synthesis. Various original architectures, shapes, and surface functionalities have been made available by this still recent approach. The synthetic pathway is based on an inward growth of dendrimers but entails certain drawbacks. The main drawback of this approach is the difficulty of producing dendrimers of a high generation. Because the growth of each generation depends on the reaction between a function present at the focal point of the previous generation dendron and those of the building block, steric inhibition may hamper such coupling for higher generations.

However, Moore and his co-workers [56] showed that by increasing the size of the building block as a function of generation, it is possible to reduce if not eliminate the problem of steric hindrance. He developed the concept of dendritic wedges including linear extensions and applied it to the convergent synthesis of phenylacetylene dendrimers.

As to the potential applications that could be derived from the dendrimers prepared by the convergent method, anyone would agree that the area of application-oriented dendrimers is still in its infancy but is bound to expand in the near future. Some high added-value applications are cited in the literature in domains such as drug delivery and molecular devices, but none has really reached the industrialization stage.

REFERENCES

1. D. Tomalia, A. Naylor, W. Goddard III, *Angew. Chem. Int. Ed. Engl.* **29,** 138 (1990).
2. K. McGrath, M. Fournier, T. Mason, D. Tirrell, *J. Am. Chem. Soc.* **114,** 727 (1992).
3. J. Fréchet, C. Hawker, K. Wooley, *J. Macromol. Sci., Pure Appl. Chem.* **A31**(11), 1627 (1994).
4. E. Buhleier, W. Wehner, F. Vögtle, *Synthesis* **155** (1978).
5. R. Denkewalter, J. Kolc, W. Lukasavage, *U.S. Patent* 4,410,688 (1983).
6. (a) D. Tomalia, H. Baker, J. Dewald, M. Hall, G. Kallos, S. Martin, J. Roeck, J. Ryder, P. Smith, *Macromolecules* **19,** 2466 (1986); (b) D. Tomalia, *Adv. Mater.* **6,** (7/8), 529 (1994).
7. (a) G. Newkome, Z. Yao, G. Baker, V. Gupta, *J. Org. Chem.* **50,** 2003 (1985); (b) C. Moorefield, G. Newkome, *Adv. Dendritic Macromol.* **1,** 1 (1994), JAI Press, Inc., Greenwich, CT.
8. (a) J.-L. Six, Y. Gnanou, *Polym. Prepr., ACS Division of Polymeric Materials* **35,** 488 (1994); (b) J.-L. Six, Y. Gnanou, *Macromol. Symp.* (1995).
9. E. Cloutet, J.-L. Six, D. Taton, Y. Gnanou, *Polym. Prepr., ACS, PMSE,* **73,** 133 (1995).

10. C. Hawker, J. Fréchet, *J. Am. Chem. Soc.* **112,** 7638 (1990).
11. T. Neenan, T. Miller, E. Kwock, H. Bair, *Adv. Dendritic Macromol.* **1,** 105 (1994), JAI Press, Inc., Greenwich, CT.
12. T. Mourey, S. Turner, M. Rubinstein, J. Fréchet, C. Hawker, K. Wooley, *Macromolecules* **25,** 2401 (1992).
13. C. Hawker, K. Wooley, J. Fréchet, *J. Am. Chem. Soc.* **115,** 4375 (1993).
14. K. Wooley, C. Hawker, J. Pochan, J. Fréchet, *Macromolecules* **26,** 1514 (1993).
15. R. Jin, T. Aida, S. Inoue, *J. Chem. Soc., Chem. Comm.* 1260 (1993).
16. G. L'Abbé, B. Haelterman, W. Dehaen, *J. Chem. Soc., Perkin. Trans.* 2203 (1994).
17. (a) J. Kremers, E. Meijer, *J. Org. Chem.* **59,** 4262 (1994); *Macromol. Symp.,* **98,** 491 (1955).
18. (a) C. Hawker, J. Fréchet, *J. Chem. Soc., Perkin. Trans.* 2459 (1992); (b) C. Hawker, J. Fréchet, *J. Am. Chem. Soc.* **114,** 8405 (1992).
19. K. Wooley, J. Fréchet, C. Hawker, *Polymer* **35,** 21, 4489 (1994).
20. (a) E. Kwock, T. Neenan, T. Miller, *Chem. Mater.* **3,** 775 (1991); (b) T. Miller, E. Kwock, T. Neenan, *Macromolecules* **25,** 3143 (1992).
21. M. Bryce, W. Devonport, A. Moore, *Angew. Chem. Int. Ed. Engl.* **33**(17), 1761 (1994).
22. W. Feast, N. Stainton, *J. Mater. Chem.* **4**(8), 1159 (1994).
23. T. Miller, T. Neenan, *Chem. Mater.* **2,** 346 (1990).
24. (a) P. Bayliff, W. Feast, D. Parker, *Polym. Bull.* **29,** 265 (1992); (b) S. Backson, P. Bayliff, W. Feast, A. Kenwright, D. Parker, R. Richards, *Macromol. Symp.* **77,** 1 (1994).
25. Y. Kim, O. Webster, *Macromolecules* **25,** 5561 (1992).
26. T. Miller, T. Neenan, R. Zayas, H. Bair, *J. Am. Chem. Soc.* **114,** 1018 (1992).
27. (a) A. Rajca, S. Utamapanya, *J. Am. Chem. Soc.* **115,** 10688 (1993); (b) A. Rajca, *Adv. Dendritic Macromol.* **1,** 133 (1994), JAI Press, Inc., Greenwich, CT.
28. Y. Liao, J. Moss, *J. Chem. Soc., Chem. Comm.* 1774 (1993).
29. (a) S. Achar, R. Puddephatt, *J. Chem. Soc., Chem. Comm.* 1895 (1994); (b) S. Achar, R. Puddephatt, *Angew. Chem. Int. Ed. Eng.* **33**(8), 847 (1994).
30. A. Morikawa, M. Kakimoto, Y. Imai, *Macromolecules* **25,** 3247 (1992).
31. A. Morikawa, M. Kakimoto, Y. Imai, *Macromolecules* **26,** 6324 (1992).
32. (a) D. Seebach, J. Lapierre, K. Skobridis, G. Greiveldinger, *Angew. Chem. Int. Ed. Engl.* **33,** 4, 440 (1994); (b) D. Seebach, J. Lapierre, G. Greiveldinger, K. Skobridis, *Helv. Chim. Acta* **77,** 1673 (1994).
33. J. Lapierre, K. Skobridis, D. Seebach, *Helv. Chimi. Acta* **76,** 2419 (1993).
34. L. Twyman, A. Beezer, J. Mitchell, *Tetrahedron Lett.* **35**(25), 4423 (1994).
35. (a) H. Chow, C. Mak, *J. Chem. Soc., Perkin. Trans.* 2223 (1994); (b) H. Chow, L. Fok, C. Mak, *Tetrahedron Lett.* **35**(21), 3547 (1994).
36. M. Damha, S. Zabarylo, *Tetrahedron Lett.* **30,** (46), 6295 (1989).
37. R. Hudson, M. Damha, *J. Am. Chem. Soc.* **115,** 2119 (1993).
38. C. Hawker, K. Wooley, J. Fréchet, *Macromol. Symp.* **77,** 11 (1994).
39. K. Wooley, C. Hawker, J. Fréchet, *Angew. Chem. Int. Ed. Engl.* **33,**(1), 82 (1994).
40. R. Spindler, J. Fréchet, *J. Chem. Soc., Perkin. Trans.* 913 (1993).

41. C. Hawker, K. Wooley, J. Fréchet, *J. Chem. Soc., Perkin. Trans. I* 1287 (1993).
42. K. Wooley, C. Hawker, J. Fréchet, *J. Am. Chem. Soc.* **115,** 11496 (1993).
43. E. Sanford, J. Fréchet, K. Wooley, C. Hawker, *Polym. Prepr., ACS Polymer Division* **34**(2), 654 (1993).
44. (a) K. Wooley, C. Hawker, J. Fréchet, *J. Chem. Soc., Perkin. Trans.* 1059 (1991); (b) C. Hawker, J. Fréchet, *Macromolecules* **23,** 4726 (1990).
45. C. Hawker, J. Fréchet, *Polymer* **33**(7), 1507 (1992).
46. I. Gitsov, K. Wooley, J. Fréchet, *Angew. Chem. Int., Ed. Engl.* **31,**(9), 1200 (1992).
47. I. Gitsov, K. Wooley, C. Hawker, P. Ivanova, J. Fréchet, *Macromolecules* **26,** 5621 (1993).
48. I. Gitsov, J. Fréchet, *Macromolecules* **27,** 7309 (1994).
49. K. Wooley, J. Fréchet, *Polym. Prepr., ACS Polym. Mater. Sci. Eng.* **67,** 90 (1992).
50. T. Kawaguchi, Z. Xu, J. Moore, *Polym. Prepr, ACS Polymer Division* **34,** 124 (1993).
51. I. Gitsov, P. Ivanova, J. Fréchet, *Macromol. Rapid Comm.* **15,** 387 (1994).
52. R. Freudenberger, W. Claussen, A. Schlüter, H. Wallmeier, *Polymer* **35,**(21), 4496 (1994).
53. W. Claussen, N. Schulte, A. Schlüter, *Macromol. Rapid Comm.* **16,** 89 (1995).
54. (a) C. Hawker, K. Wooley, J. Fréchet, *J. Chem. Soc., Chem. Comm.* 925 (1994); (b) K. Wooley, C. Hawker, J. Fréchet, F. Wudl, G. Srdanov, S. Shi, C. Li, M. Kao, *J. Am. Chem. Soc.* **115,** 9836 (1993).
55. I. Gitsov, J. Fréchet, *Macromolecules* **26,** 6536 (1993).
56. (a) J. Moore, Z. Xu, *Macromolecules* **24,** (21), 5894 (1991); (b) Z. Xu, J. Moore *Acta Polym.* **45,** 83 (1994); (c) Z. Xu, B. Kyan, J. Moore, *Adv. Dendritic Macromol.* **1,** 69 (1994), JAI Press, Inc., Greenwich, CT.

10
Dendritic Molecules by the Divergent Method

Masa-aki Kakimoto and Yoshio Imai
Tokyo Institute of Technology, Meguro-ku, Tokyo, Japan

I. INTRODUCTION

Organic chemists utilize low molecular weight molecules to produce industrially useful intermediates that in turn are used to make useful compounds of moderate-sized molecules such as pharmaceutical compounds, dyes, etc. On the other hand, macromolecular compounds of large molecular weight such as polyolefins, polyesters, and so on, are used directly by industry and consumed in thousands of tons daily. The most significant difference between the low molecular weight molecules and the polymers is that the polymers do not have a specific shape and therefore lack predictable properties unlike some low molecular weight molecules. For instance, the diameter of the polymers can be changed by changing their physical environment. This is especially true with changes in the solvent system. Chemists, to date, have not extensively studied intermediate-sized (nanoscale) molecules, which have specific shapes and functionalities along with well-defined molecular weights. These molecules can be nonlinear oligomeric units called dendrimers.

Dendrimer is the name of a fairly new family of oligomers having highly branched structures. The word was coined from a combination of dendron (Greek for *tree*) and polymer. Linear polymers typically consist of an entangled mixture of individual molecular chains each having only two terminal groups. A dendrimer, by contrast, has many branches each with its own end groups, so that each molecule contains a very high number of terminal functional groups. This gives a molecule resembling a tree. The chains of dendrimers always have regular branching, and, as a result, the density of the chains and terminal functional groups increases as the molecules grow. Star polymers, which consist of several chains emanating radially from one center point, have a geometric shape similar to the dendrimers. Because the chains in the star polymers can entangle if they are long enough, the diameter and ultimate functionality of the star polymers are not exactly decided.

Brief definitions of starburst polymers, dendrimers, and dendrons are as follows, and the reader is referred for detailed descriptions to Tomalia's excellent reviews [1]. *Starburst* is a the trademark name of poly(amide-amine)s, as shown in Scheme 1, and was presented by Tomalia. Recently, dendrimers and dendrons have been defined in general terms for a class of hyperbranched polymers. *Dendrimers* are systematically constructed to form regularly branched treelike structures. Under ideal conditions, the molecular architecture is very uniform, as shown in Figure 1. A *dendron* is one of the major subdivisions of a dendrimer. For example, Figure 1 shows the division of a dendrimer into three separate dendrons, labeled A, B, and C. Further subdivisions are also possible, with each of the three dendrons dividing into two subdendrons, each of which then divides into two subdendrons, etc. Thus, one of the major differences between linear polymers and dendrimers is this branched architecture.

The first report for the concept of Starburst dendrimers was presented by Tomalia in 1985 [2]. Newkome reported the synthesis of dendritic molecules called *arborols* at the same time [3]. As described later, these dendritic molecules were synthesized by the divergent method, in which the synthesis started from the core molecule with growth proceeding in a stepwise fashion to form large spherical molecules. This is analogous to the burst of exploding firework. In 1990, Fréchet presented a new strategy to prepare hyperbranched polymers [4]. This method was designated the convergent method, which is similar to the self-polycondensation of AB_2-type monomers, and is described in another chapter.

In this chapter the preparation of dendrimers by the divergent method is described.

II. GENERAL PROCEDURE OF THE DIVERGENT METHOD

The general synthetic route of preparing dendrimers by the divergent method is depicted in Scheme 2. The synthesis starts from a core that possesses a few ac-

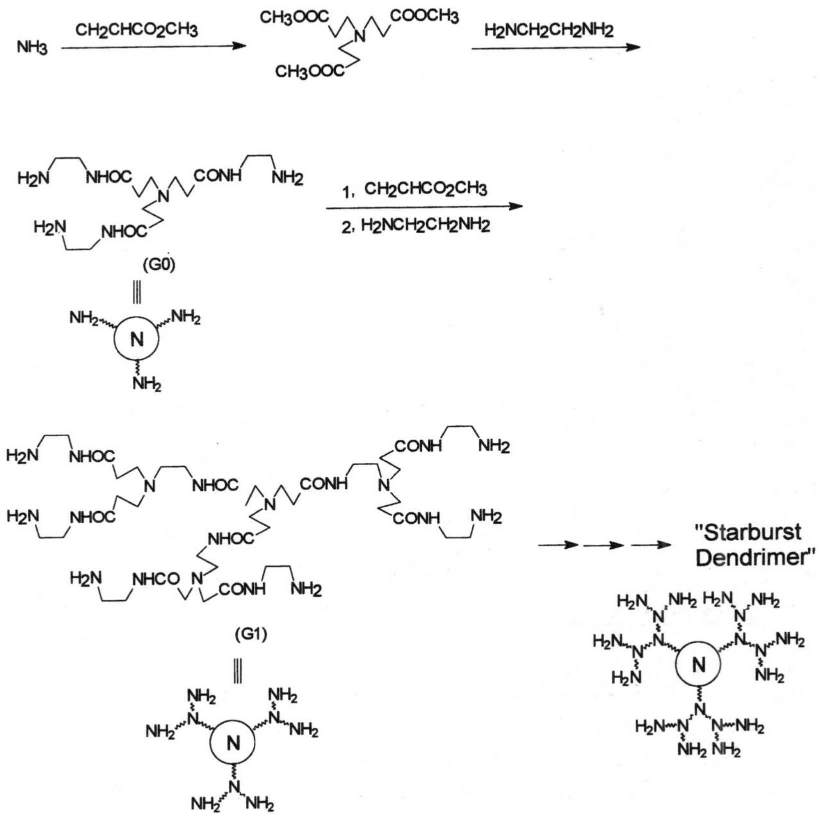

Scheme 1

tive functional groups (B). The number of the active groups (B) corresponds to the number of subdendrons that can be constructed by further reaction. In the chain extension reaction the core reacts with the building block, which contains one active functional group (A) that can react with the core active group (B), and two protected groups (Z) that can be transformed into the functional group (B) with proper treatment. After one chain extension reaction, the molecule formed is termed the first generation (G1). The protected group (Z) at the outermost position of G1 is then converted to the active functional group (B), and subsequently reacted with the building block (B-containing) molecule again to form the second generation. The dendrimers grow by repeating these chain extension and deprotection procedures until the desired molecular weight is obtained.

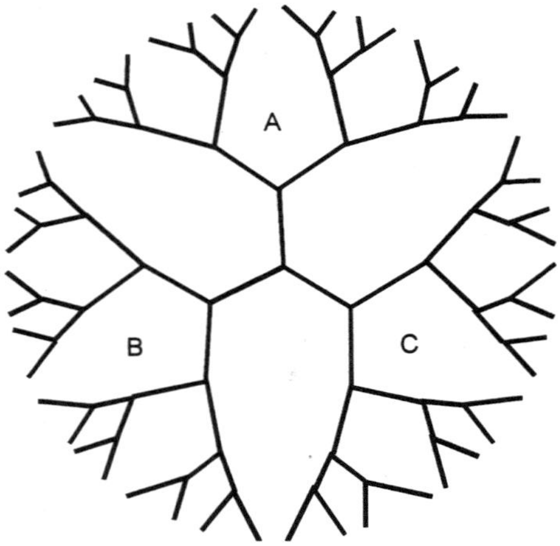

Figure 1 A dendrimer.

III. STARBURST POLYMERS (POLY(AMIDE-AMINE)S) AND POLYAMINE DENDRIMERS

Tomalia presented the synthesis of dendrimers as Starburst polymers, which consisted of poly(amide-amine)s as shown in Scheme 1 [2]. The core molecule in this case was ammonia, and the building block was methylacrylate. First, ammonia reacted with methylacrylate by a Michael-type addition. The obtained ester-terminated molecule was treated with a large excess amount of ethylenediamine to form amino groups at the terminal positions (generation 0, G0). In this reaction, the methylester is the protected form of the amine, and the treatment with ethylenediamine (ester-amide transformation) corresponds to the deprotection of the ester function. Treatment of this amine-terminated molecule again with methylacrylate results in the formation of another terminal methylester group. The molecule can be extended by repeating these operations.

As the number of generations grew, the isolated starburst polymers went from being amorphous syrups to stiff glasses. The amorphous character of this highly branched system was very different from linear crystalline homologues prepared under similar conditions. In all cases, the neat ester-terminated dendrimers were of lower viscosity than the amine-terminated products. The dendrimers were soluble in common organic solvents, and the amine-terminated ones were easily dissolved in water. The ester-terminated dendrimers exhibited surface active properties in water, giving foamy solutions.

Viscosity studies conducted in a variety of solutions showed that the

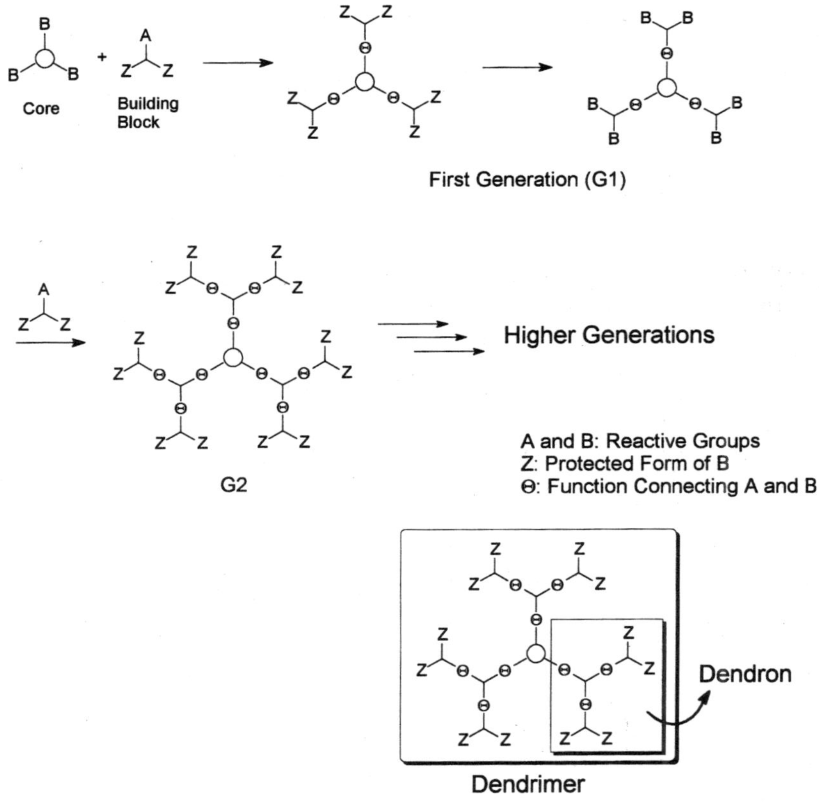

Scheme 2

amine-terminated dendrimers generally exhibit higher intrinsic viscosities than the corresponding ester-terminated ones. The Mark–Houwink shape factors (α) measured in various solvents were 0.19–0.24, indicating the expected soft, spherical structures of these dendrimers.

The Michael addition reaction of amines to conjugated double bonds is also used in the preparation of polyamine dendrimers, as shown in Scheme 3 [5]. This method is reported by Voegle and co-workers as the fundamental method for dendrimers synthesis and is termed *cascade synthesis* [6]. In this case, as shown in Scheme 3, the core and building block were 1,3-diaminopropane and acrylonitrile, respectively. After the Michael addition reaction, the terminal nitrile groups were catalytically hydrogenated to an amino group in the presence of Ni-Co catalyst. Thus, the nitrile group is the synthon (equiva-

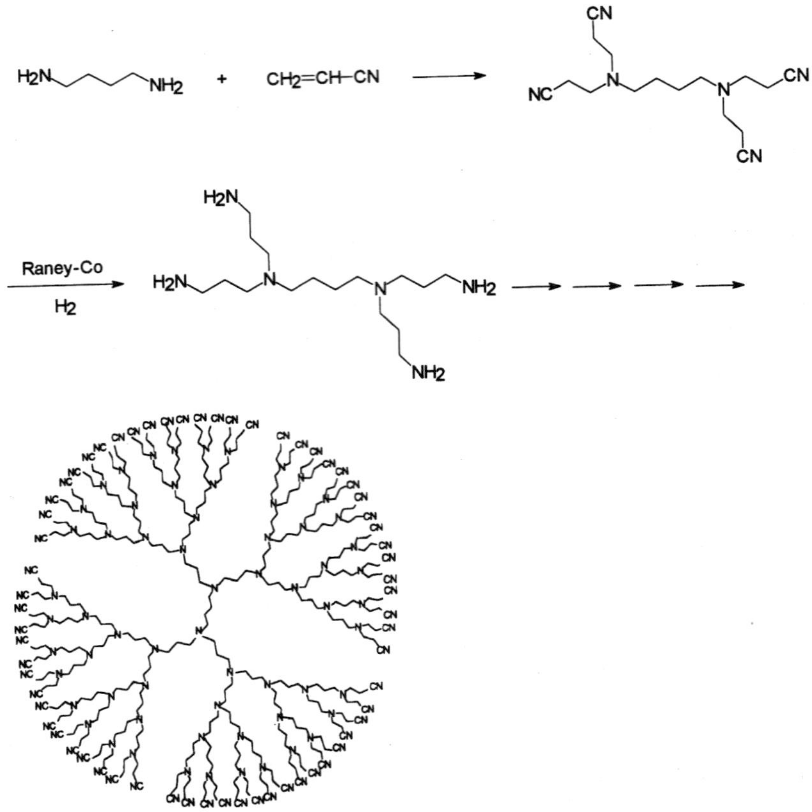

Scheme 3

lent form) of the amino group. The dendrimers grew by repeating these reactions. The dendrimers obtained were liquid, and those with terminal amine groups were readily miscible with water, while those with terminal nitrile groups dissolved in ordinary organic solvents. Since the synthetic procedure for these polyamine dendrimers is relatively simple, scaling up of this preparation sequence is expected.

IV. ARBOROLS AND CARBON-BASED DENDRIMERS

Newkome has intensively studied carbon-based dendritic molecules, which are called arborols. The first synthesis of this type of molecule is depicted in Scheme 4 [3].

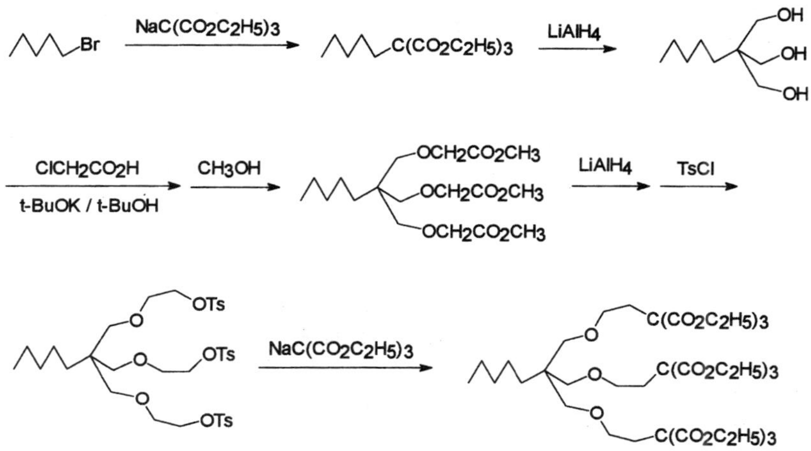

Scheme 4

In the series for synthesizing dendritic molecules, they prepared various monomers to achieve a directional divergent cascade synthesis. For instance, as shown in Scheme 5, aminotriol *1* was prepared as follows [Eq. (1)] [7]: Michael-type addition of nitromethane to acrylonitrile afforded the trisnitrile *2*, which was subsequently transformed to the triacid *3*. Alternatively, triacid *3* was obtained via a Michael-type addition of nitromethane to methyl acrylate and subsequent hydrolysis of triester *4*. Reduction of the acid moieties with diborane gave nitrotriol *5*. Further reduction with Raney nickel yielded the desired aminotriol *1*.

As initiator cores for the preparation of hydrocarbon-based dendrimers, tetrabromides *6, 7,* and *8* were examined. Tetrabromides *6* and *7* had fairly low reactivity toward the alkylation reactions because of steric hindrance caused by the neopentyl structure. Tetrabromide 8 smoothly reacted with alkylation reagents such as nitromethane and was prepared as follows [Eq. (2) in Scheme 6]: hydrogenolysis of benzyl ether *9*, which was synthesized starting from nitrotriol *10* via intermediates *11, 12,* and *13,* gave tetraol *14,* which upon treatment with hydrogen bromide gave *8*. Alternatively, as shown in Eq. (3), nitrotriol *10* was reacted with methyl iodide and base to give triether *15,* whose pendant nitro moiety was transformed to ester *16*. Reduction of ester *16* with lithium aluminum hydride gave hydroxy ether *17,* which was directly converted to tetrabromide *8*.

New divergent dendrimers were prepared by combinations of the previously mentioned functional cores and building blocks. Synthesis of Micel-

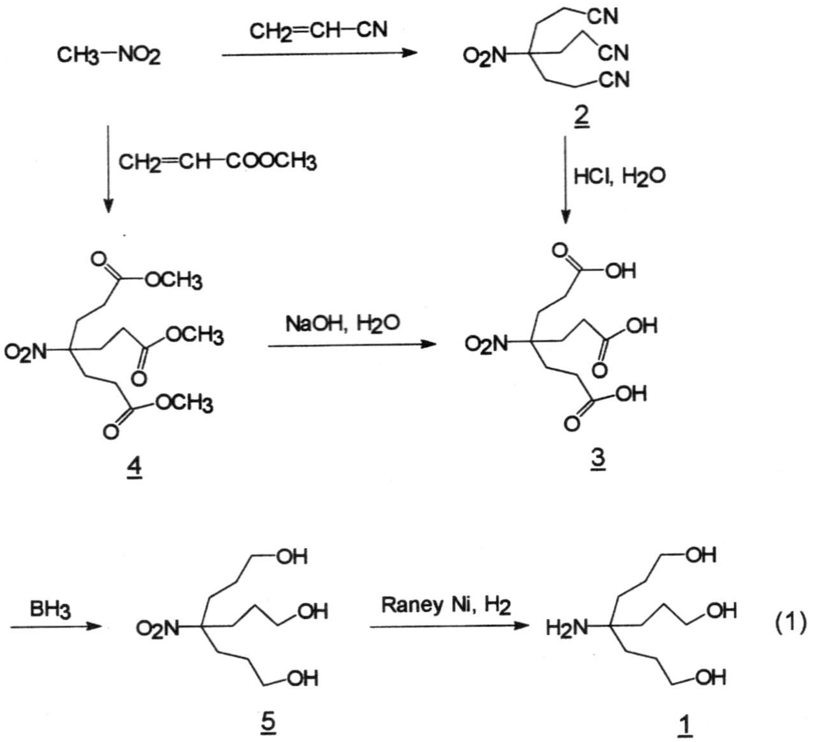

Scheme 5

lanes™ (unimolecular micelles) [8] is as follows. [Eq. (4) and (5) in Scheme 7] Alkyne *19* was synthesized from *9* via chloride *18*. Generation of the alkynide anion of *19* followed by addition of core *8* gave dodecabenzyl ether *20*. Concomitant reduction of the alkynes and hydrogenolysis of the benzyl ethers afforded nonaol *21*, which was easily transformed to the corresponding bromide; then treatment with a slight excess of the same alkynide anion yielded the 36-benzyl ether, which upon reduction and hydrogenolysis gave 36-ol *22*. Water solubility of the polyol *22* was greatly enhanced by oxidation of the terminal hydroxyl groups to carboxylic acids with $RuO_2/NaIO_4$ to afford polycarboxylic acid *23*.

This class of cascade polymers was termed Micellanes due to the obvious topological resemblance of the resulting molecules to globular micelles as well as their ability to function as a micelle. The micellar characteristics of the polytetramethylammonium salt of *23* were investigated via ultraviolet analysis

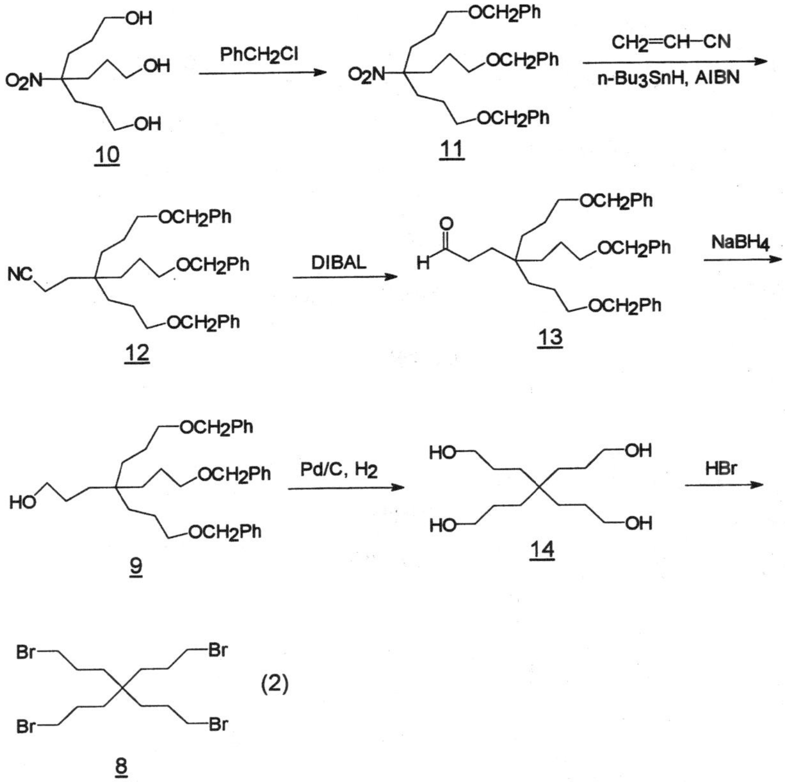

(2)

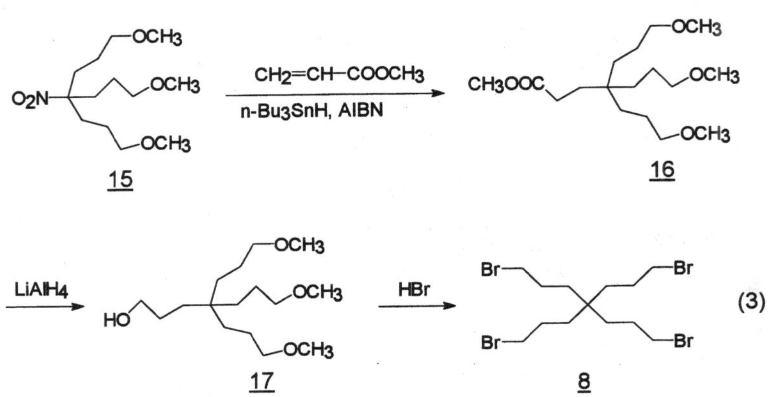

(3)

Scheme 6

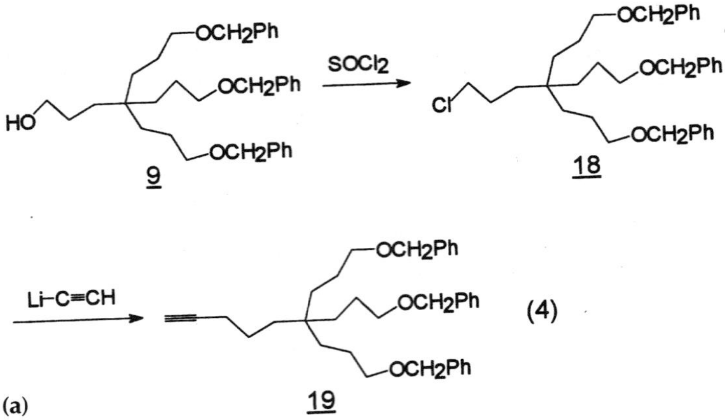

(a)

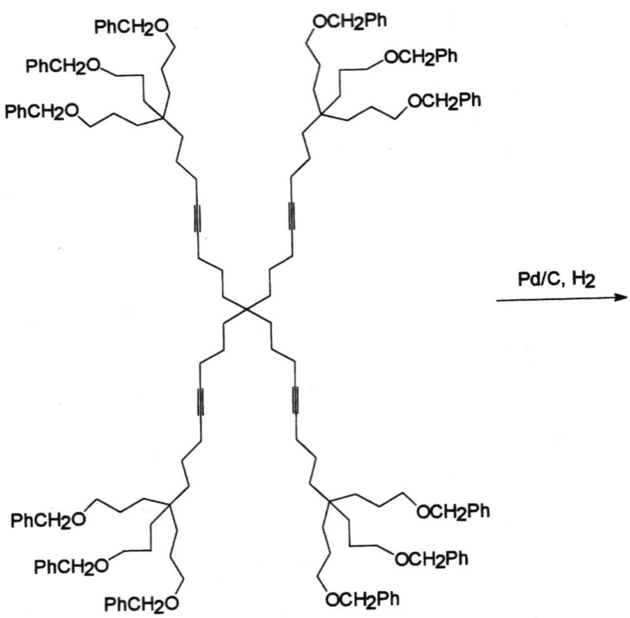

(b) 20

Scheme 7

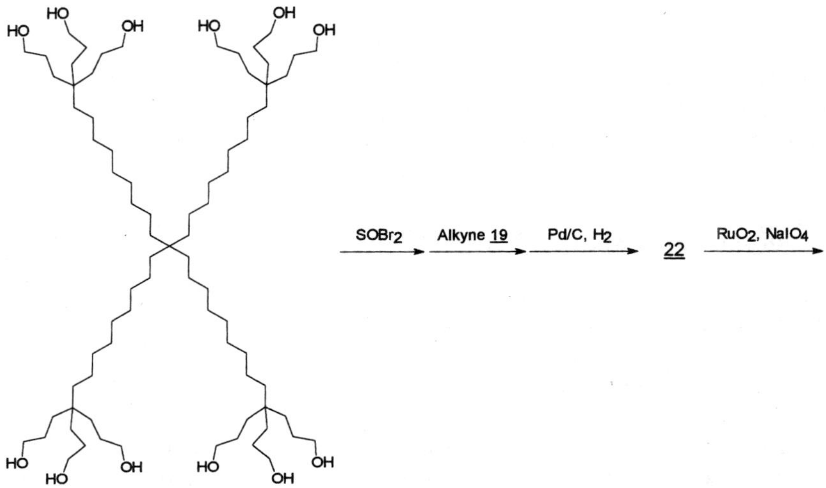

(c) 21

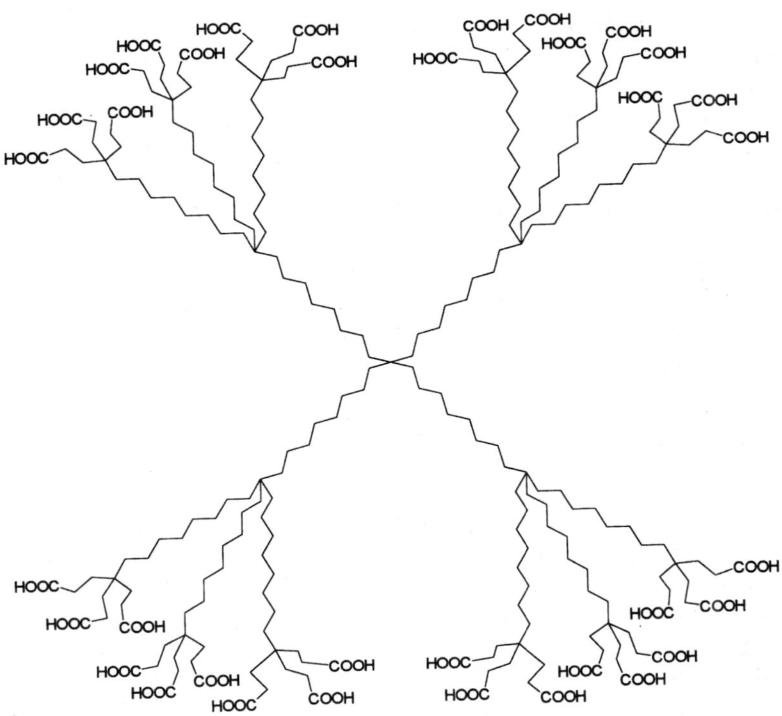

(d) 23

Scheme 7 Continued

of guest molecules such as pinacyanol chloride, phenol blue, and naphthalene combined with fluorescence lifetime decay experiments employing diphenylhexatriene as a molecular probe. The monodispersity, size, and absence of intermolecular aggregation were examined by electron microscopy.

V. POLYSILOXANE DENDRIMERS

Most of the dendrimers are constructed of a carbon-based main chain structure. Some dendrimers that possess silicon atoms in the main chain have also been reported.

As mentioned above, selection of proper synthons with active functions and high-yield coupling reactions for the chain extension is quite important in dendrimer synthesis. Siloxane bonds are usually prepared by the reaction between nucleophilic silanols and electrophilic silanes (Si^+). In the divergent synthesis of the polysiloxane dendrimers, the protection of Si^+ was employed instead of the protection of silanol [9]. Conversion from the synthon to active Si^+ species (deprotection) had to be carried out under neutral conditions, because siloxanes are easily hydrolyzed under both acidic and basic conditions. It was found that one of the most suitable synthons for Si^+ is phenylsilane, which reacts with bromine to form bromobenzene and bromosilane. Surprisingly, siloxanes are stable with bromine for short periods of time at room temperature.

Preparation of polysiloxane dendrimers by the divergent method is depicted in Scheme 8. The core G0-Ph was converted to activated core G0-Br by reaction with bromine. Next, G0-Br was reacted with diethylamine to convert the terminal bromosilane to aminosilane, which affords siloxane in higher yields by the reaction with silanols. Then, the chain was extended by the reaction with the building block B, which possessed two phenylsilanes and one silanol at its terminal positions. The obtained G1-Ph possessed six phenylsilanes at the terminal positions. G1-Ph was then successively reacted with bromine, diethylamine, and the building block to form molecules of higher generations. Repeat of these operations afforded G2-Ph and G3-Ph, which had 12 and 24 terminal phenylsilanes, respectively.

The polysiloxane dendrimers, thus obtained, were colorless oils, and were purified by gel permeation chromatography (GPC). As shown in Table 1, they had very narrow molecular weight distributions, and values of molecular weight of the purified dendrimers obtained by GPC were smaller than those of the calculated formula number. The Mark–Houwink shape factor (α) was calculated as 0.21, which was nearly identical to factors reported for Tomalia's soft, spherical Starburst dendrimers.

Figure 2 shows the relationship between the formula weight of the dendrimers and the ^{13}C-NMR relaxation time measured for the carbons of the ter-

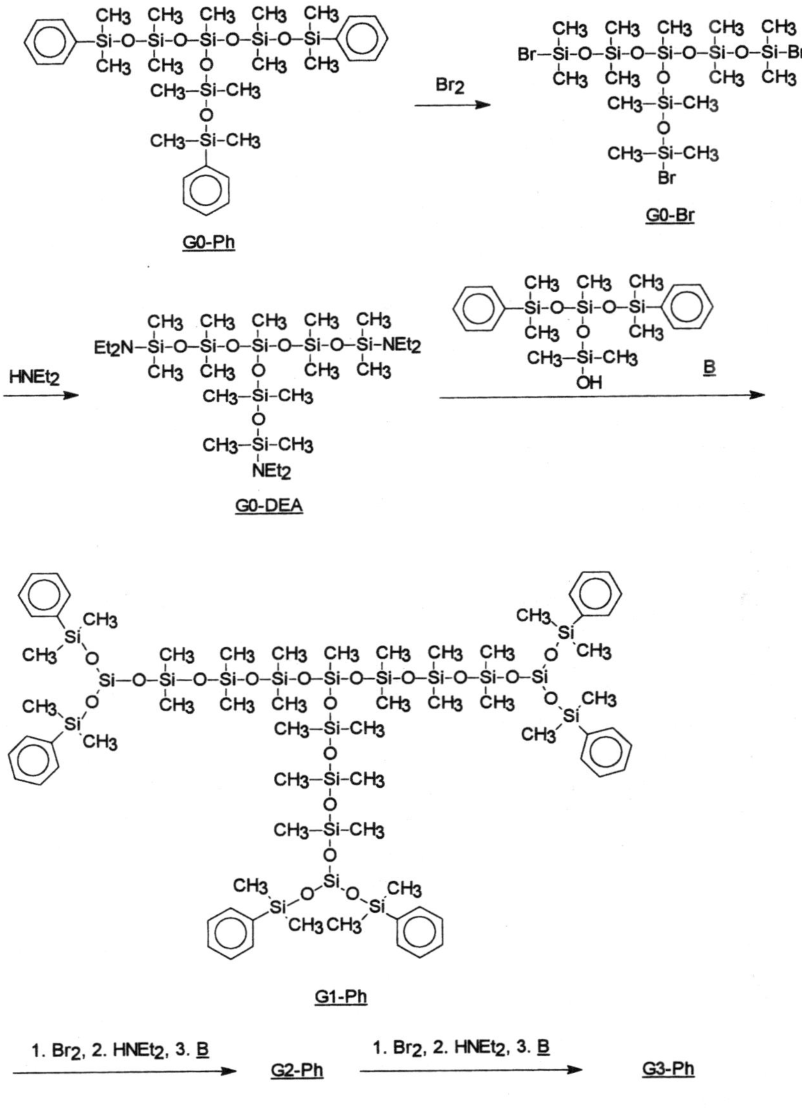
G0-Ph
Br2
G0-Br
HNEt2
G0-DEA
B
G1-Ph
1. Br2, 2. HNEt2, 3. B
G2-Ph
1. Br2, 2. HNEt2, 3. B
G3-Ph

Scheme 8

Table 1 Molecular Weight and Related Values of Polysiloxane Dendrimers

Polymer	FW^{a}	$M_n{}^{b}$	M_w/M_n	$[\eta]^{c}$	$r^{d}(nm)$
G0-Ph	718	878	1.01	0.0190	0.91
G1-Ph	1792	1613	1.06	0.0225	1.34
G2-Ph	3940	2792	1.06	0.0270	1.85
G3-Ph	8324	4819	1.14	0.0315	2.31

[a]Formula weight.
[b]Measured by gel permeation chromatography based on standard polystyrene.
[c]Intrinsic viscosity.
[d]Diameter measured on CPK molecular model.

minal phenyl-substituted silane atom. Each relaxation time in the case of G3-Ph (FW = 8324) was smaller than those of other generations. This result means that the free rotation of the bonds was restricted by intramolecular collision. In general, the inside of the dendrimers is comparatively vacant, compared to the more densely substituted outer positions of the molecule.

Kawamata's group reported an alternative synthetic route for polysiloxane dendrimers [10]. As mentioned earlier, synthesis of siloxanes requires the nucleophilic silanols and an electrophilic Si^{+} species. They used silylhydrides as the synthon of silanols. This is possible because each reactive group has the following characteristics: chlorosilanes, typical Si^{+} species, do not react with silylhydrides, and silylhydrides are hydrolyzed to silanol in the presence of palladium catalysts. Synthetic procedures are shown in Scheme 9. First, the core, possessing three silylhydrides on its terminal positions, and the building block, possessing two silylhydrides and one chlorosilane, were prepared. After the silylhydride group in the core was converted to silanol, it was reacted with the building block to form a G1 molecule that had six terminal silylhydrides. By repeating these operations, they succeeded in synthesizing G3 dendrimers having molecular weights of 15,000.

The hydrosilylation reaction is a simple and convenient method to form a carbon–silicon bond. A typical example of carbosilane dendrimers prepared by the hydrosilylation reaction is shown in Scheme 10 [11]. The core molecule, tetraallylsilane, is synthesized by the reaction of tetrachlorosilane and an allyl Grignard reagent. Hydrosilylation of trichlorosilane to the core affords a trichlorosilyl-terminated molecule. The G1 molecule was obtained by the coupling reaction with the allyl Grignard reagent. Repeating of hydrosilylation of trichlorosilane and reaction with allyl Grignard reagent produce the higher generations of carbosilane dendrimers.

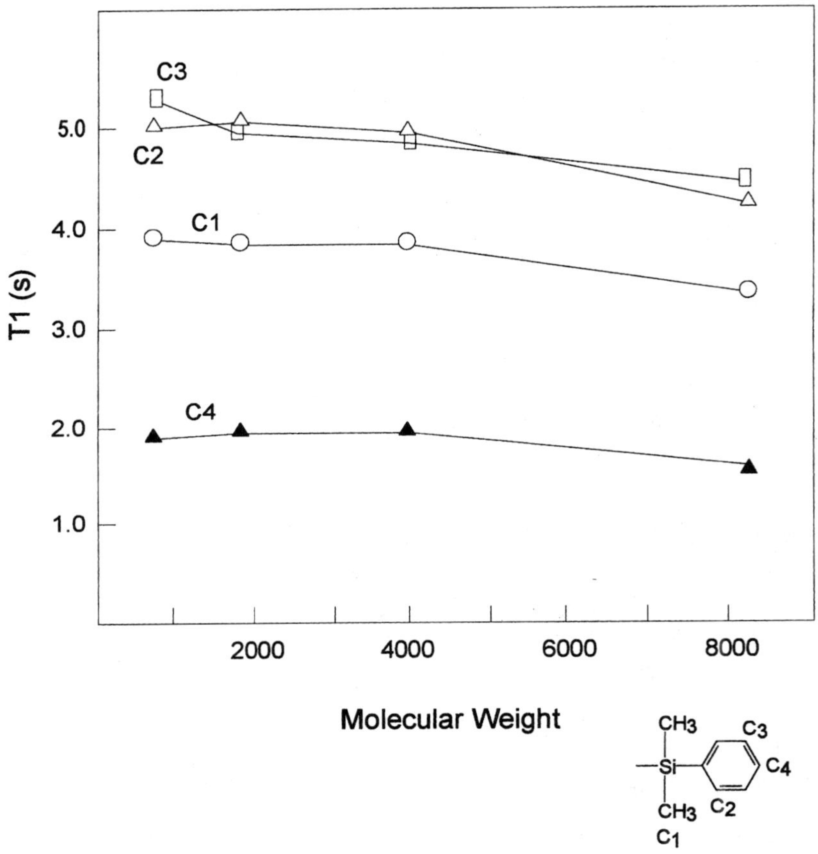

Figure 2 Relationship between the formula weight (FW) of the dendrimers and the ^{13}C-NMR relaxation time measured for the carbons of the terminal phenyl-substituted silane atom.

Using the same strategy, vinyl Grignard reagents were also used instead of allyl Grignard reagents [12]. The terminal chorosilane can be converted to a silyl hydride function by reduction using lithium aluminum hydride [13].

VI. PHOSPHONIUM CASCADE

There are also examples of phosphonium dendrimers prepared by cascade synthesis. The synthetic method of dendrimers bearing numerous phosphonium sites within the covalent structure involves the design of a suitable reagent to

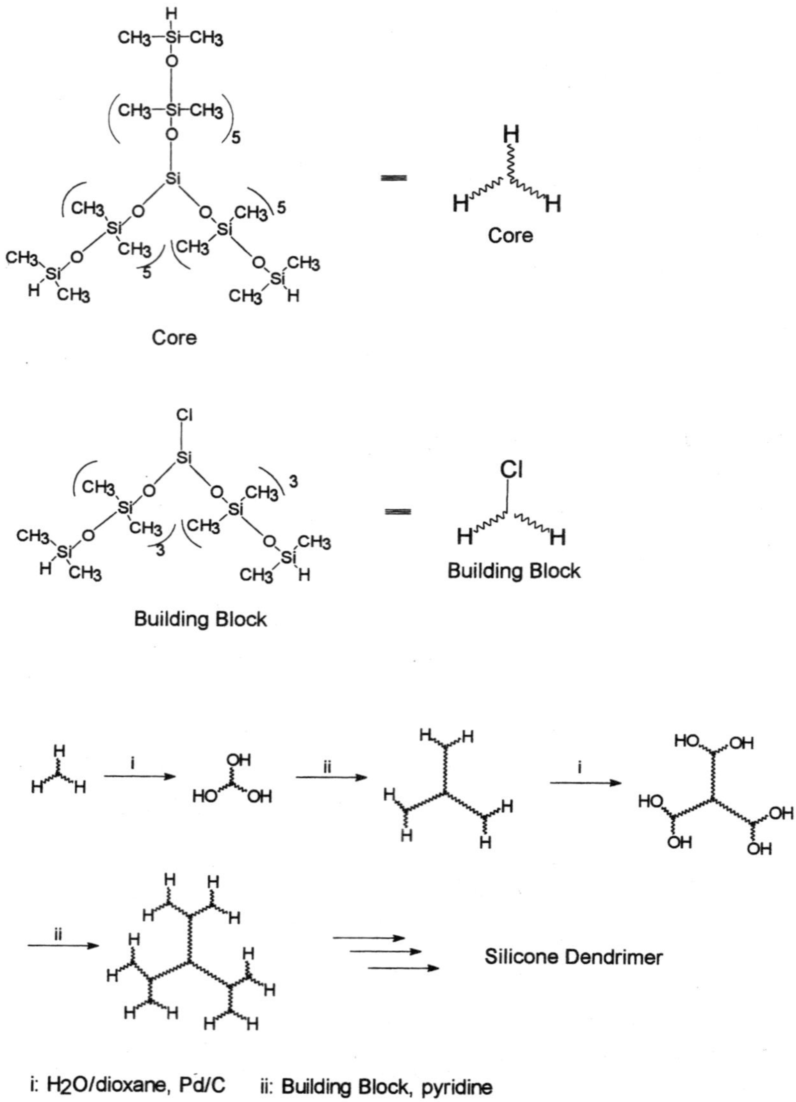
Core
Core
Building Block
Building Block
Silicone Dendrimer
i: H2O/dioxane, Pd/C ii: Building Block, pyridine

Scheme 9

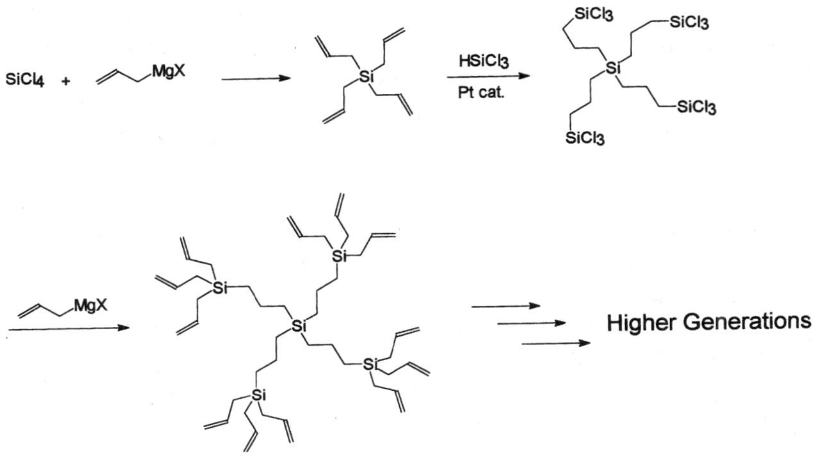

Scheme 10

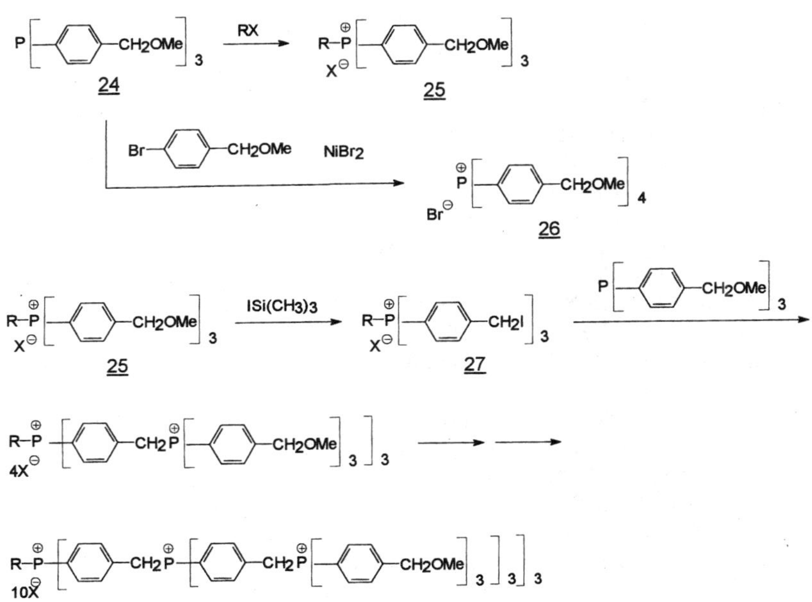

Scheme 11

serve as the repeating structural component. This component must be capable of forming a phosphonium ion species in a facile manner and must contain several symmetrically distributed masked sites, which can generate active sites identical to the original reagent. To serve as the fundamental repeating component of the phosphonium cascade species, the tertiary phosphine *24* has been prepared [14]. The phosphine center of *24* is quaternized by reaction with either alkyl halide to generate the primary core *25* for a tridirectional development of the cascade, or *p*-methoxymethylbromobenzene in the presence of nickel bromide to generate the primary core *26* for a tetradirectional development of the cascade. The approach for incorporation of successive generations of the cascade structure is illustrated in Scheme 11 for the system beginning with *25*. For the construction of further generations of the cascade structure, the benzylic ether function in *25* is efficiently cleaved using iodotrimethylsilane for an excellent yield benzylic iodide *27*. Thus, this method is ideal for the simultaneous deprotection and activation of functional groups for use in the present synthetic route.

REFERENCES

1. D. A. Tomalia, A. N. Naylor, W. A. Goddard III, *Angew. Chem., Int., Ed., Engle.*, **31,** 1571 (1992).
2. D. A. Tomalia, H. Baker, J. Dewald, M. Hall, G. Kallos, S. Martin, J. Roeck, J. Ryder, P. Smith, *Polym. J.*, **17,** 117, (1985).
3. G. R. Newkome, Z. Yao, G. R. Baker, V. K. Gupta, *J. Org. Chem.*, **50,** 2003 (1985).
4. C. Hawker and J. M. Fréchet, *J. Chem. Soc., Chem. Commun.*, 1010 (1990).
5. E. M. M. de Brabander-van den Berg, E. W. Meijer, *Angew. Chem., Int., Ed., Engle.*, **32,** 1308 (1993).
6. E. Buhlein, W. Wehner, F. Voegle, *Synthesis,* 155 (1978).
7. G. R. Newkome, C. N. Moorefield, K. J. Theriot, *J. Org. Chem.*, **53,** 5552 (1988).
8. G. R. Newkome, C. N. Moorefield, G. R. Baker, R. K. Behera, A. L. Johnson, *Angew. Chem., Int., Ed., Engle.*, **30,** 1176 (1991); G. R. Newkome, C. N. Moorefield, G. R. Baker, S. H. Grossman, *Angew. Chem., Int., Ed., Engle.*, **30,** 1178 (1991).
9. A. Morikawa, M. Kakimoto, Y. Imai, *Macromolecules,* **24,** 3469 (1991).
10. H. Uchida, Y. Kabe, K. Yoshino, A. Kawamata, T. Tsumuraya, S. Masamune, *J. Am. Chem. Soc.*, **112,** 7077 (1990).
11. A. W. van der Made, P. W. N. M. van Leeuwen, *J. Chem. Soc., Chem. Commun.*, 1400 (1992).
12. L. L. Zhou, J. Roovers, *Macromolecules,* **26,** 963 (1993).
13. D. Seyferth, D. Y. Son, *Organometallics,* **13,** 2682 (1994); K. Rengan R. Engel, *J. Chem. Soc., Chem. Commun.*, 1084 (1990); K. Rengan R. Engel, *J. Chem. Soc. Perkin Trans,* I, 987 (1991).

11

Dilute Solution Properties of Regular Star Polymers

J. Roovers
Institute for Chemical Process and Environmental Technology, National Research Council Canada, Ottawa, Ontario, Canada

I. INTRODUCTION

Flexible linear polymer chains can be represented by a random space walk of N equal steps. When the resulting average end-to-end distance R of such a walk is computed, it is observed that

$$<R^2> = \text{constant} \times N^{2\nu}$$

where ν is equal to 1/2 for the (phantom) Gaussian chain and ν is approximately equal to 3/5 (the Flory exponent) for the self-avoiding walk. In contrast with ordinary three-dimensional objects, the average segment density of the polymer coils, N/R^3, is not constant but decreases with increasing number of steps in the polymer chain. Therefore, when properties of polymer chains are studied as a function of their molecular weight, the concomitant variation of the segment density is always implicitly present. There appears to be only one method to control the segment density independently of the molecular weight of the polymer, and this is by introducing long-chain branching. However, branching modifies the connectivity of the chain segments.

The regular star polymer has the simplest branching architecture because the star architecture can be characterized by a single parameter f, the functionality of the branch point. The f arms of a regular star have equal molecular weights. Furthermore, the effect of branching on the properties of polymers is strongest when the branches are arranged in the regular star architecture. For example, comb and graft polymers have properties that reduce to star polymer properties in the limit of an infinitely small backbone or of infinitely large branches. The instantaneous shape of star polymers becomes more spherical with increasing functionality, and stars with large functionality have properties approaching globular polymers.

Star polymers are excellent models for block copolymers dissolved in selective solvents and for end-associating polymers that form polymeric micelles [1,2]. These polymers, with their dense core and swollen corona, are of great technical importance but are difficult to study because their degree of association is highly variable. Knowledge of the properties of stars is also important for the design of the properties of star block copolymers, which are an important class of industrial materials. The conformation of the arms of a star is also the limiting conformation of polymers attached to a surface with infinite curvature ($1/R \rightarrow \infty$). Therefore, the properties of star polymers are related to the properties of polymer-coated colloidal particles.

Fortunately, powerful experimental methods are available to prepare regular star polymers with full control over their arm molecular weight and functionality. The properties of regular star polymers have been reviewed [3,4]. Other reviews, in the context of branched polymers, are also available [5–7]. Specialized topics on branched polymers in comparison with linear polymers have also been covered [8–12].

This review deals with the static and dynamic properties of regular star polymers at zero concentration and with the dependence of these properties on the polymer concentration in dilute solution. We stress the experimental results and make comparisons with theoretical models and computer simulations.

II. STATIC PROPERTIES

A. The Gaussian Chain

Star polymers are distinguished from linear polymers by their smaller size and correspondingly higher segment density. The mean square radius of gyration of a polymer is defined as the mean square distance of the N segments of the chain from their center of mass:

$$<R_G^2> = \left(\frac{1}{N}\right) \sum_i^N <S_i^2>$$

When the chain segments obey random-flight statistics and $N >> 1$, the preceding sum can be replaced by [13,14]

$$<R^2_{G,0}> = \frac{b^2}{N^2} \sum_{j=1}^{N} j\,(N - \mathrm{j}) = \frac{b^2}{2N^2} \sum_{i}^{N} \sum_{j}^{N} (j - i)$$

with b the length of the chain segment and the zero subscript indicating the Gaussian chain. For linear polymers $<R^2_{G,0}> = b^2N/6$.

For star polymers with f arms each having y segments a distinction must be made between the case where the ith and jth segments are found on the same arm

$$\sum_{i} \sum_{j} (j - i) = y^3/3$$

and where i and j are on different arms

$$\sum_{i} \sum_{j} (i + j) = y^3$$

The first sum occurs f times and the second sum occurs $f(f - 1)$ times. For a star

$$<R^2_{G,0}> = \frac{b^2}{2(fy)^2} \left(\frac{fy^3}{3} + f(f - 1)y^3 \right)$$

and replacing fy by N gives

$$<R^2_{G,0}> = \frac{b^2N}{6} \left(\frac{3f - 2}{f^2} \right) = <R^2_{G,0}>_{\mathrm{lin}} \left(\frac{3f - 2}{f^2} \right) \tag{1}$$

Equation 1 compares the mean square radius of gyration of a regular star with f arms with the radius of gyration of the linear polymer having the same number of segments [14]. The Zimm–Stockmayer branching factor

$$g_{ZS} = \frac{3f - 2}{f^2} \tag{2}$$

has values between one ($f = 1$ and 2) and zero ($f \rightarrow \infty$). At large f, g_{ZS} varies inversely with f. The segment density distribution in a regular star polymer in which segment–segment distances conform to the Gaussian distribution has also been calculated [15]. As will be shown in this review, Eq. (2) plays a benchmark role in the description of the properties of star polymers, but it must be kept in mind that it relates to an unrealistic phantom chain. The effect of chain length distributions in the arms of the star on the branching parameter has also been considered [10,14,16].

B. Scaling Model of Star Polymers

The scaling model of Daoud and Cotton has had a strong influence, because it succeeds in accounting for observed deviations from Eqs. (1) and (2) in stars with many arms [17–19]. The Daoud–Cotton scaling model starts from the reasonable assumption that the local segment density is high near the center of a star with many arms and decreases with the radial distance, r, from the center [17]. The effect of the high segment concentration on the chain conformation is treated by means of the "blob" model used originally to describe polymer chains in semidilute solution [9]. A blob is characterized by a correlation length, ξ, over which the chain conforms to a self-avoiding random walk without interactions with other chains. In semidilute solution $\xi \sim c^{-3/4}$ [9].

The segment density in the blob is given by

$$\phi(r) = \frac{n(r)a^3}{\xi^3(r)} \tag{3}$$

where n is the number of segments with volume a^3 in the blob volume ξ^3. We have

$$\xi(r) \sim an^{1/2}(r)\alpha(r)$$

and

$$\alpha^5(r) \approx n^{1/2}(r)v = n^{1/2}(r)\,(1 - 2\chi)$$

The local expansion coefficient, $\alpha(r)$, is related to the Flory polymer–solvent interaction parameter χ. Introducing these relations into Eq. (3) yields

$$\phi(r) \sim n(r)^{-4/5}\, v^{-3/5} \tag{4a}$$

$$\xi(r) \sim an(r)^{3/5}\, v^{1/5} \tag{4b}$$

$$\alpha(r) \sim n(r)^{1/10}\, v^{1/5} \tag{4c}$$

which show that the segment density, blob length, and the local chain expansion coefficient depend on the radial position in the star.

In order to determine how these properties scale with the functionality of the star, it is assumed that the blobs of each arm are restricted to and cover one conical sector of space and that the blob dimension is determined by the relation

$$4\pi r^2 \approx f\xi^2(r) \tag{5}$$

The segment density in the spherical shell between r and $r + \xi$ is then given by

$$\phi(r) = \frac{f{\cdot}n(r)a^3}{r^2\xi(r)} \tag{6}$$

Combining Eqs. (5) and (6) with Eqs. (4a) to (4c) yields

$$n(r) \sim \left(\frac{r}{a}\right)^{5/3} f^{-5/6}\, v^{-1/3} \tag{7a}$$

$$\phi(r) \sim \left(\frac{r}{a}\right)^{-4/3} f^{2/3}\, v^{-1/3} \tag{7b}$$

$$\xi(r) \sim rf^{-1/2} \tag{7c}$$

$$\alpha(r) \sim \left(\frac{r}{a}\right)^{1/6} f^{-1/12}\, v^{1/6} \tag{7d}$$

Note that Eq. (7c) agrees with the hypothesis in Eq. (5). Equations (7a) to (7d) show how n, ϕ, ξ, and α vary with r, f, and the solvent quality v.

Daoud and Cotton realized that the segment density in stars becomes so high that the semidilute solution condition $\alpha(r) = 1$ is met for radial distances $r < r_1$. From Eq. (7d)

$$r_1 \sim af^{1/2}\, v^{-1} \tag{8}$$

For $r < r_1$

$$\xi(r) \sim an^{1/2}$$

and

$$\phi(r) \sim n^{-1/2}$$

With the use of Eq. (5) we introduce the dependence on f and obtain

$$n(r) \sim \left(\frac{r}{a}\right)^{2} f^{-1} \tag{9a}$$

$$\phi(r) \sim \left(\frac{r}{a}\right)^{-1} f^{1/2} \tag{9b}$$

$$\xi(r) \sim rf^{-1/2} \tag{9c}$$

Note that $\xi(r)$ has the same dependence on r and f as in the swollen outer region of the star. Equations (9a) to (9c) are expected to be applicable to the whole star polymers in a Θ-solvent.

Furthermore, in a smaller sphere around the star center, defined by $r < r_2$, the segment density is constant and of order unity. From Eq. (9b)

$$r_2 \sim af^{1/2}$$
$$\xi(r) \sim a$$

In this model the star polymer is divided into three regions. The segment density is constant of order unity near the center; further out there is a region with unswollen polymer [Eq. (9b)] and the corona is characterized by swollen arms [Eq. (7b)].

The radius of the star polymer is obtained by integrating the segment densities in the three regions of the star

$$(Nf)a^3 = \int_0^{r_2} r^2 \, dr + \int_{r_2}^{r_1} \left(\frac{r}{a}\right)^{-1} f^{1/2} r^2 \, dr + \int_{r_1}^{R} \left(\frac{r}{a}\right)^{-4/3} f^{2/3} v^{-1/3} r^2 \, dr$$

which yields

$$R \sim a\left[Nf + \frac{1}{10}\,\frac{f^{3/2}}{v^2} + \frac{f^{3/2}}{6}\right]^{3/5} v^{1/5} f^{-2/3} \tag{10}$$

For stars with long arms in a good solvent, the radius can be approximated by

$$R \sim aN^{3/5} f^{1/5} v^{1/5} \tag{11}$$

Equation (11) shows that the dimension of the star has the same dependence on molecular weight, $N^{3/5}$, as the linear polymer. However, the stars are smaller than the linear polymer of the same molecular weight, $R_{\text{lin}} \sim a(Nf)^{3/5} v^{1/5}$, but larger than the linear polymer with the molecular weight of a single arm $R \sim aN^{3/5} v^{1/5}$.

For stars with short arms or stars in a θ-solvent the radius is predicted to scale as

$$R \sim aN^{1/2} f^{1/4} \tag{12}$$

which again shows that the star dimension is larger than the dimension of a single arm but smaller than the dimension of the linear polymer with the same molecular weight. The scaling predictions are summarized in Table 1. In constructing Table 1, it has been assumed that radius R may be replaced by R_G. It should be noted that the dependence of $g = R_G^2/(R_G^2)_{\text{lin}}$ on f is different in the scaling model than the f^{-1} dependence given in Eq. (2) for the case of the Gaussian chain.

The scaling model gives insights and trends but is not ready for numerical comparison with experimental data. For this, Gast [20] specified that ξ is the blob diameter so that Eq. (5) can be written as

$$4\pi r^2 = f\pi \left(\frac{\xi(r)}{2}\right)^2 \tag{13}$$

Table 1 Scaling of Star Dimensions

Region	*Condition*	$R_G{}^a$	$g_G{}^b$
Hard core	$N << f^{1/2}$	$(Nf)^{1/3}a$	$N^{-1/3}f^{-1/3}$
Unperturbed	$f^{1/2}v^{-2} >> N >> f^{1/2}$	$N^{1/2}f^{1/4}a$	$f^{-1/2c}$
Self-avoiding	$n >> f^{1/2}v^{-2}$	$N^{3/5}f^{1/5}a$	$f^{-4/5c}$

[a]Radius of gyration.
[b]$g_G = \langle R_G^2\rangle_{star}/\langle R_G^2\rangle_{lin}$
[c]$g_G = f^{1-3v}$
Source: Ref. 97.

and the number of segments per blob is given by the more general

$$n_s = \left(\frac{\xi}{a_s}\right)^{1/v} \tag{14}$$

where a_s is the Kuhn step length and $3/5 > v > 1/2$ depending on the solvent quality. The number segment density is then given by

$$\rho(r) = \frac{(4r/f^{1/2}a_s)^{1/v}}{\pi/6(4r/f^{1/2})^3}$$

or

$$\rho(r) = A(f)\left(\frac{r}{a_s}\right)^{1/v} r^{-3} \tag{15}$$

where

$$A(f) = \frac{3 \cdot 4^{1/v}}{32\pi} f^{(3v-1)/2v}$$

contains the dependence on the star functionality. It should be noted that Eq. (15) is in agreement with Eq. (7b) ($v = 3/5$) and with Eq. (9b) ($v = 1/2$).

Integration of the number segment density over the whole volume of the star gives the radius according to

$$R = \left[\frac{8N_s f^{(1-v)/2v}}{3 \cdot 4^{1/v}\, v}\; a_s{}^{1/v}\right]^v \tag{16}$$

where N_s is the number of segments with step length a_s *per arm*. Values of R calculated with Eq. (16) compare reasonably well with the radii of starlike micelles of poly(styrene-*b*-ethyleneoxide) in cyclopentane, a θ-solvent for polystyrene [20]. Equation (16) was also applied to 64- and 128-arm polybutadiene stars in a good solvent. Agreement to within 20% was obtained with experimental values [21]. The difference can easily be accounted for by the approximations made in Eq. (15) and in deriving Eq. (16).

C. Radius of Gyration

Because of the fundamental nature of the size of polymer coils, considerable effort has been devoted to the experimental measurement of the size of regular star polymers both in good solvents and under θ-conditions. The size of a polymer coil is obtained from the angular dependence of scattered radiation at low angles. The angular dependence of the intensity of light (or neutrons) scattered by a polymer solution is analyzed in terms of the product of the form factor, $P(Q)$, and the structure factor, $S(Q)$:

$$I(c,Q) = I(0,0) \cdot P(c,Q) \cdot S(c,Q) \tag{17}$$

The form factor or intramolecular scattering function is due to segment–segment correlations within a single polymer coil and is observed in (very) dilute solution where interference of the scattering intensity from other coils is negligible, i.e., when $S(Q) \to 1$. $Q = (4\pi/\lambda)\sin(\theta/2)$ is the scattering vector that depends on the wavelength in the medium, λ, and θ is the angle between incident and scattered radiation. Q ranges typically between 4.10^{-4} and 5×10^{-3} Å^{-1} when light-scattering measurements are performed with monochromatic light with a wavelength between 3660 Å (UV) and 6328 Å in solutions with a refractive index between $n = 1.35$ and $n = 1.65$ and with θ between 15° and 150°. Q varies typically from 4.10^{-3} to 0.15 Å^{-1} in small-angle neutron scattering (SANS) experiments.

The form factor of a regular star polymer has been derived under the assumption of the Gaussian segment distribution [22,23]:

$$P(u) = \frac{2}{\mu} + \frac{\mathrm{f}}{\mu^2}\left[(f-3) - 2(\mathrm{f}-2)\exp\left(\frac{-\mu}{f}\right) + (f-1)\exp\left(\frac{-2\mu}{f}\right)\right] \tag{18}$$

where $\mu = Q^2 R_G^2$. Here, R_G^2. is the mean-square radius of the linear polymer. In terms of $x = QR_{G,\mathrm{arm}}$ [10,24,25]

$$P(Q) = \frac{2}{fx^4}\left[x^2 - (1 - \exp(-x^2)) + \frac{(f-1)}{2}[1 - \exp(-x^2)]^2\right] \tag{19}$$

At low values of Q, in the Guinier range, the form factor reduces to

$$\lim_{Q\to 0} P(Q) = \exp(-\tfrac{1}{3}R_G^2 Q^2) \tag{20}$$

and R_G^2 is obtained from the slope of a plot of ln $P(Q)$ against Q^2. Expansion of the exponential in Eq. (20) yields

$$\lim_{Q\to 0}\left[\frac{1}{P(Q)}\right] = 1 + \frac{R_G^2}{3} Q^2 - \cdots \tag{21}$$

which is the form used in the classic Zimm plot [26]. Berry has shown that a plot of $P(Q)^{-1/2}$ against Q^2 is often more nearly linear and leads to more accurate values of R_G^2 [27].

It is clear from Eqs. (20) and (21) that experimental conditions for determining R_G are optimal when the experiment is performed at $Q{\cdot}R_G \approx 1$ [15]. For light scattering this requires that 1000 Å > $R_G \geq$ 200 Å. In the case of SANS, values of $R_G <$ 100 Å are most accurate. When R_G is large, experimental values of $QR_G >> 1$, and plots of $P(Q)^{-1}$ against Q^2 show the strong upward curvature expected from Eqs. (18) or (19). In those cases the Guinier plot underestimates and the Zimm plot overestimates the value of R_G. In the case of star polymers with a narrow molecular weight distribution, the curvature in $P(Q)$ against Q^2 plots increases with increasing f and the low $Q{\cdot}R_G$ requirement becomes more stringent. See Figure 1.

Ullman has described a method to make first-order corrections based on the Q dependence of scattering of Gaussian chains [28]. A linearization scheme based on the scattering function of Gaussian chains has also been proposed [29]. In order to establish the initial slope, experimental values of I/I_0 are multiplied by the theoretical $P(\mu)(1 + \mu/3)$ [Eq. (18)], where μ is obtained with an estimated value of R_G^2 If the correct value of R_G^2 has been used, $(I/I_0)P(\mu)$ $(1 + \mu/3)$ is a linear function of $g_{zs}\,\mu$ with a slope of 1/3.

It was pointed out by Burchard that a Kratky plot of Eq. (19), i.e., a plot of $x^2\,P(Q)$ against x has a maximum when f is greater than 4 [30]. The position of the maximum lies between $x_{max} = 2.522$ for $f = 4$ and $x_{max} = 1.94$ for $f \rightarrow \infty$. The values of $x^2P(Q)$ at the maximum also depend on f. In principle this allows for the determination of f when it is unknown. Alternatively, from the position of the maximum R_G can be estimated. It is useful for the analysis of SANS data [31] that have a large Q range. In light scattering x_{max} is seldom reached experimentally [4].

After the discussion of the methods for extracting the radius of gyration of branched polymers from experimental scattering data, it will have become clear that results have to be treated with some caution. One has to be aware that z-average values of $<R_G^2>$ are obtained. In the ratio $<R_G^2>/M_w$ the molecular weight is an additional source of error. Finally, in the ratio, $g_G = <R_G^2>_{star}/<R_G^2>_{lin}$, the denominator itself has a margin or error. It is therefore felt that good values of $g_G(f)$ can only be obtained when narrow molecular weight star and linear polymers are studied simultaneously with the same equipment and procedures and when several samples of each functionality are available over the widest possible molecular weight range.

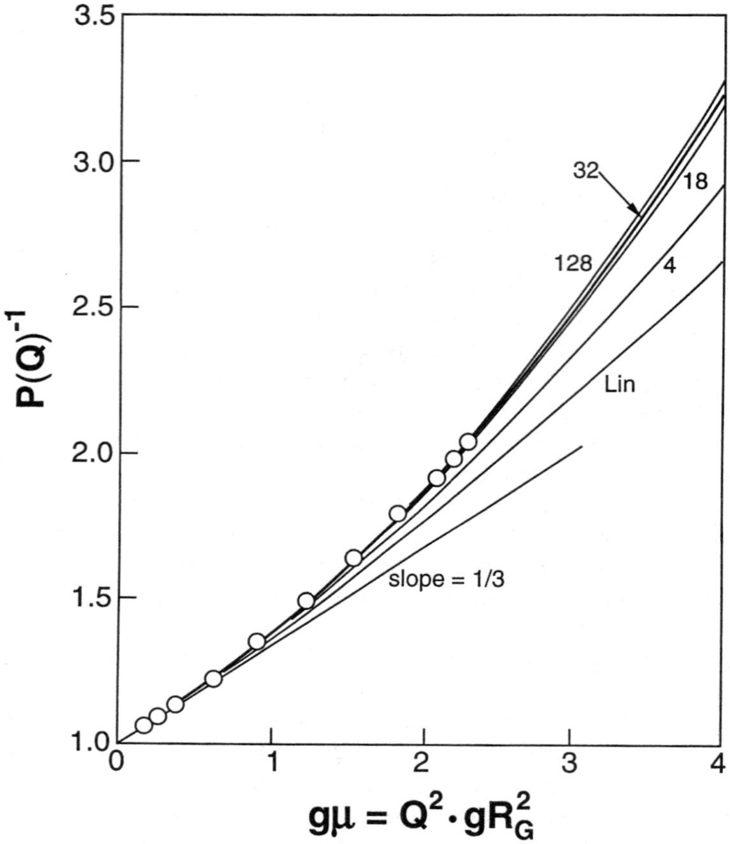

Figure 1 Zero-concentration intramolecular scattering function or form factor of a 32-arm polybutadiene star obtained by light scattering in cyclohexane with $g <R_G^2> = 1.47 \times 10^5$ Å^2. Lines are theoretical curves for star polymers with different functionality under Gaussian chain assumption: Eq. (18). All curves and the experimental data have the asymptote with slope 1/3.

The results of several studies of the radius of gyration of polybutadiene stars with $f = 4$, 18, 32, 64 and 128 are shown in Figure 2. The values of $<R_G^2>$ were obtained either by light scattering [21,32–35] or by SANS [36]. It is noted that the molecular weight range in the good solvent, cyclohexane at 25°C, is considerably larger than in the θ-solvent, dioxane at 26.5°C. This is partly due to the lack of SANS data for low molecular weight samples in dioxane. Figure 2 provides good experimental evidence that the molecular weight dependence

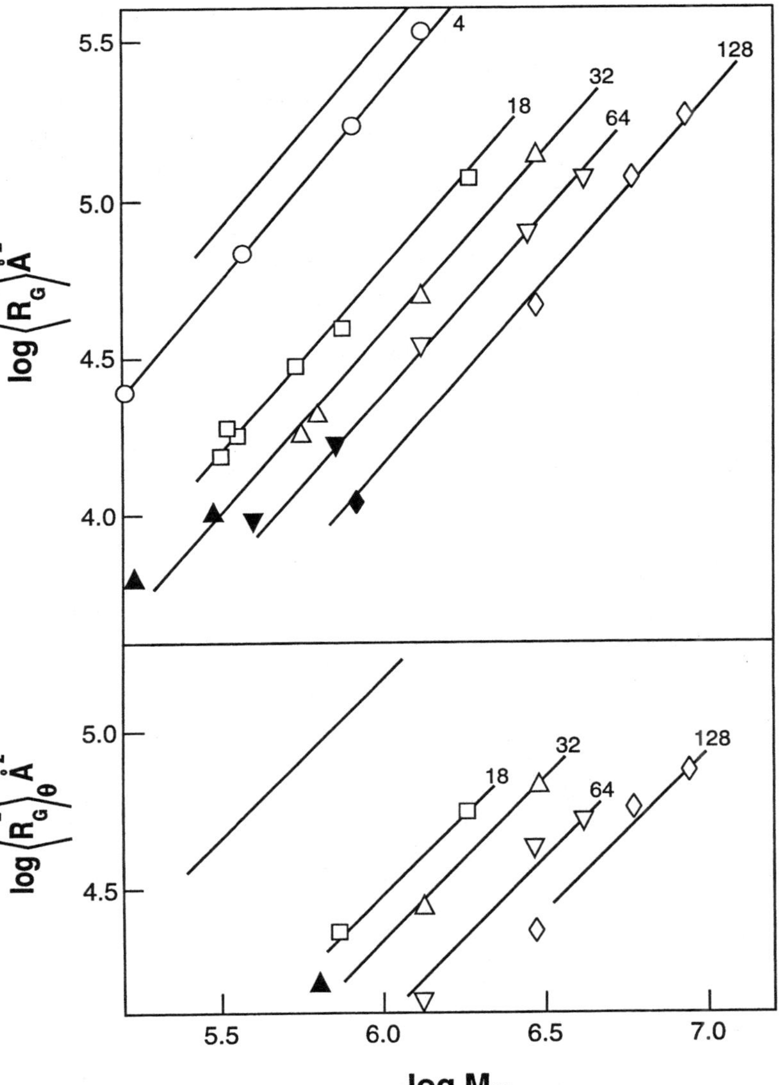

Figure 2 Double-logarithmic plot of the mean-square radius of gyration against molecular weight for polybutadiene stars. The functionality of the stars is indicated. Top: good solvent: cyclohexane 25°C. Bottom: θ-solvent: dioxane 26.5°C. Open symbols: light scattering. Full symbols: SANS. Full lines without symbols summarize the linear polymer data.

of $<R_G^2>$ has the same exponent for the stars as for the linear polymer. This is certainly true in the limit of high molecular weight in the good solvent. In the θ-solvent the experimental scatter in the data makes the evidence for a common power law less well documented. Similar excellent data sets are available for polystyrene with from 3 to 12 arms in cyclohexane [37–39] at the θ-temperature and in toluene, a good solvent [32,37–40]. In the case of polyisoprene, data are available for stars with 4, 6, 8, 12, and 18 arms in dioxane at the θ-temperature [41–43] and for stars with 8, 12, and 18 arms in cyclohexane [43]. A number of authors have also reported on individual samples. These results are more difficult to audit but many appear to be in reasonable agreement with the existing body of data.

Therefore, in the limit of high molecular weight

$$<R_G^2> \propto M^{2\nu}$$

where ν depends on the solvent quality only. From data of the type presented in Figure 2 it is straightforward to derive the experimental branching (or shrinkage) factor for regular stars

$$g_G(f) = \frac{<R_G^2>_{\text{star}}}{<R_G^2>_{\text{lin}}} \tag{22}$$

at equal molecular weight. This is in accordance with the original definition [14]. Some authors [44] have preferred to consider

$$g_G^{\text{a}}(f) = \frac{<R_G^2>_{\text{star}}}{<R_G^2>_{\text{arm}}}$$

because the arm material is often available for direct comparison. Note that $g_G^a(f) = f^{2\nu}\, g_G(f) > 1.0$.

Values of $g_{G,\theta}$ and g_G are summarized in Table 2a and 2b, respectively, and shown in Figure 3 as a function of the star functionality. The Gaussian chain relation, Eq. (2), is included for comparison. It can be concluded that $g(f)$ is a universal function for flexible star polymers.

At small f, $g_{G,\theta}$ is indistinguishable from g_{zs} [Eq. (2)]. When $f > 10$, $g_{G,\theta} > g_{zs}$ and in the limit of large f

$$g_{G,\theta} = 1.5 f^{-0.67}$$

The latter relation is intermediate between the $f^{-0.50}$ prediction of Daoud–Cotton [Eq. (12)] and the Zimm–Stockmayer prediction [Eq. (2)]. In good solvents $g_G \approx g_{ZS}$ up to $f = 8$ and larger than g_{ZS} for $f > 10$. In the limit of large f

$$g_G = 1.8 f^{-0.78} \tag{23}$$

Table 2 Values of $g_{G,\theta}$ and g_G

	Star polymer (no. of Samples)			*Refs.*		
f	*Polystyrene*	*Polyisoprene*	*Polybutadiene*	*PS*	*PI*	*PBd*
	a. θ-solvents: $g_{G,\theta} = <R_G^2>_{\theta,\text{star}}/<R_G^2>_{\theta,\text{lin}}$					
3	0.82(4)			37		
4	$0.63_3(5)$[a]	0.65(3)		38	41	
6	$0.45_8(3)$	0.46(3)		38	41	
8		$0.41_5(2)$,$0.43_8(4)$			42,43	
12	0.28(1),$0.27_6(2)$	0.33(2),$0.36_6(4)$		39,37	42,43	
18	$0.22_8(1)$	$0.29_4(3)$	0.20(2)	39	43	34
32			$0.15_1(3)$			33
64			$0.09_2(3)$			21
128			$0.06_0(3)$			21
	b. Good Solvents: $g_G = <R_G^2>_{\text{star}}/<R_G^2>_{\text{lin}}$					
3	$0.78_5(4)$			37		
4	$0.62_4(3)$,$0.60_6(3)$		$0.67_6(5)$	40,32		32
6	$0.44_9(1)$,0.46_5			40,32		
8		$0.34_4(6)$			43	
12	$0.24_3(1)$,$0.25_4(7)$	$0.24_7(5)$		39,37	43	
18	$0.20_3(1)$	$0.17_9(6)$	0.19(2), $0.18_5(4)$	39	43	34,35
32			$0.11_5(6)$			33
64			$0.06_8(5)$			21
128			$0.04_3(4)$			21

[a]Includes one sample of G.C. Berry, *J. Polym. Sci.*, A-2, (9), 687 (1971).

which agrees well with the Daoud–Cotton model, which yields $f^{-0.8}$ (Table 1) for $\nu = 3/5$ or $f^{0.765}$ for $\nu = 0.588$. It should be noted that while $g_G \approx g_{G,\theta}$ for stars with small f, $g_G < g_{G,\theta}$ when $f > 10$.

A closer inspection of Figure 2 suggests that $<R_G^2>$, and therefore g_G, values of low molecular weight star polymers are higher than the limiting value at high molecular weight. This phenomenon has been studied in detail by Huber et al. for 12-arm star polystyrene [45]. It appears more easily observable with polystyrene than with polydiene stars. A Monte Carlo simulation supported the experimental observation [46]. The scaling model implicitly predicts that low molecular weight star polymers will have relatively larger dimensions because

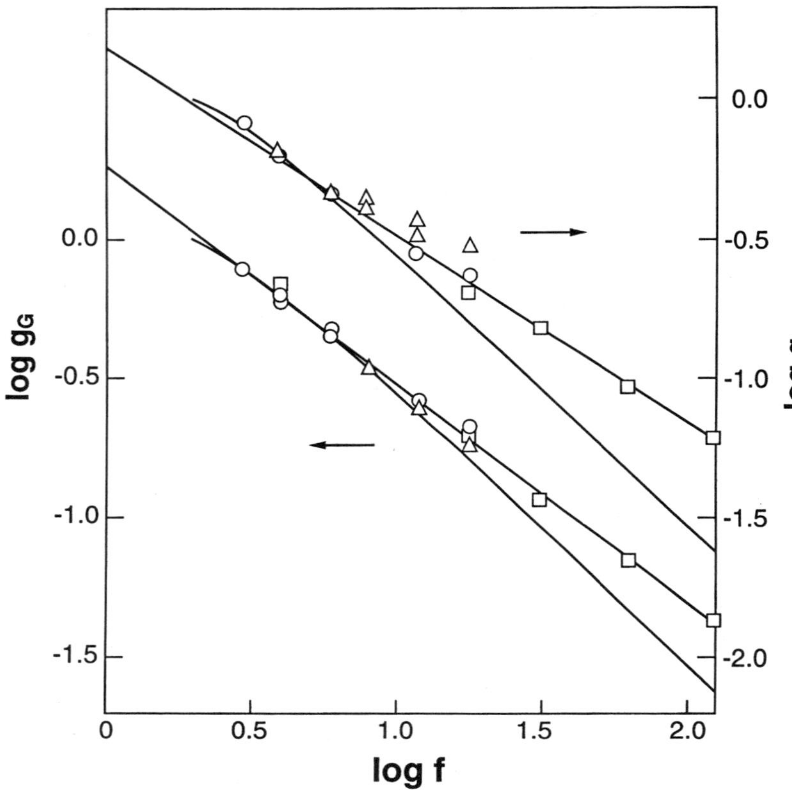

Figure 3 Double-logarithmic plot of the dependence of $g_G = \langle R_G^2 \rangle_{star} / \langle R_G^2 \rangle_{lin}$ on the functionality of the star. Top line: θ-solvent. Bottom line: good solvent. Symbols: circle, polystyrene; triangle, polyisoprene; square, polybutadiene. Unmarked curves are for the theoretical Gaussian chain, Eq. (2).

the chain expansion is largest near the core and tapers off with the radial distance from the core. Detailed studies of the dimensions of partially labeled 12-arm star polyisoprenes in a good solvent have clearly shown that most of the chain expansion occurs in the interior part of the star. The central part of the star containing 50% of the mass has R_G = 50 Å, while the whole star polymer has R_G = 70 Å and the 50% by mass in the outer shell has R_G = 78 Å [47].

D. Comparison with Other Theories

Freed and his co-workers and Vlahos and Kosmas have used renormalization group methods to derive values of $g(f)$ in an effort to understand the factors that affect the excluded volume of stars [48–50]. The results, to first order in $\epsilon = 4 - d$, where d is the dimensionality of the space, are shown in the third column of Table 3. These results [48] agree with experimental results at least to $f = 6$. The fourth column of Table 3 gives the values of g derived by Douglas and Freed [49] according to

$$g_G = g_{ZS} \times \frac{1 + a_G(\text{star})}{1 + a_G(\text{lin})}$$

where $a_G = (3K_f/32) - 1/4$ with K_f the virial coefficient in

$$<R_G^2> = \left(\frac{Nb^2}{6}\right)(1 + K_f z - \cdots)$$

Theoretical values of K_f are given on page 119 of reference 8. This theory combines renormalization group methods with the two-parameter theory. Agreement with experiment is good up to $f = 12$.

The fifth column of Table 3 lists values of g_G calculated following Yamakawa–Burchard [8,45]. In general,

$$g_G = g_{ZS}\left(\frac{\alpha^2_{G,\text{star}}}{\alpha^2_{G,\text{lin}}}\right) \tag{24}$$

where α_G^2 is calculated from

$$\alpha_G^5 - \alpha_G^3 = K_f z = K_f\, k'\, M^{1/2}$$

At constant molecular weight, the expansion of the star from its Gaussian coil size is larger than the expansion of the linear polymer because K_f increases with f [8]. For linear polybutadiene $k' = 0.003359$ is found experimentally. With M set at 10^7 (approximating the high molecular weight limit) values of α_G^2 are calculated for each f. It can be seen from Table 3 that the results of the Yamakawa–Burchard theory agree remarkably well with experimental results up to $f = 64$. It appears, therefore, that application of a simplified two-parameter theory reproduces star polymer dimensions to high f in the good solvent limit.

The failure of the Zimm—Stockmayer Gaussian chain model and the lack of agreement with the Daoud–Cotton exponent at the θ-point have been pointed out. It is generally recognized that this is due to an important contribution of three-body interactions in stars with many arms. Candau and coworkers [51] introduced three-body interactions to account for certain devia-

Table 3 Comparison of $g_G(f)$ with Different Theories

f	$g_G{}^a$	$g_G{}^b$	$g_G{}^c$	$g_G{}^d$	$g_{G,\theta}{}^e$	$\alpha_{G,\theta}{}^{2,f}$
3	0.78_5	0.759	0.778	0.787	0.82	
4	0.62	0.619	0.631	0.639	0.64	≈1.0
5		0.526	0.525	0.537		
6	0.45	0.462	0.453	0.464	0.46	
8	0.34	0.378	0.354	0.369	0.42	
10		0.326	0.292	0.306		
12	0.25	0.291	0.248	0.263	0.28	
18	0.19	0.231	0.173	0.190	0.21	1.265
24		0.199	0.135	0.150		
32	0.11_5		0.105	0.118	0.151	1.645
64	0.06_8		0.059	0.068	0.092	1.983
128	0.04_3		0.032_5	0.038	0.060	2.573

[a]Experimental values in good solvent
[b]Ref. 48, eq. 5.9, with ν = 3/5. Slightly higher values are obtained with ν = 0.588.
[c]Ref. 49, see also Ref. 32.
[d]Yamakawa–Burchard theory, Eq. (24).
[e]Experimental values at θ-temperature.
[f]Square expansion coefficient at the Flory θ-temperature of polybutadiene (dioxane at 26.5°C).

tions observed in stars near the Flory θ-temperature [51]. At $T = \theta$, the expansion of the star due to three-body interactions is given by

$$\alpha^8_{G,\theta} - \alpha^6_{G,\theta} = \tfrac{2}{3} C'_{M,\ell} \left(\frac{1}{g^3_{ZS}} \right) \tag{25}$$

where

$$C'_{M,\ell} = \frac{3^{5/2}}{(2\pi)^3} \frac{\bar{v}^3}{N_A^2 V_1} \left(\frac{\langle R_G^2 \rangle_\theta}{M} \right)^{-3} \frac{1}{2} \left(\frac{1}{3} - \chi_2 \right)$$

When applied to polybutadienes with $f = 18$ to $f = 128$, Eq. (25) does not lead to a constant value of $(1/3 - \chi_2)$. Boothroyd et al. [52] have proposed a modification that uses the elastic free energy of the star rather than using the elastic free energy of a linear chain. This changes Eq. (25) to

$$\alpha^8_{G,\theta} - \alpha^6_{G,\theta} = \frac{2}{7} \frac{A \cdot G}{g^3_{ZS}} \tag{26}$$

where $A = \tfrac{2}{3} C'_{M,l}$ and $G = (15f - 14)/(3f - 2)^2$. A plot of experimental values

$\alpha^2_{G,\theta}$, given in the last column of Table 3, against $G/(\alpha^{\theta}_{\Sigma\epsilon\theta} \cdot g_{ZS})^3$ yields a straight line with $A = 0.0933$ and $(1/3 - \chi_2) = 0.39$ for polybutadiene in dioxane. Although the latter value is too large to be realistic, the calculation lends strong support for three-body interactions as the main contribution to the expansion of star polymers from the Gaussian chain dimension at the Flory θ-point. Boothroyd applied this model to polyethylene stars with $f = 3$ to $f = 18$ (which shows much smaller expansions than the polybutadiene stars) and came to the same conclusion [52]. An analysis of experimental polyisoprene star polymer data by means of Eq. (26) led to $1/3 - \chi_2 = 2.4$ [53].

Batoulis and Kremer have shown that three-body interactions can account for the small deviations of star dimensions from the Gaussian dimension between $f > 5$ and $f = 12$ [54]. They also reported a $f^{-1/2}$ dependence of $g_{G,\theta}$ in agreement with scaling theory. A theoretical study of the compensation between attractive two-body and repulsive three-body interactions provides rather good agreement with the experimentally obtained θ-temperature dimensions for 12- and 18-arm polystyrene stars [55].

Separate comparisons have been made between the Gaussian chain [Zimm–Stockmayer, Eq. (2)] and the θ-solvent results, on the one hand, and the same Gaussian chain and good solvent data, on the other hand. Direct comparison between the dimensions at the θ-point and in the good solvent should be avoided because the expansions are governed by different phenomena in each case.

E. Comparison with Simulations

The results of numerical computations of the static dimensions ratio of regular star and linear polymers are summarized in Table 4a, and 4b for the θ-condition [56–61] and the good solvent [44,57,62–65], respectively. The numerical methods occupy a position intermediate between theoretical models and experiments. They have their own limitations, problems, and uncertainties that are usually discussed in the original publications. For lattice models the definition of the θ-state [56,57,59] as well as the definition of the self-avoiding walk [62–64] (good solvent) have to be verified, along with any possible dependence on the type of lattice. For off-lattice models the interaction potentials are varied to simulate θ and good solvent conditions. Often, the number of segments per arm in the simulated chain is small, and extrapolation to large chains is required. This is usually done by extrapolation of $N^{-1} \to 0$ but this extrapolation is sometimes problematic [58,63]. Molecular dynamics simulations have also been used to study star polymers [44]. Both molecular dynamics and Monte Carlo simulations require long computer times due to the slow relax-

Table 4 Comparison of Star Dimension with Computations

a. θ-Solvent

f	$g_{G,\theta}{}^{a}$	$g_{G,\theta}{}^{b}$	$g_{G,\theta}{}^{c}$	$g_{G,\theta}{}^{d}$	$g_{G,\theta}{}^{e}$	$g_{G,\theta}{}^{f}$	$g_{G,\theta}{}^{h}$
3	0.83			0.81	0.79	0.74	0.82
4	0.67	0.59 ± 0.06		0.66	0.68	0.65	0.64
5					0.55	0.54	
6	0.50	0.44 ± 0.05	0.49	0.50	0.48	0.45	0.46
8	0.41			0.41	0.39	0.40	0.42
10				0.35		0.34	
12			0.33	0.31	0.28		0.28
18			0.21				0.21

[a]θ-condition taken as $<R_G^2> = A(f)\,N^{1.0}$. Includes $g(9) = 0.35_5$. Ref. 56.
[b]Quoted values are for $a' = 0$, i.e, the random flight model. Ref. 57.
[c]Off-lattice with Lennard–Jones potential. Ref. 58.
[d]θ-condition approximated by $\xi = -0.5$. Ref 59.
[e]Ref. 60.
[f]Includes $g(20) = 0.20$. Ref. 61.
[h]Experimental values.

b. Good Solvent

f	$g_G{}^{a}$	$g_G{}^{b}$	$g_G{}^{c}$	$g_G{}^{d}$	$g_G{}^{e}$	$g_G{}^{f}$
3	0.75				0.766	0.785
4	0.59	0.62_3 ± 0.04			0.611	0.62
5	0.51				0.508	
6	0.42	0.45 ± 0.03	0.44	0.448	0.439	0.45
8						0.34
10				0.287		
12			0.233			0.25
18			0.168			0.19

[a]Good solvent based on $<R_G^2> = A(f)\,N^{1.176}$. Ref. 62.
[b]Ref. 57.
[c]Off-lattice with Lennard–Jones potential with $\varepsilon/kT = 0.1$. Ref. 63.
[d]Includes $g(20) = 0.156$, $g(30) = 0.114$, $g(40) = 0.0897$ and $g(50) = 0.0754$ recalculated from $g_G{}^{a} = <R_G^2>/<R_G^2>_{\mathrm{arm}}$ with $2\nu = 1.18$. Ref. 44.
[e]Ref. 65.
[f]Experimental values.

ation processes even for dilute star polymers. Recent, very accurate, Monte Carlo lattice calculations give impressive results for small f [65].

As can be seen from Table 4a, agreement of simulation results with experimental data is usually within experimental error for stars with up to $f = 18$, despite the fact that the θ-state is quite difficult to simulate. In Table 4b, the same good agreement with experiment is observed. In particular it has been noted that $g_G = g_{zs}$ and simultaneously satisfies the Daoud–Cotton requirement that $g_G \propto f^{-0.80}$ [44,64,65]. It is obvious that these results are fortuitous over a small range $4 < f < 18$. It has been shown that such assertion is due to allowance for inaccuracies [64], while experimental results in Figure 3 clearly show that the true Daoud–Cotton asymptote is obtained only when $f > 18$. The excellent agreement of the dimensions of stars with 30, 40, and 50 arms obtained by molecular dynamics simulations [44] with the experimental values for polybutadiene stars with $f > 18$ should be stressed. The simultaneous good agreement of these results with the Daoud–Cotton model is a guarantee for the essential correctness of the present understanding of star polymer dimensions in good solvents.

F. The Segment Density Distribution

The form factors of star polymers originally obtained from SANS experiments over a wide range of QR_G values have often been cited to be in good agreement with a Gaussian chain model, Eq. (18). This is probably the case for stars with low functionalities up to $QR_G \approx 5$ in good solvents and to higher QR_G values in θ-solvents [46,52,66,67]. A more recent example of the experimental form factor of a 18-arm star in a good solvent is shown in Figure 4. The data are represented in the generalized Kratky form. Agreement between the experimental and calculated curves are excellent to $QR_G \approx 3$. However, Eq. (18) does not reproduce the $Q^{-5/3}$ dependence expected in a good solvent at high Q.

A detailed analysis of the form factor over the whole Q range was undertaken by Dozier et al. [68]. They analyzed results on 8- and 12-arm star polymers in good solvent and θ-solvents and concluded that the form factor of stars consists of two regions. In the low Q range, the form factor is characteristic of the whole molecule and is described by a Guinier relation in terms of the radius of gyration, $P(Q) = \exp[-(1/3)\, Q^2 R_G^2]$, Eq. (20). At high values of Q, on the order of the inverse of the size of the outermost blob, $\xi \approx R_G / f^{1/2}$ [see Eq. (5)], the form factor reflects the dilute solution within these blobs where the pair correlation function is given by

$$g(r) = \left(\frac{r}{a_s}\right)^{1/\nu} r^{-3}$$

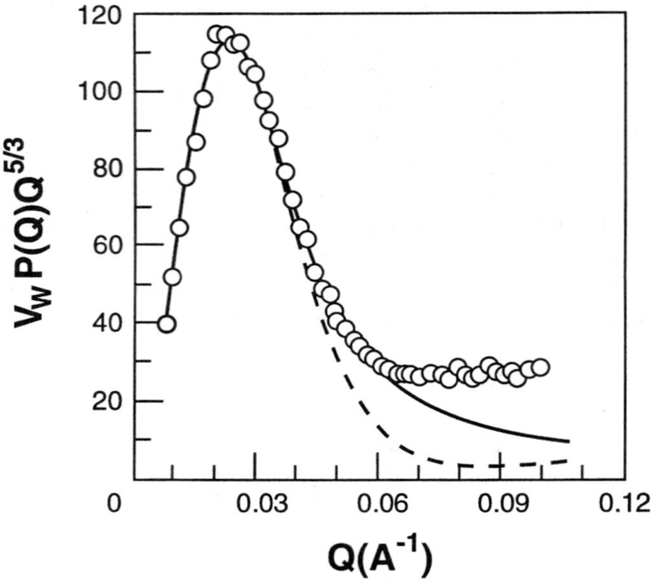

Figure 4 Generalized Kratky representation of the form factor of an 18-arm polyisoprene in methylcyclohexane-d_{14} (R_G = 74 Å). The solid line is a fit to the data with Eq. (19) (Gaussian conformation). The dashed line is a fit for a swollen star according to Eq. (28). (From Ref. 31.)

for values of $r \leq \xi$. Compare with Eq. (15). These large blobs contain the major fraction of the star polymer mass. The total form factor is approximated by the sum of contributions in the two regions. This approximation is justified on the basis of the widely different distances probed in the two regions.

The method of Dozier has also been applied to a series of stars with from 8 to 128 arms [31]. The experimental data were fitted to [31,68]

$$V_w P(Q) = V_w \exp[-\tfrac{1}{3} Q^2 R_G^2] + \frac{\alpha}{Q\xi} \frac{\sin[u \tan^{-1}(Q\xi)]}{[1 + Q^2\xi^2)^{u/2}} \tag{27}$$

V_w is the molar volume of the polymer chain where $u = 1/(\nu - 1)$ and $\alpha = \bar{v}_p N f^{-1/2}$. Comparison of Eq. (27) with experiment is shown in Figure 5. The agreement is very good except for the 128-arm star where the calculated

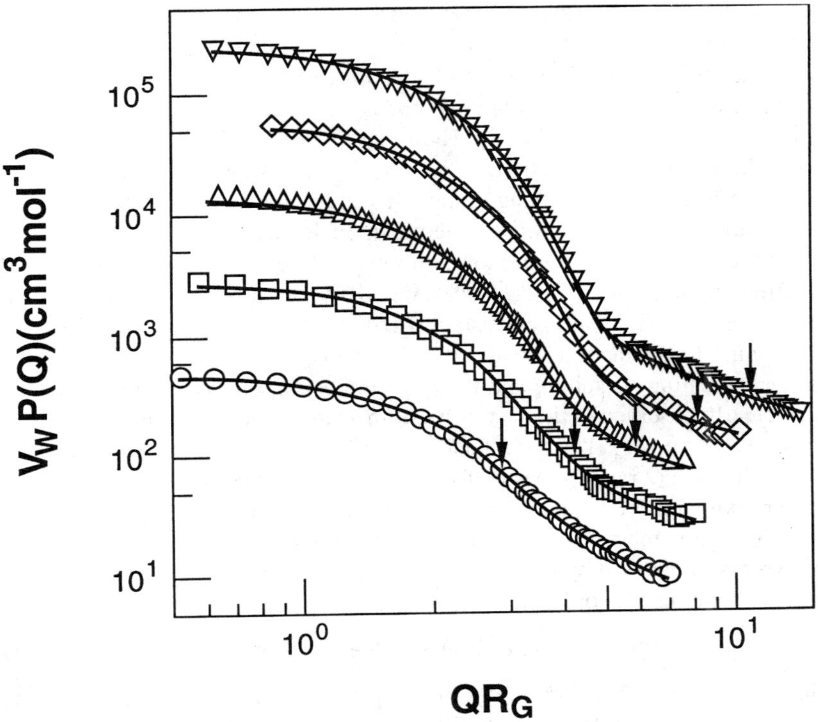

Figure 5 Zero-concentration form factors of star polymers with different functionalities. Data obtained in a good solvent, methylcyclohexane-d_{14} with SANS. From bottom to top: $f = 8$ (polyisoprene), $f = 18$ (polyisoprene), $f = 32$ (polybutadiene), $f = 64$ (polybutadiene), and $f = 128$ (polybutadiene). The data are offset vertically for clarity. The solid line represents Eq. (27). Arrows indicate $Q = R_G^{-1} f^{1/2}$, the onset of the asymptotic regime in which scattering is caused by the swollen blobs in the corona. (From Ref. 31.)

and experimental drop-off between the high and low Q-ranges are slightly different.

The second term on the right-hand side of Eq. (27) dominates the form factor when $QR_G \geq f^{1/2}$. This point is indicated for each polymer by arrows in Figure 5. In the high Q-range, three parameters can be varied to obtain agreement between experiment and Eq. (27): $\alpha = \bar{v}_p N f^{-1/2}$ (N: number of segments per arm; $\bar{v}_p$: partial specific volume of polymer), $\xi \propto R_G / f^{1/2}$, and $u = 1/(\nu - 1)$.

It is found that values of α and ξ that lead to good agreement for different stars with different f and N obey the relations given by their definitions. There is some uncertainty about the values of ν and u that are applicable. In the good solvent, ν is equal to 3/5 (see Figure 4) but experimental results on stars are also compatible with $\nu = 2/3$ [31,68]. The latter value has also been observed in molecular dynamics simulations of stars with many arms [69]. A value of $\nu = 2/3$ is consistent with stretching of the arms and is expected in small stars with many arms. Resolution of this problem is impeded by the small Q range over which the scattering within a blob is observed (see Figure 5). There is good experimental evidence that $P(Q)$ scales as $Q^{-5/3}$ for very large 18-arm stars in a good solvent [68]. This is consistent with the SAW configuration of the arms of the star. The distinction between the $Q^{-3/2}$ and $Q^{-5/3}$ dependence of $P(Q)$, the former indicating stretching and the latter indicating ordinary SAW, has been quite clearly observed in the scattering from a single protonated arm in 12-arm star polystyrene [67]. The form factor of a 12-arm polystyrene star in a θ-solvent in the high Q-range has been found to scale as Q^{-2} consistent with $\nu = 1/2$, or (approximately) Gaussian segment distribution in the star [68].

In Figure 4 the experimental results are also compared with the theoretical expression derived by Alessandrini and Carignano [70] by renormalization group techniques. Accordingly,

$$P(y) = f(y)\{1 + A_1 [\exp(-A_2 y) - 1] + A_3 y + A_4 y^2\} \tag{28}$$

where $y = \frac{3}{4} f/(3f - 2)\, Q^2 R_G^2$ and $f(y)$ is Eq. (19). Values of A_1, A_2, A_3, and A_4 are given for a star with $3 \leq f \leq 18$ [70]. It can be seen that the theoretical curve decreases more rapidly than the experimental one. The $Q^{-5/3}$ dependence at high Q is well reproduced.

A further point should be made with regard to the ratio of the scattering intensities in the low and high Q range, as shown in Figure 5. At low Q values the scattering intensity is due to the coherent scattering of the whole molecules and is proportional to $[n_b(\xi/a_s)^{1/\nu}]^2$ where $n_b = fR_G/\xi$, the number of scattering blobs and $(\xi/a_s)^{1/\nu}$ [see Eq. (14)], the number of segments per blob. At high values of Q where $Q{\cdot}\xi = 1$, the intensity of scattering is due to the incoherent superposition of coherent scattering from within each blob. This is given by $n_b[(\xi/a_s)^{1/\nu}]^2$. Since $R_G/\xi = f^{1/2}$, the ratio of scattering intensity at low and high Q values is proportional to $f^{3/2}$. This has in fact been observed experimentally [31]. This analysis shows clearly that the Daoud–Cotton model correctly describes the internal structures of star polymers with many arms. The existence of a (small) interior region with stretched chains is more conclusively derived from the radii of gyration than from the experimental form factors.

Furthermore, in Figure 5 it can be seen that the form factor of the 128-

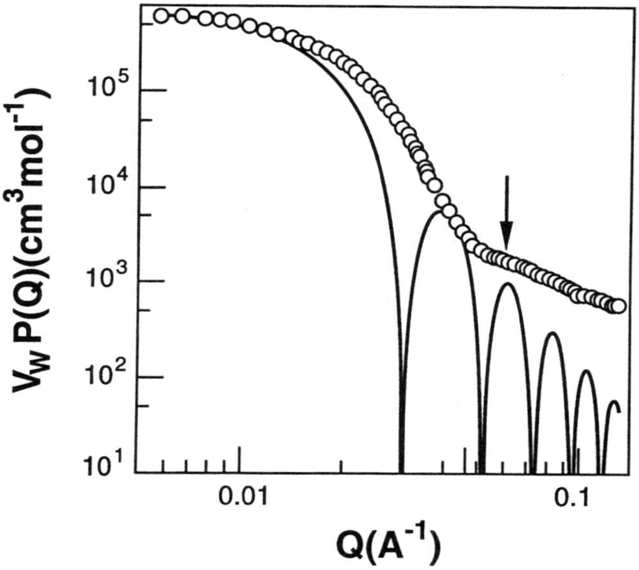

Figure 6 Comparison of the form factor of a star polymer with 128 arms with the form factor of an equal-density hard sphere calculated by Eq. (29). The radius of the sphere was taken as $R = (5/3)^{1/2} R_{G,\text{star}}$. The arrow marks the secondary maximum in the star polymer data. (From Ref. 31.)

arm star polymer has a small secondary maximum at $QR_G \approx 6$. In the limit of very high functionality the form factor of a star is expected to resemble that of an equal-density hard sphere [31]. This comparison is made in Figure 6. The hard-sphere form factor is calculated from

$$P(Q) = (3/v^3)\,(\sin v - v \cos v)^2 \tag{29}$$

where $v = QR = (5/3)^{1/2}\, QR_G$. As expected, the star form factor is considerably more smeared than the form factor of an equal-density hard sphere, but it is interesting to note that the small secondary maximum in the form factor of the star is consistent with the third maximum of the form factor of the hard sphere.

Two pictures of the conformation of star polymers have been proposed [17,71]. Daoud and Cotton [17] proposed that stars consists of a series of concentric shells each containing *f* blobs that regularly increase in size $\xi\,(r)$ as the radial distance from the center increases. The other model [71] has only a small number of interior shells with increasing thickness ξ, but the main fraction of the star is contained in f swollen blobs with a diameter of the order of the radius of gyration R_G. A version showing the $\xi \propto R_G \cdot f^{1/2}$ of the latter model is shown in Figure 7. The $\xi \propto R_G \cdot f^{1/2}$ dependence is based on recent SANS evidence [31].

III. HYDRODYNAMIC PROPERTIES

A. Intrinsic Viscosity

The experimental ratios of the intrinsic viscosities of star and linear polymers of equal molecular weight, $g_{[\eta]} = [\eta]_{star}/[\eta]_{lin.}$, are given in Table 5a and 5b as a function of f for the θ-solvent and good solvent, respectively.

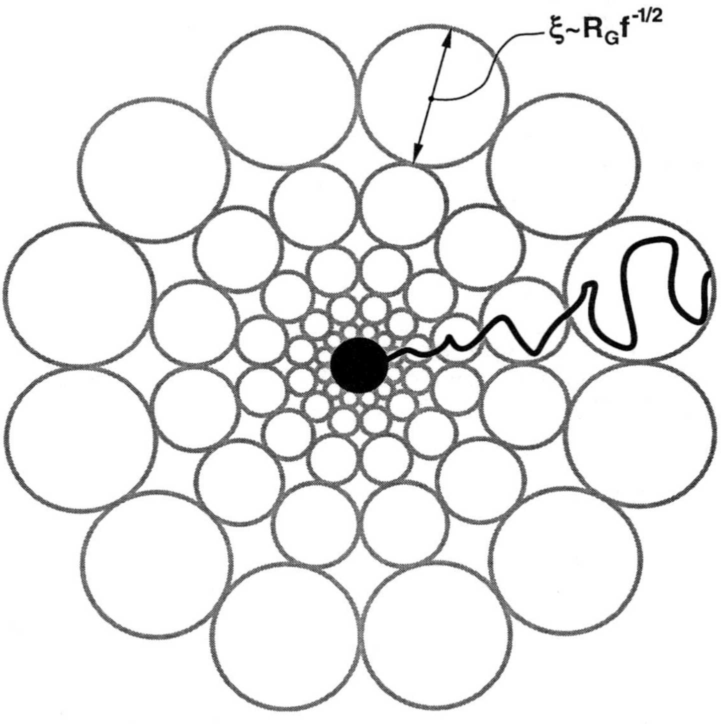

Figure 7 Schematic two-dimensional representation of the blob model of a regular star polymer in dilute solution.

As was shown before for g_G, each value of $g_{[\eta]}$ is based on a set of measurements of stars with the same functionality. An example of actual intrinsic viscosity molecular weight plots for different functionality of the star is shown in Figure 8. Polystyrene results are given in Ref. 37 and polyisoprene results are given in Ref. 43. The quality of the intrinsic viscosity data is generally superior to the radii of gyration data due to the well-known simplicity and accuracy with which intrinsic viscosities are measured. Moreover, intrinsic viscosity measurements have no upper molecular weight limit. A lower molecular weight limit, however, occurs in good solvent. For stars with $f \geq 6$, the dependence of $g_{[\eta]}$ on functionality is given by

$$\log g_{[\eta],\theta} = 0.35 - 0.69 \log f \tag{30}$$

and in good solvents

$$\log g_{[\eta]} = 0.36 - 0.80 \log f \tag{31}$$

When $f < 6$, values of $g_{[\eta]}$ are slightly less than predicted by Eqs. (30) and (31). It is shown in Figure 9 that $g_{[\eta]}$ always lies between g_{ZS} and $g_{ZS}^{1/2}$. Furthermore from Tables 5a and 5b it can be seen that $g_{[\eta]} < g_{[\eta],\theta}$. Relations of the type

$$g_{[\eta]} \propto g_G^m$$

have been sought. However, since data over a large range of functionalities have become available, it has become clear that no constant exponent can be adopted to fit all results satisfactorily.

Theoretical calculation of the intrinsic viscosity of regular stars goes back to 1959 [78]. In the free-draining case, the Rouse model for linear polymers applied to stars yields [79]

$$g_{[\eta]} = g_{ZS}$$

This relation finds application in the melt and concentrated solution of small stars [12].

In the non-free-draining case, in dilute solution, the theory of Kirkwood and Riseman with preaveraged hydrodynamic interactions is applied to star polymers. Zimm and Kilb obtained [78]

$$g_{ZK} = \left(\frac{2}{f}\right)^{3/2} \frac{[(f-1)0.390 + 0.196]}{0.586} \tag{32}$$

where the $(f-1)$ term originates from the $(f-1)$ times degenerate odd eigenvalues sum: $\sum_{k,\text{odd}} (1/\lambda_{f,k}) = 0.390$ and $\lambda_{f,k}$ has been approximated by $\lambda_{2,k}$. The sum of the even eigenvalues is $\sum_{k,\text{even}} (1/\lambda_{f,k}) = \sum_{k,\text{even}} (1/\lambda_{2,k}) = 0.196$. For the linear polymer the sum of all eigenvalues is just the sum of the odd and even eigenvalues. Zimm and Kilb observed that

Table 5 Experimental Ratios of Intrinsic Viscosities of Star and Linear Polymers of Equal Molecular Weight

a. Flory θ-temperature: $g_{[\eta],\theta} = [\eta]_{\theta,\text{star}}/[\eta]_{\theta,\text{lin}}$

	Polymer			*Refs.*		
f	*Polystyrene*	*Polyisoprene*	*Polybutadiene*	*PS*	*PI*	*PBd*
3	0.81_0,0.86		0.85_3	73,37		74
4	0.76_1,0.74_3	0.77_2	0.75_9	72,77	41	74
5		0.73_5		38	43	
6	0.631	0.62_5		38	41	
8		0.54_7			43	
12	0.41,0.42	0.43_3		39,37	43	
18	0.35	0.30_5	0.28_4	39	43	34
32			0.19_8			33
64			0.127			21
128			0.080_5			21

b. Good Solvent: $g_{[\eta]} = [\eta]_{\text{star}}/[\eta]_{\text{lin}}$

	Polymer				*Refs.*		
			Polybutadiene				
f	*Polystyrene*	*Polyisoprene*	*Cyclohexane*	*Toluene*	*PS*	*PI*	*PBd*
3	0.83_8,0.83_7[a]			0.82_1	37,73		74
4	0.72_6,0.70_0[a]	0.73_3		0.70_9,0.72_9	72,77	41	74,75,76
5		0.60				43	
6	0.57_0	0.58_9			38	41	
8		0.43_3				43	
12	0.35,0.34_5	0.320			39,37	43	
18	0.26	0.23_2	0.23_4	0.22_5	39	43	35,34
32			0.15_4	0.14_4			33
64			0.090	0.082_7			21
128			0.049_7	0.044_7			21

[a] In benzene.

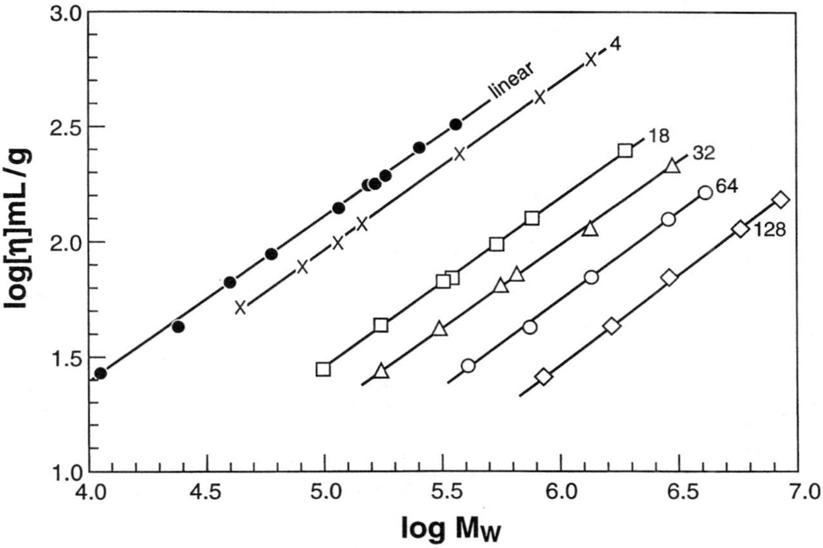

Figure 8 Double-logarithmic plot of intrinsic viscosity against molecular weight for polybutadiene stars with different functionality. Solvent: toluene, 35°C. (From Ref. 31.)

$$g_{ZK} \approx g_{ZS}^{1/2} \tag{33}$$

and these calculations were well supported by intrinsic viscosity results on four- and eight-arm stars available at that time [80]. It is now understood that the agreement of Eqs. (32) and (33) with experiment is in part due to the polydispersity of the arms in stars prepared by step-growth polymerization.

Osaki and Schrag have showed that the dependence of $g_{[\eta]}$ on the strength of the hydrodynamic interaction parameter causes values of $g_{[\eta]}$ to be intermediate between g_{ZS} and g_{ZK} [81]. However, they had already observed that experimental values of $g_{[\eta]}$ for stars with $f = 9$ and $f = 13$ are considerably lower than predicted by Eq. (32), and they fully realized that this cannot be due to increased free-draining behavior in more dense star polymer coils. The experimental work and model calculations of Osaki and Schrag on the frequency dependence of the complex moduli of dilute solutions of star polymers have been summarized [11,12].

The origin of the major discrepancy between the experimental results

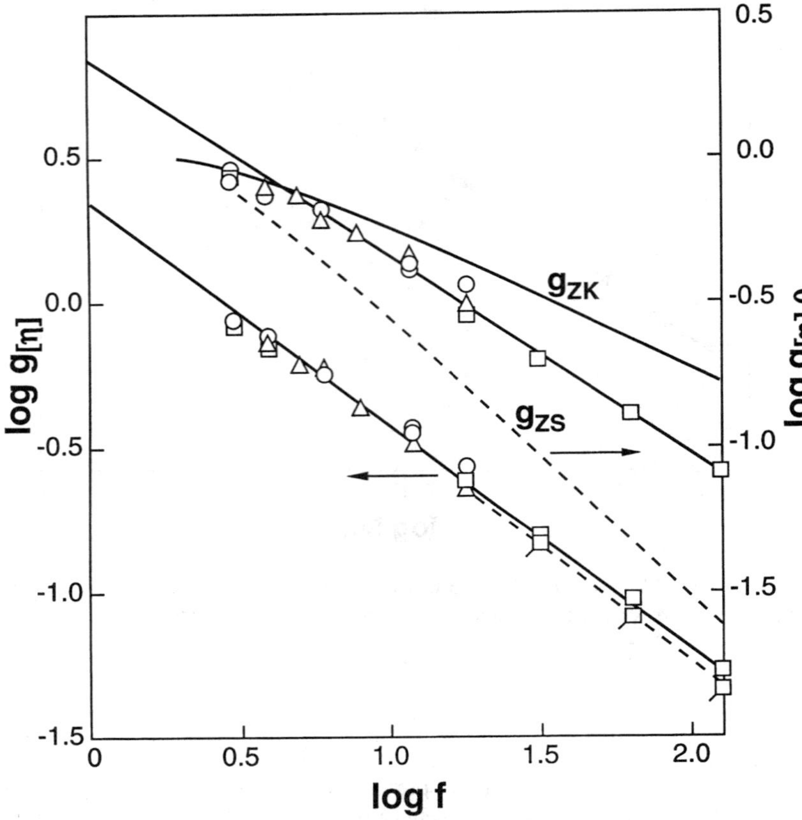

Figure 9 Dependence of $g_{[\eta]} = [\eta]_{star}/[\eta]_{lin}$ on the functionality of the star. Top: data obtained at the θ-temperature. Solid line marked g_{ZK} is Eq. (33). Dashed line marked g_{ZS} is Eq. (2). Bottom: data obtained in good solvents. Symbols; ⊡: polybutadiene stars in toluene (dashed line); other symbols as in Figure 3.

and values predicted by Eq. (32) was elucidated by Zimm [82]. He compared the hydrodynamic properties of computer-generated stars and linear chains with and without preaveraging of the hydrodynamic interactions. In all cases, the non-preaveraged values of $g_{[\eta]}$ are lower than the preaveraged g_{ZK} values and closer to the experimental results. Differences due to excluded volume effects could not be discerned, although experimental results clearly show that

$g_{[\eta]} < g_{[\eta],\theta}$. This may be due to smaller apparent coil expansion, changes in the persistence length, and increased partial free draining in the good solvent.

Considerable further computational work on star polymers has been undertaken by Freire and his collaborators [63,83–87]. This is summarized in Table 6. The latest development, simulation on chains with thermodynamic interactions and including non-preaveraged hydrodynamic interactions, modulated by Brownian motion, yields values of $g_{[\eta]}$ that are in agreement with available experimental results within the uncertainty limits. It seems therefore that all relevant factors contributing to the intrinsic viscosity of stars have been accounted for and that improvements in the simulations on larger stars with more segments and larger samples will lead to greater accuracy that can no longer be checked against experimental results. A clear example is the recent result of the simulation on a 12-arm star giving $g_{[\eta],\theta} = 0.42_5$ and $g_{[\eta]} = 0.36$, which are both in excellent agreement with experiment [87].

Douglas and Freed have developed a semi-empirical theory to obtain the intrinsic viscosity of a star polymer in a good solvent from its value in a θ-solvent [49,89]. Accordingly,

$$g_{[\eta]} = g_{[\eta],\theta} \frac{1 - 0.276 - 0.015\,(f-2)}{1 - 0.276} \tag{34}$$

which is the numerical form of

Table 6 Summary of $g_{[\eta]}$ Computations

	f				
Type	*4*	*6*	*12*	*18*	*Refs.*
MC preaveraged	0.81_4	0.70	0.52	0.43	78,83
MC non-preaveraged[a]	0.73 ± 0.056	0.56 ± 0.038			82
MC non-preaveraged		0.58 ± 0.02	0.38 ± 0.03	0.24 ± 0.03	84
MC θ-solvent		0.59 ± 0.03	0.39 ± 0.06	0.28 ± 0.04	85
MC good solvent		0.58_4 ± 0.008	0.373 ± 0.005	0.221 ± 0.002	63
MC good solvent[b]	0.733	0.571			88
MC BD[c]		0.532	0.374	0.276	86
Experimental $g_{[\eta],\theta}$	0.77–0.74	0.63	0.42	0.28–0.35	
Experimental $g_{[\eta]}$	0.70–0.74	0.58	0.32–35	0.22–0.26	

[a]Monte Carlo mean of $a' = 0$ and $a' = 0.6$ results.
[b]Average of $a = 1/2$ and $a = 1/4$.
[c]Monte Carlo with Brownian dynamics.

$$g_{[\eta]} = g_{[\eta],\theta} \frac{1 + \alpha_{[\eta]}(\text{branched})}{1 + \alpha_{[\eta]}(\text{linear})} \tag{35}$$

with $\alpha_{[\eta]} = (3/32)C_{[\eta]} - (3/8)$ and $C_{[\eta]}$ is the coefficient of expansion

$$[\eta] = [\eta]_\theta(1 + C_{[\eta]}\, z - \cdots)$$

of the linear polymer. $C_{[\eta]} = -0.276$ [8]. The problem with the approach is that $\alpha_{[\eta]}$ (branched) has been obtained from a comparison of experimental data of $g_{[\eta],\theta}$ and $g_{[\eta]}$ and then fed into Eq. (35) to produce the $-0.015\ (f-2)$ term in Eq. (34). In Ref. 90 the authors show that this approach gives reasonably good estimates of $g_{[\eta]}$ for stars with up to 18 arms. Extension to higher functionalities leads to unphysical results, however. This is most certainly due to the fact that ternary interactions become very important in the θ-state, and these are not accounted for in a two-parameter model.

It is worthwhile to mention that comparisons of the intrinsic viscosity of a star polymer have also been made with the intrinsic viscosity of a single arm or with the double arm or span molecular weight [43]. In these cases

$$\frac{[\eta]}{[\eta]_{\text{arm}}} = g_{[\eta]} \cdot f^{\text{a}}$$

and

$$\frac{[\eta]}{[\eta]_{\text{span}}} = g_{[\eta]} \cdot \left(\frac{f}{2}\right)^{\text{a}}$$

where "a" is the exponent in the intrinsic viscosity–molecular weight relationship of the linear polymer. It has been shown that a plot of $[\eta]/[\eta]_{\text{arm}}$ or $[\eta]/[\eta]_{\text{span}}$ as a function of f goes through a (broad) maximum at $f = 6$ (good solvent) and $f = 5$ (θ-solvent) [7,43].

B. Intrinsic Translational Friction Coefficient

The intrinsic translational friction coefficient $[f]$

$$[f] = \frac{f_0}{\eta_s} \tag{36}$$

where the zero subscript means that the friction coefficient has been obtained after extrapolation of experimental results to zero concentration and η_s is the solvent viscosity. The friction coefficient of a polymer is obtained from the measurement of its sedimentation velocity s_0 after extrapolation to zero concentration,

$$f_0 = \frac{M_w\,(1 - \bar{v}_p\,\rho_s)}{s_0^0\,N_\text{A}}$$

(with $\overline{v}_p$ the partial specific volume of the polymer in the solution and ρ_s the solvent density.) The term s_0^0 has a superscript as well to indicate the initial sedimentation coefficient. The friction coefficient can also be obtained from

$$f_0 = \frac{kT}{D_0} \tag{37}$$

(Einstein's relation). The measurement of the self-diffusion coefficient, D_0, is now mostly performed by dynamic light scattering and it is implicit that the data have been obtained at low $Q^2 <R_G^2>$ and have been extrapolated to the zero scattering vector and to zero concentration.

The ratio of the friction coefficient of a star and linear polymer with the same molecular weight has traditionally been denoted by h [15]. For internal consistency we will use g_H in recognition that this ratio also represents the ratio of hydrodynamic radii. The experimental results are tabulated in Table 7. It can be seen that g_H is relative insensitivity to functionality between $f = 3$ and $f = 128$.

The experimental results are shown in Figure 10. In the case of high functionality stars

Table 7 $g_H = [f]_{star} / [f]_{lin}$

	Polystyrene		*Polyisoprene*[a]	*Polybutadiene*		*Refs.*		
f	*θ-Solvent*	*Good*	*good*	*θ-Solvent*	*Good*	*PS*	*PI*	*PBd*
3	0.93_3[b]	0.93_3[b]				45		
4	0.94_2	0.93_6	0.91			72	43	
6	0.89	0.86				40		
8			0.81				43	
12	0.81	0.70	0.75			39	43	
	0.80	0.73				37		
	0.78[c]	0.74[c]				45		
18	0.76[d]	0.68[d]	0.64	0.72	0.67	39	43	35
	0.71_2[d]	0.58_3[d]				45		
32				0.63_1	0.59			33
64				0.54_9	0.49_7			21
128				0.47_2	0.41_3			21

[a]Linear polyisoprene data in Ref. 95.
[b]Average of two determinations.
[c]Average of four larger samples.
[d]One sample.

$$g_{H,\theta} = 1.35\, f^{-0.215}$$

and

$$g_H = 1.35\, f^{-0.244}$$

Theoretical work on the calculation of the friction coefficient of stars was initiated by Stockmayer and Fixman [15]. For the free-draining case g_H is equal to $g_G^{1/2}$. However, in dilute solution, friction between solvent and polymer segments is (hydrodynamically) screened and

$$[f] = \xi \sum_{\ell=1}^{N} \psi_\ell < N\xi$$

where ξ is the friction on one segment and

$$\psi_\ell = 1 - \frac{\xi}{6\pi\eta_s} \sum_{s\neq\ell} \langle r_{s\ell}^{-1} \rangle\, \psi_s$$

The latter equation indicates that screening is produced more effectively by nearest neighbors and dies away as $r_{s\ell}$ increases. Calculation of $\sum \langle r_{s\ell}^{-1} \rangle$ for stars with a Gaussian segment distribution with use of the Kirkwood–Riseman preaveraging of the hydrodynamic interactions and the omission of the effect of Brownian motion on the hydrodynamic interactions yields [15]

$$g_{SF} = f^{1/2}\, [2 - f + \sqrt{2}\,(f - 1)]^{-1} \tag{38}$$

This theoretical result is compared with the experimental results in Figure 10. It is clear that $g_{SF} > g_G^{1/2}$ but still g_{SF} overestimates the effect of branching on the friction coefficient of stars. The agreement is rather poor even for stars with low f. At high values of f numerical agreement is perhaps not expected, but g_H is also less sensitive to branching than the limiting form of Eq. (38), which predicts $g_{SF} \propto f^{-1/2}$. Nevertheless, much of later theoretical work takes Eq. (38) as the benchmark in the same way as Eq. (2) serves as the benchmark for the static dimensions of stars.

The first indication that the preaveraging technique has a substantial effect on the calculation of the friction coefficients of star polymers is found in the calculation of the Q-dependence of D_0, i.e., the contributions of internal modes [90,91]. Zimm showed that preaveraging of the hydrodynamic interactions in the Kirkwood–Riseman model was largely to blame for the discrepancy between g_{SF} and experimental values of g_H [57]. Further refinements of the Monte Carlo simulations with thermodynamic interactions to represent θ-solvent conditions or good solvent conditions combined with non-preaveraged hydrodynamic interactions lead to close agreement between calculated and

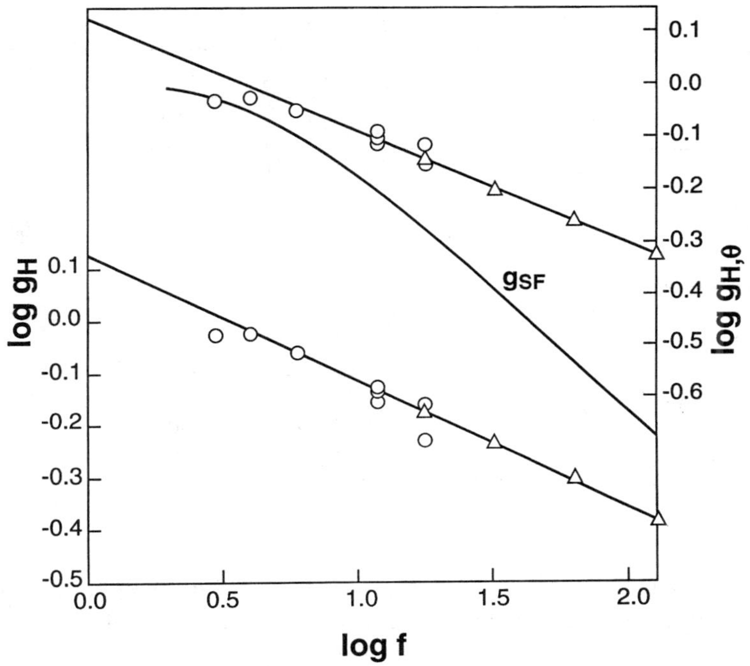

Figure 10 Dependence of $g_H = [f]_{star}/[f]_{lin}$ on the functionality of the star. Top: data obtained at the θ-temperature. Line marked g_{SF} is Eq. 38. Bottom: data obtained in good solvents. Symbols as in Figure 3.

experimental values of g_H for stars with up to 18 arms [63,84,85]. The results of computer simulations have been collected in Table 8.

C. Relation Between Hydrodynamic and Static Properties: The Equivalent Sphere Notation

The relation between the hydrodynamic properties and the static dimensions of the polymer coil have attracted much interest because they reveal two "universal constants" Φ′ and P' according to [92]:

$$[\eta] = \frac{\Phi' \langle R_G^2 \rangle^{3/2}}{M}$$

$$[f] = P' \langle R_G^2 \rangle^{1/2}$$

Table 8 Summary of g_H Computations

Type	*f* = *4*	*6*	*12*	*18*	*Refs.*
SF[a]	0.892	0.798	0.623	0.528	15
MC preaveraged:					
θ-solvent	0.925	0.845	0.73	0.65	57,84
good solvent	0.913	0.832			
MC non-preaveraged:					
θ-solvent	0.962	0.892	0.82	0.77	57,85
good solvent	0.940	0.873			
MC θ-solvent		0.89 ± 0.02	0.82 ± 0.03	0.78_5 ± 0.007	63
MC good solvent		0.89_2 ± 0.009	0.785 ± 0.007	0.682 ± 0.001	63
Experimental $g_{H,\theta}$	0.94_2	0.89	0.80	0.72	
Experimental g_H	0.93_6	0.86	0.74	0.68	

[a]Stockmayer–Fixman. Eq. (38).

Subsequent detailed studies of linear polymers have shown that both constants depend slightly on molecular weight, solvent quality, and drainage effects [93]. Φ' and P' also depend on branching. Rather than describing the dependence of Φ' and P' on the functionality of the star polymers, we will consider the ratios R_H/R_G and R_V/R_G where R_H is the equivalent sphere radius of the intrinsic friction coefficient given by Stokes' law

$$[f] = 6\pi R_H \tag{39}$$

and R_V is the equivalent sphere radius of the intrinsic viscosity defined by Einstein's relation

$$[\eta]M = \left(\frac{10\pi}{3}\right) N_A R_V^3 \tag{40}$$

This approach has the advantage of comparing lengths and has become popular since measurements of the diffusion coefficient and radius of gyration can be made on the same solutions with the same instrumentation [91,94].

The dependence of R_H/R_G and R_V/R_G on functionality of the stars has been studied in detail in good solvents and θ-solvents by several groups [35,40,43,45]. It is worthwhile first to summarize the available experimental ratios for the linear polymer ($f = 2$). In θ-solvents, R_V/R_G is found to be 0.85, 0.84, and 0.83 for polystyrene [40], polybutadiene [35], and polyisoprene [95], respectively, and this value has been confirmed by Zimm [57] and Freire [58] by means of computer simulations. In good solvents, experimental results yield R_V/R_G equal to 0.79 (PS in toluene), 0.84 (PBd in cyclohexane), and 0.79 (PI in cyclohexane) [93]. It is noted that values of R_V/R_G in a good solvent are slightly lower than in the Θ-solvent. These values have been included in Figure 11 at $f = 2$. The experimental ratio R_H/R_G of linear polymers is equal to 0.78 (PS) and 0.74 (PBd) for the Θ-condition and 0.76 (PS, toluene) and 0.72 (PBd, cyclohexane) for the good solvent. Note that these values are considerably higher than the theoretical value 0.65 [8,96]. For the hard sphere R_V/R_G is identical with R_H/R_G and equal to $(5/3)^{1/2}$ [8]. The hard sphere limit is a natural limit for regular stars with many arms. However, this limit may not be attained in practice for stars with high functionality because of their inhomogeneous segment distribution or the adoption by the polymer of a nonspherical shape.

Figure 11 summarizes the dependence of the ratio R_V/R_G on the star functionality. The data are taken from several listings [35,43,45] and individual

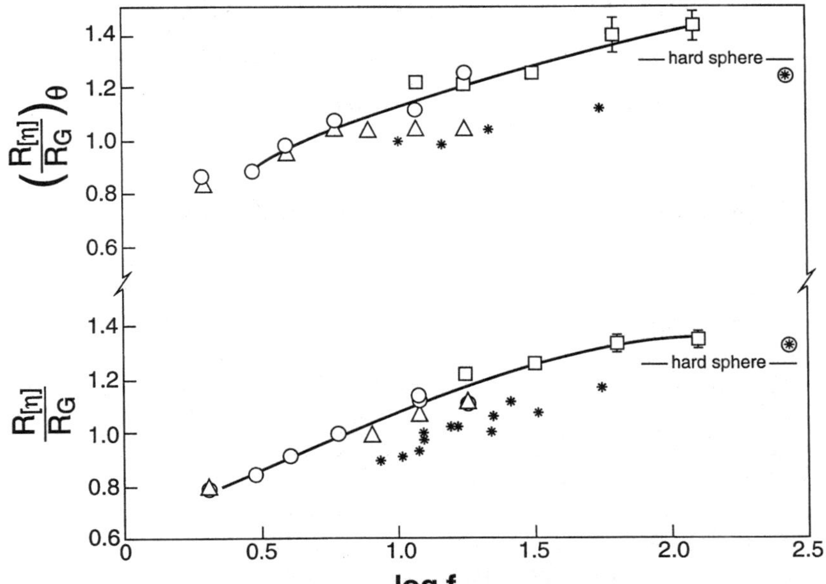

Figure 11 Ratio of viscosimetric radius and radius of gyration as a function of the functionality of the star. Top: data obtained at the Θ-temperature. Bottom: data obtained in good solvents. Symbols: *, divinylbenzene-coupled star polymers—Ref. 43. ⊛, multiarm stars—Ref. 97. Other symbols as in Figure 3.

papers [21,33]. Data for polyisoprenes stars made by the divinylbenzene coupling method [43] and multiarm star polymers with 270 arms [97] are included to show that some, presently unidentified, effects tend to lower the experimental R_V/R_G ratio in special star polymers. The low R_V/R_G values of the divinylbenzene stars may be caused by the larger z-average R_G values of somewhat polydisperse samples. The 270-arm stars are surmised to be aspherical [97].

Figure 11 clearly shows that R_V/R_G increases from less than unity to the hard sphere limit as the number of arms in the star increases. The values of R_V/R_G for the θ case are always slightly higher than for the good solvent. Note also that the ratio R_V/R_G for 64- and 128-arm stars is larger than the hard sphere limit. Plots of R_H/R_G as a function of the star functionality are essentially identical to the R_V/R_G plot in Figure 11. For linear polymers R_V/R_H has been found to vary between 1.13 [40,43] and 1.03 [35] but for stars with many arms $R_H = R_V$ within the limits of experimental error [21,33,40].

IV. CONCENTRATION-DEPENDENT PROPERTIES

A. The Second Virial Coefficient in Good Solvents

The osmotic second virial coefficient is defined by

$$\frac{\pi}{\pi_0} = 1 + A_2M_nc + A_3M_nc^2 \tag{41}$$

or, if the measurement has been made by a scattering experiment,

$$\frac{[Kc/\Delta R]}{[Kc/\Delta R]_0} = 1 + 2A_2M_wc + 3A_3M_wc^2 \tag{42}$$

The determination of the second virial coefficient requires that measurements be made in the correct concentration range. In general, the concentration should be as low as possible, taking into account the necessary accuracy of the measurements. Indeed, in most cases assumptions are made about the third virial coefficient in order to determine A_2. It is customary to derive A_2 in good solvents from the more linear $(\pi/\pi_0)^{1/2}$ versus c plots, which implies that $A_3 = \frac{1}{4}A_2^2M$ or from $(Kc/\Delta R)^{1/2}$ versus c plots, which implies that $A_3 = \frac{1}{3}A_2^2M$ [92]. Recent detailed studies have shown that both relations between the third and second virial coefficients are approximately appropriate for linear polymers [98], but this is not necessarily true for branched polymers.

In this section the dependence of the second virial coefficient on the functionality of regular star polymers in good solvent is considered. Typical results of the second virial coefficient as a function of molecular weight are shown in Figure 12 for different star functionalities. It can be seen that A_2 of stars with

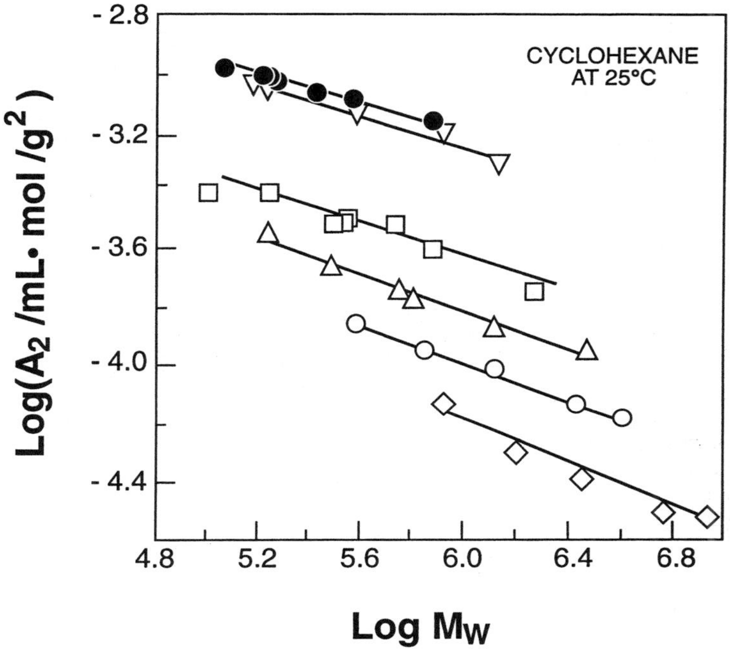

Figure 12 Double-logarithmic plot of the second virial coefficient against molecular weight for polybutadiene stars. From top to bottom: ●, linear; ▽, 4-arm; □, 18-arm; △, 32-arm; ○, 64-arm; and ◇, 128-arm stars. (From Ref. 31.)

low functionality and linear polymers have the same molecular weight dependence. However, stars with high functionality have a stronger (negative) molecular weight dependence. It follows that the ratio

$$g_A = \frac{(A_2)_{\text{star}}}{(A_2)_{\text{lin}}} \tag{43}$$

is a constant for low functionality stars but decreases with increasing molecular weight for high functionality stars. Values of g_A have been collected in Table 9. A range of values is given for 32, 64, and 128-arm polybutadienes. The dependence of g_A on star functionality is shown in a double logarithmic plot in Figure 13.

The large scatter seen in Table 9 and Figure 13 testifies to the experimental challenge of accurately measuring A_2 of star and linear polymers. For $f \geq 8$

Table 9 $g_A = (A_2)_{star}/(A_2)_{lin}$

			Polymer			*Refs.*		
f	*TP*[a]	g_A*[b]	*PS*	*PI*	*PBd*	*PS*	*PI*	*PBd*
3		0.96_8	1.13[c]			37,45		
4	0.84	0.923	0.94 ± 0.03		0.94 ± 0.04	32,72		32
6	0.69	0.816	0.86 ± 0.02			32,38		
8		0.668		0.76 ± 0.1[d]			43	
12	0.43	0.336	0.51;0.70[c]	0.63 ± 0.06[d]		39,45	43	
18	0.31		0.39	0.56 ± 0.03[d]	0.37 ± 0.03	39	43	32
32	0.18_5				0.20–0.25			33
64	0.09_5				0.13–0.15			21
128	0.04_7				0.06_5–0.09_5			21

[a] g_A by two-parameter (TP) theory. Eq. (45)
[b] Renormalization group modification of TP theory. Eq. (46)
[c] Calculated with $(A_2)_{lin} = 7.1 \times 10^{-3}\, M^{-0.242}$ [32].
[d] Calculated with $(A_2)_{lin} = 1.464 \times 10^{-2}\, M_w^{-0.247}$ from light-scattering measurements [95].

$$g_A = 3.89 f^{-0.80} \tag{44}$$

Equation (44) can be compared with the corresponding relation for g_G [Eq. (23)]. The almost identical exponent is fortuitous.

There are two theoretical models for the dependence of g_A on the functionality of the star. The first is the Yamakawa extension of the two-parameter theory (TP) to large values of z (= $k'M^{1/2}$) [8]. Accordingly

$$g_A = \frac{[h_0(\bar{z})]_{star}}{[h_0(\bar{z})]_{lin}} \tag{45}$$

where $h_0(\bar{z})$ in closed form is given by

$$h_0(\bar{z}) = (\tfrac{1}{2}C(f)\bar{z})\ \ln(1 + 2C(f)\bar{z})$$

and $\bar{z} = z/\alpha_s^3$ and $C(f)$ is the coefficient of the first term in the z-expansion of the virial coefficient $A_2 = (N_A n^2 \beta / 2M^2)(1 - C(f)z + \cdots)$. Values of $C(f)$ of regular stars are given by [8]

$$C(f) = \frac{2.133}{f^{1/2}}\,[1.343 + 0.4532(f-1) + 0.1032(f-1)^2]$$

The high molecular weight limit values of g_A according to the two parameter theory are given in Table 9. It can be seen that the TP values of g_A overestimate the effect of functionality on the second virial coefficient, as already noted by Huber [45]. However, TP values of g_A depend on molecular weight ($\bar{z}$), as

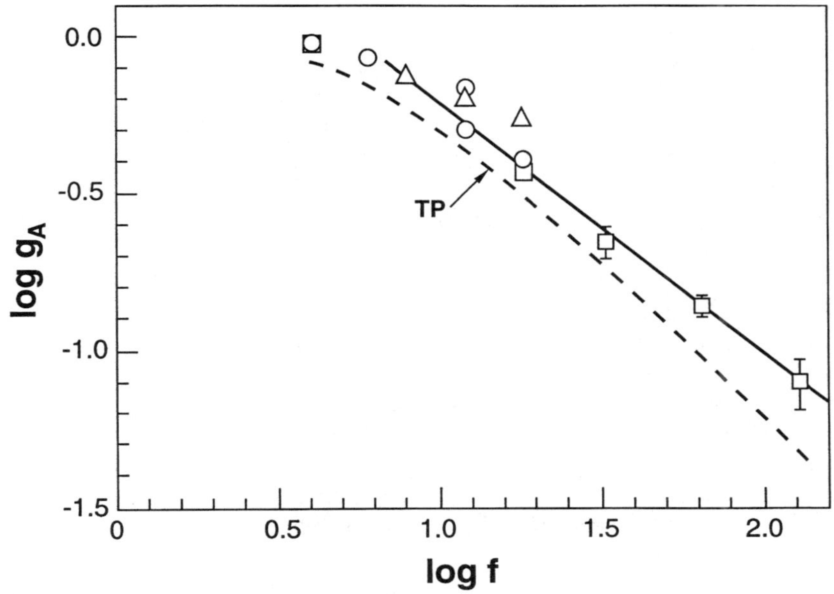

Figure 13 Double-logarithmic plot of $g_A = (A_2)_{star}/(A_2)_{lin}$ against the functionality of the star. The dashed line represents the two-parameter theory: Eq. (45). g_A decreases with molecular weight for 32-, 64-, and 128-arm stars. The error bars indicate the experimentally observed range of g_A values.

shown in Figure IV.7 in [8]. Agreement between the theory and experimental results is in fact better than shown in Table 9 when the molecular weight dependence is taken into account, especially for stars with 12, 18, and 32 arms. For stars with 64 and 128 arms the TP theory underestimates g_A by 25 and 40%, respectively. There are strong parallels between the applicability of the TP theory to g_A and g_G (see Table 3, column 5).

Douglas and Freed introduced renormalization group techniques into the TP theory [32,49]. Accordingly,

$$g_A^* = \frac{1 + a_h(\text{star})}{1 + a_h(\text{linear})} \tag{46}$$

where $a_h = (3/32)C_h + 1/2$. Values of g_A^* compare very well with experimental results for stars with four and six arms but decrease rapidly below the experimental values for stars with $f > 6$.

The effect of branching can also be studied by means of the dependence of the interpenetration function on the functionality of the star:

$$\Psi = \frac{A_2M^2}{(4\pi^{3/2}N_A)R_G^3} \tag{47}$$

Values of A_2 and R_G are obtained simultaneously from (light) scattering measurements, and data on the linear parent polymer are not needed. In general, Ψ is molecular weight dependent. Because mostly high molecular weight light-scattering data are available we will assume that we are dealing with the high molecular weight asymptotic values of Ψ. In good solvents $\Psi_{\text{linear}} = 0.26$ [8,49]. Values of Ψ as a function of the functionality of the regular star are given in Table 10. Uncertainties in the measured values of A_2, M, and R_G have erased detection of any possible dependence of Ψ on the molecular weight of stars at constant functionality. Therefore, only average values of Ψ are quoted in Table 10. It can be seen that Ψ increases with f. It is somewhat surprising that the hard sphere value $\Psi = 1.61$ is exceeded by the 64- and 128-arm stars. This is probably due to the nonuniformity of the segment density distribution in the stars, as was noted previously for the ratios R_V/R_G and R_H/R_G. If the segment density is higher in the center, R_G will be relatively smaller than the overall radius and if A_2M^2 is proportional to the overall volume of the star, then Ψ may exceed the hard-sphere limit.

The second virial coefficient has also been expressed in terms of the radius of the thermodynamic equivalent sphere:

$$R_T = \left(\frac{3}{16\pi N_A} \text{A}_2M^2\right)^{1/3} \tag{48}$$

Table 10 $\Psi = A_2M^2/4\pi^{3/2}N_A\text{R}_G^3$

	Polymer			*Refs.*		
f	*PS*	*PI*	*PBd*	*PS*	*PI*	*PBd*
3	0.42,0.35			37,45		
4	0.53		0.45	32,72		32
6	0.70			32,38		
8		0.77			43	
12	0.62	1.04		37	43	
18	1.06,1.08	1.51	1.20	37,39	43	32
32			1.6			33
64			1.9			21
128			2.1			21

It can be shown from Eqs. 47 and 48 that

$$\frac{R_T}{R_G} = \left[\frac{3\pi^{1/2}}{4}\Psi\right]^{1/3} \tag{49}$$

Therefore, the dependence of R_T/R_G and Ψ on the functionality of the star reveals the same properties and what has been said about Ψ applies equally to R_T/R_G.

Recently, some measurements of the third virial coefficients of linear and star polymers have been published. In the case of linear polymers the ratio A_3/A_2M^2 has been found to increase from about 0.25 to 0.45 with increasing molecular weight [98,99]. However, others claim that A_3/A_2M^2 is independent of molecular weight [100]. Only a few experimental results on star polymers are available. For one high molecular weight 128-arm polybutadiene star good experimental evidence gives $A_3/A_2M^2 = 0.61 \pm 0.02$ in good agreement with the hard-sphere value (5/8) [101]. This result is supported by less extensive measurements on one 32- and 64-arm star [101]. Burchard indicated that $A_3/A_2M^2 = 0.54$ for a 12-arm star [102]. At present it appears that A_3/A_2M^2 increases with f from the linear to the hard-sphere limit, but more extensive data are needed to better describe the dependence.

B. The $A_2 = 0$ Temperature: θ_{A_2}

For star polymers the temperature at which $A_2 = 0$ is lower than the Flory θ-temperature of the linear polymer and depends on the molecular weight and the functionality [103]. Several studies have confirmed this phenomenon for polystyrene stars [37,38] and polyisoprene stars [41–43]. All studies have assumed that the lower θ_{A_2} temperature of the star polymer is exclusively due to their branched structure and is not affected by the increased weight fraction of foreign chemical groups present in the multiple chain ends or in the central coupling unit of the star. There is experimental evidence that such groups affect the observable θ_{A_2} [104]. Furthermore, in the discussion it should be kept in mind that determination of θ_{A_2} usually has a 1 to 2 K accuracy.

Decreases of 10 to 20 K in θ_{A_2} from the Flory θ-temperature have been observed in polystyrene stars in cyclohexane. In the case of polyisoprene stars, the θ depressions in dioxane are small and can only be observed with 8-arm, 12-arm, and 18-arm stars. The θ_{A_2} depression depends also on the type of θ-solvent used [41]. Polybutadiene stars in dioxane have very small θ_{A_2} depressions. The largest value, $\theta_{A_2} - \theta = -4$ K has been observed for low molecular weight 64- and 128-arm stars [21]. Boothroyd observed a 12 K lower θ_{A_2} for a polyethylene star in biphenyl-d_{10}. [52].

A compilation of available results for polystyrene stars is shown in Figure 14. Figure 14 includes results on a three-arm star with MW = 6500 and on a 15-arm star with MW = 3.10^6. It can be seen that $\theta - \theta_{A_2}$ increases with decreasing MW and increasing *f*. Results on polyisoprene stars with 8, 12, and 18 arms lead to a similar qualitative conclusion.

Candau, Rempp, and Benoit (CRB) have developed a theory that ascribes the θ_{A_2} temperature lowering in branched polymers to increased three-body contacts in polymers with increased segment density [51]. The effects of ternary contacts are balanced by increased attractive binary contacts at a lower temperature than the Flory θ-temperature.

Approximately

$$\frac{\theta}{\theta_{A_2}} - 1 = \frac{3^{3/2}}{2^4} \frac{C'_M}{C_M \psi_1} \frac{1}{M^{1/2} \alpha^3 g^{3/2}} H(\mathrm{K}') \tag{50}$$

where C_M and C'_M contain polymer-dependent constants. Moreover C'_M depends on $(1/3 - \chi_2)$, which describes the strength of the ternary interactions [see Eq. (25)]. Bauer has argued that setting $H(\mathrm{K}') = 1$ leads to results that are closer to the exact solution [42]. It is further assumed that the value of ψ_1, derived from $dA_2/d(1 - \theta/T)$ at θ or at θ_{A_2}, is independent of branching. This has been shown for 4- and 6-arm polystyrene stars [38] and for 8- and 12-arm polyisoprenes [42]. Evaluation of the experimental θ_{A_2} data by means of Eq. (50) leads to $(1/3 - \chi_2) \approx 0.045$ for polystyrene stars in cyclohexane (see Figure 14). Deviations are observed at very low and very high molecular weights. For polyisoprene in dioxane, values of $(1/3 - \chi_2)$ are of the same order of magnitude but seem to increase slightly with *f* from $f = 4$ to $f = 18$ [38].

Equation (50) is not expected to be applicable to star polymers with $f \geq 12$ because higher than ternary contacts become increasingly important. Furthermore, the temperature dependence of A_2 near θ_{A_2} differs [21] from that found in linear polymers, and ψ_1 can no longer be assumed constant.

Ganazzoli and Allegra proposed that the θ-temperature depression is of the general form

$$\theta(N) = \theta_\infty - \phi\chi(f)\, N^{-1/2}$$

with

$$\chi(f) = f^{1/2}\, [f(2 - 2^{1/2}) + 2^{1/2} - 4]$$

and ϕ contains polymer-specific constants [105]. They have criticized the CRB approach because it considers a star polymer as a constant-density cloud of segments with no connectivity. Consequently, the three-body interactions decrease

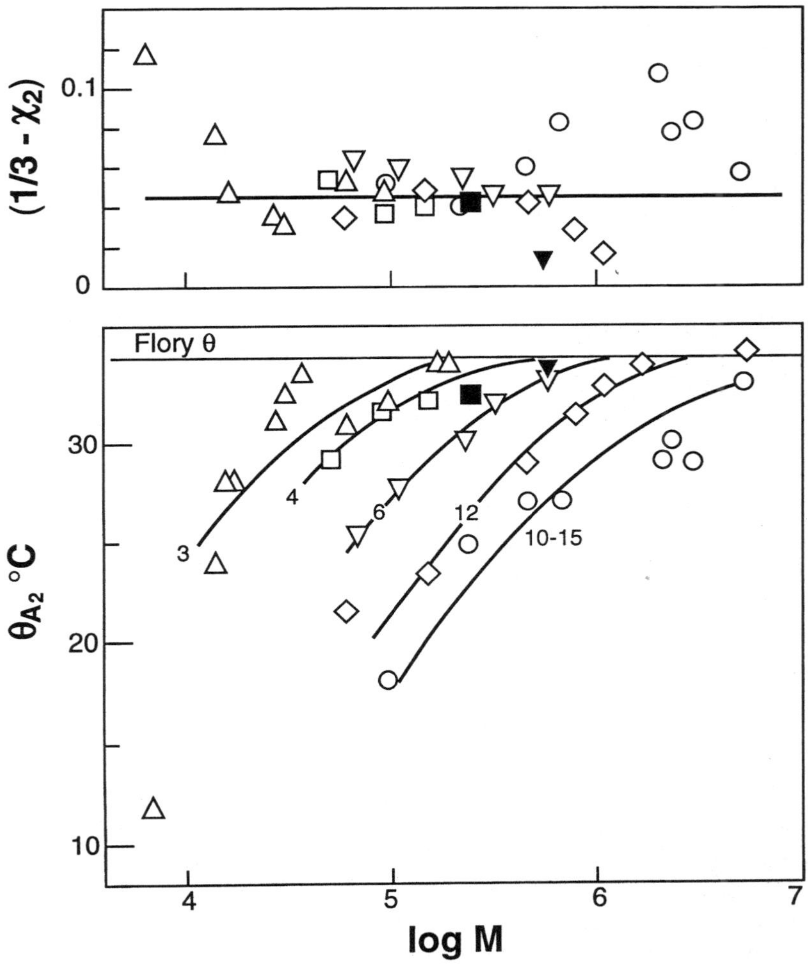

Figure 14 θ_{A_2}-temperature depression. Bottom: experimental values of θ_{A_2} plotted against molecular weight for polystyrene stars in cyclohexane. The curves for each functionality are a guide to the eye. Top: calculated values of $1/3 - \chi_2$. See Eq. (50) in text. Symbols: △, 3-arm stars [37]; □, 4-arm stars [38]; ■, 4-arm star quoted in [51]; ▽, 6-arm stars [38]; ▼, 6-arm star quoted in [51]; ◇, 12-arm stars [37]; and ○, divinylbenzene stars quoted in [51].

with increasing chain length in the CRB theory rather than being parallel with the two-body interactions.

C. The Huggins Coefficient

The concentration dependence of the viscosity of a polymer solution can be expressed by a virial-like expression

$$\eta_{sp} = \frac{\eta}{\eta_s} - 1 = [\eta]c + k_2c^2 + k_3c^3 + \cdots \tag{51}$$

where the intrinsic viscosity is the first virial coefficient. Huggins observed that $k_2/[\eta]^2 = k_H$ is approximately independent of the molecular weight of the polymer. In the case of flexible polymers small increases of k_H are observed at low molecular weight. However, the Huggins coefficient depends on the solvent quality and on branching.

It should be mentioned that intrinsic viscosities and Huggins coefficients are obtained from viscosity measurements with low shear rate (or capillary) viscometers on polymer solutions for which $0.1 < \eta_{sp} < 0.5$. In the case of linear polymers and stars with a small number of arms a plot of η_{sp}/c against c or a plot of $(\ln \eta_{rel})/c$ against c yields straight lines according to

$$\frac{\eta_{sp}}{c} = [\eta] + k_H[\eta]^2c \tag{52}$$

and

$$\frac{\ln \eta_{rel}}{c} = [\eta] + k_K[\eta]^2c \tag{53}$$

where k_K is called the Kraemer constant. However, for stars polymers with high functionality of the branch point, Eq. (52) is insufficient and a third term must be added to the right-hand side to account for upward curvature [33]. Equation (53), however, is still appropriate. The procedure for data analysis is then to derive $[\eta]$ and k_K with Eq. (53). The intrinsic viscosity is then used to analyze the data according to

$$\frac{(\eta_{sp}/c) - [\eta]}{c} = k_H[\eta]^2 + k'_H[\eta]^3c \tag{54}$$

from which k_H and k'_H are obtained. See Figure 15. The internal consistency of the procedure is checked against the requirement that $k_H - k_K = 0.5$. Curvature in η_{sp}/c versus c plots is also observed in 64- and 128-arm star data obtained in a θ-solvent [21].

Values of k_H for different functionalities of regular stars have been col-

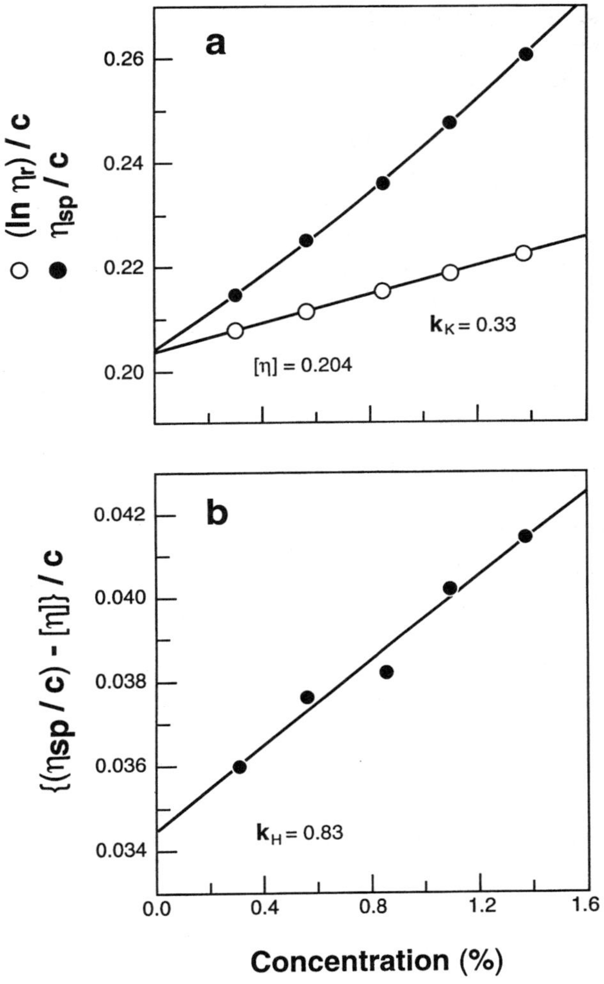

Figure 15 Evaluation of intrinsic viscosity and Huggins coefficient for 128-arm star polybutadiene in dioxane at 26.5°C. Top: application of Eq. (52) and (53). Bottom: application of Eq. (54).

lected in Table 11a and 11b for the θ-solvent and good solvent case, respectively. For linear polymers k_H varies between 0.5 and 0.6 in θ-solvents and $k_H \approx 0.3$ in good solvents. The problems involved in the theoretical calculation of k_H for a flexible polymer have been outlined [8]. The values of k_H increase with increasing f from the value for linear polymers and reach values slightly less than unity for 64- and 128-arm stars. The high f limit appears identical in θ- and good solvents. This limit is quite near the best theoretical result for the case of the hard sphere (0.99) [106]. Values of k'_H show considerable scatter due to the high degree of experimental accuracy required when using Eq. (54). In good solvents, $k'_H = 0.25 \pm 0.2, 0.6 \pm 0.1, 0.8 \pm 0.2$, and 0.9 ± 0.2 for 18-, 32-, 64- and 128-arm polybutadienes. These values are probably independent of molecular weight. This is in agreement with results that show that η_{rel} scales with the product $[\eta]c$ over a wide range of concentrations from the dilute to the semidilute [107]. Under θ conditions k'_H is equal to 0.8 ± 0.2 for 64- and 128-arm stars.

D. Concentration Dependence of the Friction Coefficient

The concentration dependence of the translational friction coefficient is given by

$$f = f_0 (1 + k_s c) \tag{55}$$

Experimentally, k_s is determined from the concentration dependence of the sedimentation coefficient according to

$$\frac{1}{S^0} = \frac{1}{S_0^0} (1 + k_s\, c_{av})$$

or, more recently, from the concentration dependence of the diffusion coefficient measured by dynamic light scattering

$$D = D_0 (1 + k_D c)$$

where, following Yamakawa [8], k_D contains a thermodynamic and a friction term

$$k_D = 2A_2M - k_s - \bar{v}_p$$

Because k_s depends on molecular weight, it is customary to evaluate k_0 in

$$k_s = k_0 \left[\left(\frac{4\pi}{3} \right) N_A \left(\frac{R_H^{\,3}}{\mathrm{M}} \right) \right] \tag{56}$$

which, as the Huggins coefficient of intrinsic viscosity, is independent of

Table 11 Huggins Coefficients

a. θ-solvent

	Polymer			Refs.		
f	*Polystyrene*	*Polyisoprene*	*Polybutadiene*	*PS*	*PI*	*PBd*
2	0.55	(0.63)	0.60		110	109
3						
4	0.59	0.57	0.63	109	110	109
6	0.63	0.65		109	110	
12	0.37–1.68[a]			108		
18			0.90			109
32			0.83			33
64			0.83			21
128			0.90			21

b. Good Solvents

	Polymer							
		Polyisoprene		Polybutadiene		Refs.		
f	*Polystyrene*	*Toluene*	C_6H_{12}	*Toluene*	C_6H_{12}	*PS*	*PI*	*PBd*
2	0.34	0.27	0.34	0.36		109	95	109
			0.32				111	
3			0.36				43	
4	0.41	0.34	0.36	0.41		109	43	109
6	0.48	0.37				109		
8		0.52	0.60				43	
12		0.69	0.63				43	
18		0.74	0.77	0.74	0.73		43	109
32				0.77	0.77			33
64				0.92	0.91			21
128				0.86	0.94			21

[a]At the θ_{A_2} = temperature of each sample; k_H decreases with increasing molecular weight.

molecular weight but depends on the solvent quality and possibly also on branching.

Values of k_0 obtained on linear polymers and on regular star polymers have been collected in Table 12. The polystyrene results are based on sedimentation experiments. The polybutadiene results are from dynamic light scat-

Table 12 Dependence of k_0 in Eq. (56) on Functionality

	θ-solvent		*Good solvent*		*Refs.*	
f	*PS*	*PBd*	*PS*	*PBd*	*PS*	*PBd*
2	2.1 ± 0.5	2.0 ± 0.4	4.7 ± 0.5	5.0 ± 0.6	108	35
4	2.6 ± 1		5.5 ± 0.5		108	
6	3.2 ± 0.4[a]		6.9[a]		108	
18		1.7 ± 0.4		7.2 ± 2		35
32		2.0 ± 0.5		7.3 ± 1		33
64		3.2 ± 1.5		9.0 ± 1.5		21
128		2.5 ± 1		9.0 ± 2		21

[a]Two samples.

tering. Except for the six-arm stars all values are the average of five or more samples. The error margin encompasses all results.

The values of k_0 at the θ-temperature do not vary in a systematic way with the functionality of the star and are within experimental error equal to k_0 for linear polymer. Pyun and Fixman [112] calculate $k_0 = 2.23$ at the θ-temperature for linear polymers, in good agreement with experimental results. In good solvents, values of k_0 increase from the value of linear polymers (≈5) to about $k_0 = 9$ for 64- and 128-arm stars. The latter is slightly higher than the hard sphere value $k_0 = 7.16$ [112]. This has also been observed for some highly branched comb polymers [97]. The increase in k_0 with increasing arm functionality is indicative of the decreased ability of the spheres to interpenetrate each other.

V. TIME-DEPENDENT PROPERTIES

We have dealt with the long-time hydrodynamic properties of star polymers in dilute solution in Sections III. A and III.B. The intrinsic viscosity is the sum of products of moduli and relaxation times. In term of frequency (time^{-1})

$$[\eta] = \lim_{\omega \to 0} \frac{[G'']}{\omega \eta_s}$$

where ω is the radian frequency, η_s is the solvent viscosity and

$$[G''] = \lim_{c \to 0} \frac{G'' - \omega \eta_s}{c}$$

with G'' the loss part of the complex modulus [11,12]. At low frequency $[G'']$ is a linear function of ω and [η] is the zero-shear intrinsic viscosity. However, at high frequency (or short times) deviations from linearity are observed and $[\eta]$ decreases with increasing frequency. For comparison with theoretical models, the reduced moduli

$$[G'']_R = \frac{[G'']M}{RT}$$

$$[G']_R = \frac{[G']M}{RT}$$

are evaluated.

Ferry, Schrag, and their collaborators have compared the frequency dependence of the moduli of dilute solutions of star and linear polymers after extrapolation to zero concentration [113,114]. Values of $[G'']_R$ and $[G']_R$ are plotted double logarithmically against $\omega\tau_{01}$. The longest relaxation time, τ_{01}, is given by $\eta_s[\eta]M/RTS_1$ and S_1 for star polymers is given by

$$S_1 = (f-1)\sum_{p,\mathrm{odd}} \frac{\tau_{0p}}{\tau_{01}} + \sum_{p,\mathrm{even}} \frac{\tau_{0p}}{\tau_{01}} \tag{57}$$

Equation (57) shows the same degenerate form of relaxation times as Eq. (30). The characteristic relaxation times are derived from the exact Zimm–Kilb eigenvalues [11,81]. The matching of the experimental and theoretical dependence of the moduli on frequency has only one adjustable parameter, i.e., the hydrodynamic interaction parameter, h^*. The free-draining case was evaluated by Ham [79].

An example of the moduli-frequency dependence for the case of a nine-arm star polystyrene is shown in Figure 16. For comparison $\omega\tau_{01}$ has been multiplied by S_1 to realize identical values of $[G'']_\mathrm{R}$ for linear and star polymers at low frequency. It can be seen that star polymers have relatively higher values of $[G'']_R$ and $[G']_R$ than linear polymers at high frequency. Also, the transition from low to high frequency occurs over a narrower frequency range. The low value of $[G']$ of the star at low frequency compared to the linear polymer is to be noted. Unfortunately, in order to fit the θ-temperature data an unrealistically high hydrodynamic interaction parameter $h^* = 0.40$ had to be used. This problem does not occur in good solvents.

The limiting high-frequency behavior, several decades higher in frequency than those shown in Figure 16, has also been studied [115]. A second Newtonian high frequency viscosity can be observed in the polymer solution. This secondary Newtonian viscosity is about two to eight times smaller than the low-frequency (zero-shear) viscosity.

The frequency-dependent response of star polymer solutions has also been studied by oscillatory flow birefringence [116]. The magnitude and the phase angle of the complex mechano-optic factor are obtained. Theoretical cal-

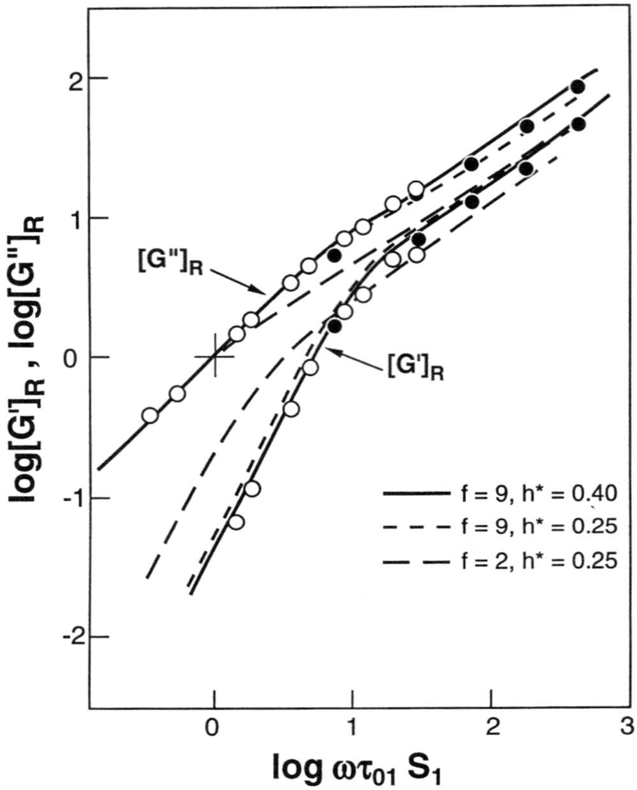

Figure 16 Logarithmic plot of the reduced intrinsic shear moduli against reduced frequency for a nine-arm star polymer in two θ-solvents. Symbols: open circles, decalin; full circles, dioctylphthalate. Curves: exact eigenvalue evaluation of Zimm–Kilb theory with $f = 9$, $N_b = 100$ and $h^* = 0.40$ and 0.25. For comparison, the moduli curves of a linear polymer $f = 2$, $N_b = 100$, and $h^* = 0.25$ are also shown. (From Ref. 114.)

culations are based on the same Zimm–Kilb set of relaxation times. It is gratifying to see that the experiments are able to identify qualitatively the presence of linear and three-arm material in a four-arm star polystyrene [116]. Extensive model calculations have been done recently on star and other branched polymers [117,118]. These are required in the process of matching theoretical with experimental data and to enhance the analytical capabilities of frequency-dependent properties [119].

The relaxation dynamics of star polymers has been studied most successfully by means of dynamic scattering experiments. The QR range in light scattering is too small, and deviations from translational motion due to internal motions have not been observable in star polymers solutions. Calculations of the first Q-dependent coefficient

$$\frac{\Gamma}{Q^2} = D_0\,(1 + C \cdot Q^2 R_G^2 + \cdots)$$

have been made for stars [91]. Quasi-elastic neutron scattering (QENS) allows probing the internal dynamics of polymers because with cold neutrons the accessible momentum and energy transfer correspond to the intramolecular distances and relaxation times of the polymer. At each scattering angle, corresponding to a segment distance, $QR_G << R_G$, an initial relaxation rate is obtained. The dependence of the reduced relaxation rate Γ/Q^3 is shown as a function of QR_G in Figure 17. The reduced rate is expected to be a constant at high QR_G in agreement with the Zimm internal relaxation modes [47,67,120]. At $QR \leq 1$, Γ/Q^3 is expected to increase with decreasing Q due to the observa-

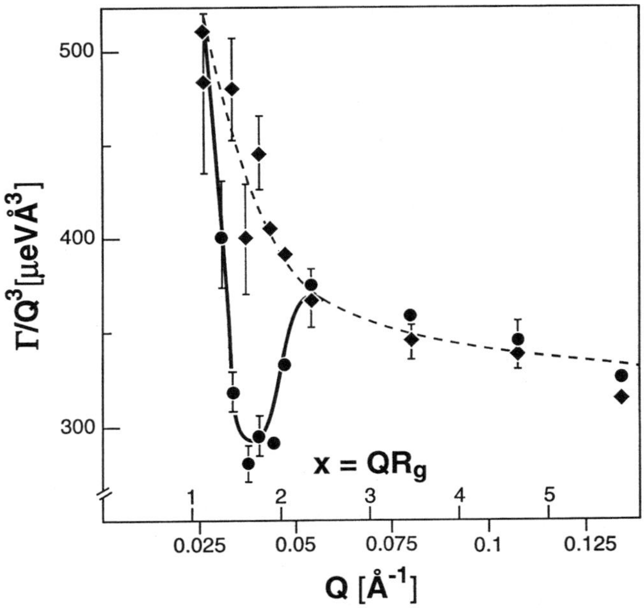

Figure 17 Reduced relaxation rates Γ/Q^3 for a star polymer (•) and for a one-arm labeled star (♦) as a function of Q. The curves are guides to the eye. The position of the minimum in the relaxation rate of the whole star corresponds to the position of the maximum in the Kratky plot of the form factor. (From Ref. 67.)

tion of the Q^2-dependent translational diffusion process. This can be seen in Figure 17 in the case of the single arm of a star. The Q-dependence of Γ/Q^3 of the single arm resembles that of a linear polymer or four-arm star [120] although the transition from Q^2 to Q^3 dependence of Γ is broader for the single arm [67]. In the case of the star polymer Γ/Q^3 versus Q is characterized by a sharp minimum at about $QR_G \approx 1.5$. It was established that the position of the minimum corresponds with the position of the maximum in the Kratky plot of static scattering [47,67,120] (see Figure 4). The slowing down of relaxation processes for the highly populated segment–segment distances of the order of R_G is due to interarm processes, as shown by the absence of such slowing down in the single arm. A quantitative analysis of the minimum in Figure 17 has been given [47]. It is to be noted that at each Q only initial rates of relaxation are presently measured because the experimental measurement time $\leq$ 16 ns. In some cases, the relaxation process does not correspond to a single exponential and slowing down of the relaxation at longer times is observed [47]. This has been tentatively assigned to the effect of arm–arm entanglement, which has been suggested earlier on the basis of molecular dynamics simulations of stars with many arms [69]. However, arm–arm entanglements are presently not incorporated in the blob model of star polymers.

Guenza and Perico have used their optimized Rouse–Zimm local dynamics model to show that the observed slowing down originates with the chain segments near the center of the star. Segments far from the star center have relaxation rates identical to those of linear polymers [121–123]. Recently, Carignano and Alessandrini have also reproduced the slowing down of relaxation rates by means of renormalization group methods [124].

VI. CONCLUSIONS

The static dimensions of regular star polymers with few arms can be well described by the Gaussian chain model in θ-solvents and by classic chain expansion in good solvents. The segment density distribution and the overall static dimensions of high functionality polymers in good solvents is in essential agreement with the scaling model. The experimental results have shown that the outer blob of swollen arms is of the size of $R_G f^{-1/2}$ and accounts for the bulk of the star polymer mass. The segment density distribution and dimensions in the θ-condition have not received sufficient attention and are therefore not known in great detail. Modification of the scaling model by replacing $\nu = 3/5$ to $\nu = 1/2$ is clearly unsatisfactory. In the θ-solvent three-body interactions play a major role. Their incorporation into a two-parameter theory explains some available data, e.g. on the θ-temperature depression. However, when f is large, increased deviations are expected and are not yet fully accounted for.

It has been clearly shown that the hydrodynamic properties of star polymers cannot be compared with those of linear polymers within the Kirkwood–Riseman model with preaveraged hydrodynamic interactions. In order to match calculated with experimental results, non-preaveraging of hydrodynamic interactions becomes increasingly more important as f increases. It is interesting to note that non-preaveraging increases g_H but decreases $g_{[\eta]}$ from the calculated preaveraged values.

Many star properties, especially when $f \geq 18$, resemble the properties of the constant-density hard sphere with the same radius of gyration. This comparison can only be made for the zero concentration limit and in very dilute solution because the star polymers have a soft segment density near the surface that allows for mutual interpenetration.

In stars with very small arms the mass and volume of the central coupling unit may have to be taken into account. A method has been proposed [126]. Also, the polymer chain segments close to the central coupling unit may have conformations quite different from the random walk and thereby strongly affect properties of stars with small arms. Therefore, star polymers prepared with polydivinylbenzene cores, starlike comb polymers, and polymacromonomers are expected to have properties slightly different from the regular stars described in this review. It will also be interesting to compare the properties of the regular stars with those of dendrimers and arborescent polymers when these become available.

ACKNOWLEDGMENT

The author thanks Dr. G. Grest and Dr. C. Vlahos for comments on parts of the early draft.

REFERENCES

1. Y. Gallot, M. Leng, and H. Benoit, *J. Chim. Phys.*, **59,** 1093 (1962).
2. A. Halperin, *Macromolecules,* **20,** 2943 (1987); **22,** 3806 (1989).
3. B. J. Bauer and L. J. Fetters, *Rubber Chem. Technol.,* **51,** 406 (1978).
4. S. Bywater, *Adv. Polym. Sci.,* **30,** 90 (1979).
5. P. A. Small, *Adv. Polym. Sci.,* **18,** 1 (1975).
6. J. Roovers, *Encylopedia of Polym. Sci. Eng.,* 2nd ed, Vol. 2. John Wiley, New York, 1983, p. 478.
7. J. Roovers, *Polymeric Materials Encyclopedia,* Vol. 1. J. C. Salamone, ed. CRC Press, Baton Rouge, 1996, p. 850.
8. H. Yamakawa, *Modern Theory of Polymer Solutions,* Harper and Row, New York, 1971.

9. P. G. de Gennes, *Scaling Concepts in Polymer Physics,* Cornell University Press, Ithaca, N.Y., 1979.
10. W. Burchard, *Adv. Polym. Sci.,* **48,** 1 (1983).
11. K. Osaki, *Adv. Polym. Sci.,* **12,** 1 (1973).
12. J. D. Ferry, *Viscoelastic Properties of Polymers,* 3rd ed. John Wiley, New York, 1980.
13. H. Kramers, *J. Chem. Phys.,* **14,** 415 (1946).
14. B. H. Zimm and W. H. Stockmayer, *J. Chem. Phys.,* **17,** 1301 (1949).
15. W. H. Stockmayer and M. Fixman, *Annals N.Y. Acad. Sci.,* **57,** 334 (1953).
16. W. Burchard, *Macromolecules,* **7,** 835 (1974).
17. M. Daoud and J. P. Cotton, *J. Phys.* (*Les Ulis*), **43,** 531 (1982).
18. T. M. Birshtein and E. B. Zhulina, *Polymer,* **25,** 1453 (1984).
19. T. M. Birshtein, E. B. Zhulina, and O. V. Borisov, *Polymer,* **27,** 1078 (1986).
20. L. J. M. Vagberg, K. A. Cogan, and A. P. Gast, *Macromolecules,* **24,** 1670 (1991).
21. J. Roovers, L.-L. Zhou, P. M. Toporowski, M. van der Zwan, H. Iatrou, and N. Hadjichristidis, *Macromolecules,* **26,** 4324 (1993).
22. H. Benoit, *J. Polym. Sci.,* **11,** 507 (1953).
23. Yu Ya Kol'bovskii, *Polym. Sci. USSR,* **3,** 1, (1962); **3,** 326 (1962).
24. H. Benoit and G. Hadziioannou, *Macromolecules,* **21,** 1449 (1988).
25. J. L. Alessandrini and M. A. Carignano, *Macromolecules,* **25,** 1157 (1992).
26. P. Outer, C. I. Carr, and B. H. Zimm, *J. Chem. Phys.,* **18,** 1086 (1950).
27. G. C. Berry, *J. Chem. Phys.,* **44,** 4550 (1966).
28. R. Ullman, *J. Polym. Sci.: Polym. Phys. Ed.,* **23,** 1477 (1985).
29. P. M. Toporowski and J. Roovers, *Macromolecules,* **11,** 365 (1978).
30. W. Burchard, *Macromolecules,* **10,** 919 (1977).
31. L. Willner, O. Jucknischke, D. Richter, J. Roovers, L.-L. Zhou, P. M. Toporowski, L. J. Fetters, J. S. Huang, M.-Y. Lin, and N. Hadjichristidis, *Macromolecules,* **27,** 3821 (1994).
32. J. F. Douglas, J. Roovers and K. F. Freed, Macromolecules, **23,** 4168 (1990).
33. L.-L. Zhou, N. Hadjichristidis, P. M. Toporowski, and J. Roovers, *Rubber Chem. Technol.,* **65,** 303 (1992).
34. P. M. Toporowski and J. Roovers, *J. Polym. Sci.: Part A: Polym. Chem.,* **24,** 3009 (1986).
35. J. Roovers and J. E. Martin, *J. Polym. Sci.: Part B: Polym. Phys.,* **27,** 2513 (1989).
36. O. Jucknischke, Ph. D. Thesis, Munster University, Germany (1995).
37. N. Khasat, R. W. Pennisi, N. Hadjichristidis, and L. J. Fetters, *Macromolecules,* **21,** 1100 (1988).
38. J. Roovers and S. Bywater, *Macromolecules,* **7,** 443 (1974).
39. J. Roovers, N. Hadjichristidis, and L. J. Fetters, *Macromolecules,* **16,** 214 (1983).
40. J. Roovers and P. M. Toporowski, *J. Polym. Sci.: Polym. Phys. Ed.,* **18,** 1907 (1980).
41. N. Hadjichristidis and J. Roovers, *J. Polym. Sci.: Polym. Phys. Ed.,* **12,** 2521 (1974).

42. B. J. Bauer, N. Hadjichristidis, L. J. Fetters, and J. Roovers, *J. Am. Chem. Soc.,* **102,** 2410 (1980).
43. B. J. Bauer, L. J. Fetters, W. W. Graessley, N. Hadjichristidis, and G. F. Quack, *Macromolecules,* **22,** 2337 (1989).
44. G. S. Grest, K. Kremer, and T. A. Witten, *Macromolecules,* **20,** 1376 (1987).
45. K. Huber, W. Burchard, and L. J. Fetters, *Macromolecules,* **17,** 541 (1984).
46. K. Huber, W. Burchard, S. Bantle, and L. J. Fetters, *Polymer,* **28,** 1990 (1987); **28,** 1997 (1987).
47. D. Richter, B. Farago, L. J. Fetters, J. S. Huang, and B. Ewen, *Macromolecules,* **23,** 1845 (1990).
48. A. Miyake and K. F. Freed, *Macromolecules,* **17,** 678 (1984).
49. J. F. Douglas and K. F. Freed, *Macromolecules,* **17,** 2344 (1984).
50. C. H. Vlahos and M. K. Kosmas, *Polymer,* **25,** 1607 (1984).
51. F. Candau, P. Rempp, and H. Benoit, *Macromolecules,* **5,** 627 (1972).
52. A. T. Boothroyd, G. L. Squires, A. R. Rennie, J. C. Horton, and A. M. B. G. de Vallêra, *Macromolecules,* **22,** 3130 (1989); A. T. Boothroyd, *Europhys. Lett.,* **9,** 237 (1989).
53. A. T. Boothroyd and L. J. Fetters, *Macromolecules,* **24,** 5215 (1991).
54. J. Batoulis and K. Kremer, *Europhys. Lett.,* **7,** 683 (1988).
55. F. Ganazzoli, *Macromolecules,* **25,** 7357 (1992).
56. J. Mazur and F. McCrackin, *Macromolecules,* **10,** 326 (1977).
57. B. H. Zimm, *Macromolecules,* **17,** 795, (1984); **17,** 2441 (1984).
58. J. J. Freire, J. Pla, A. Rey, and R. Prats, *Macromolecules,* **19,** 452 (1986).
59. A. Kolinski and A. Sikorski, *J. Polym. Sci.: Polym. Chem. Ed.,* **20,** 3147 (1982).
60. J. Batoulis and K. Kremer, *Europhys. Lett.,* **7,** 683 (1988).
61. G. Grest, *Macromolecules,* **27,** 3493 (1994).
62. S. G. Whittington, J. E. G. Lipson, M. K. Wilkinson, and D. S. Gaunt, *Macromolecules,* **19,** 1241 (1986).
63. A. Rey, J. J. Freire, and J. Garcia de la Torre, *Macromolecules,* **20,** 342 (1987).
64. A. J. Barrett and D. L. Tremain, *Macromolecules,* **20,** 1687 (1987).
65. J. Batoulis and K. Kremer, *Macromolecules,* **22,** 4277 (1989).
66. J. C. Horton, G. L. Squires, A. T. Boothroyd, L. J. Fetters, A. R. Rennie, C. J. Glinka, and R. P. Robinson, *Macromolecules,* **22,** 681 (1989).
67. D. Richter, B. Farago, J. S. Huang, L. J. Fetters, B. Ewen, *Macromolecules,* **22,** 468 (1989).
68. W. D. Dozier, J. S. Huang, and L. J. Fetters, *Macromolecules,* **24,** 2810 (1991).
69. G. S. Grest, K. Kremer, S. Milner, and T. A. Witten, *Macromolecules,* **22,** 1904 (1989).
70. J. L. Alessandrini and M. A. Carignano, *Macromolecules,* **25,** 1157 (1992).
71. M. Adam and D. Lairez, *Fractals,* **1,** 149 (1993).
72. J. Roovers and S. Bywater, *Macromolecules,* **5,** 384 (1972).
73. J. Herz, M. Hert, and C. Strazielle, *Makromol. Chem.,* **160,** 213 (1972).
74. N. Hadjichristidis and J. Roovers, *Polymer,* **26,** 1087 (1985).

75. J. Roovers, *Polymer,* **26,** 1091 (1985).
76. J. Roovers, *Macromolecules,* **20,** 148 (1987).
77. C. Strazielle and J. Herz, *Eur. Polym. J.,* **13,** 223 (1977).
78. B. H. Zimm and R. W. Kilb, *J. Polym. Sci.,* **37,** 19 (1959).
79. J. S. Ham, *J. Chem. Phys.,* **26,** 625 (1957).
80. J. R. Schaefgen and P. J. Flory, *J. Am. Chem. Soc.,* **70,** 2709 (1948).
81. K. Osaki and J. L. Scrag, *J. Polym. Sci.: Polym. Phys. Ed.,* **11,** 549 (1973).
82. B. H. Zimm, *Macromolecules,* **17,** 795, 2441 (1984).
83. R. Prats, J. Pla, and J. J. Freire, *Macromolecules,* **16,** 1701 (1984).
84. J. J. Freire, A. Rey, and J. Garcia de la Torre, *Macromolecules,* **17,** 1815 (1984).
85. J. J. Freire, A. Rey, and J. Garcia de la Torre, *Macromolecules,* **19,** 457 (1986).
86. A. Rey, J. J. Freire, and J. Garcia de la Torre, *Macromolecules,* **23,** 3948 (1990).
87. A. Rey, J. J. Freire, M. Bishop, and J. H. R. Clarke, *Macromolecules,* **25,** 1311 (1992).
88. M. K. Wilkinson, D. S. Gaunt, J. E. G. Lipson, and S. G. Whittington, *Macromolecules,* **21,** 1818 (1988).
89. J. F. Douglas, J. Roovers, and K. F. Freed, *Macromolecules,* **23,** 4168 (1990).
90. W. Burchard, M. Schmidt, and W. H. Stockmayer, *Macromolecules,* **13,** 580 (1980).
91. W. Burchard, K. Kajiwara, D. Nerger, and W. H. Stockmayer, *Macromolecules,* **17,** 222 (1984).
92. P. J. Flory, *Principles of Polymer Chemistry,* Cornell University Press, Ithaca, N.Y., 1953.
93. K. F. Freed, S.-Q. Wang, J. Roovers, and J. F. Douglas, *Macromolecules,* **21,** 2219 (1988).
94. W. Burchard, M. Schmidt, and W. H. Stockmayer, *Macromolecules,* **13,** 1265 (1980).
95. N. S. Davidson, L. J. Fetters, W. G. Funk, N. Hadjichristidis, and W. W. Graessley, *Macromolecules,* **20,** 2614 (1987).
96. J. Oono, *J. Chem. Phys.,* **79,** 4629 (1983).
97. J. Roovers, P. M. Toporowski, and J. E. Martin, *Macromolecules,* **22,** 1897 (1989).
98. Y. Nakamura, T. Norisuye, and T. Teramoto, *J. Polym. Sci.: Part B: Polym. Phys.,* **29,** 153 (1991).
99. Y. Nakamura, K. Akasake, K. Katayama, T. Norinuye, and A. Teramoto, *Macromolecules,* **25,** 1134 (1992).
100. K. R. Kniewski and W.-H. Kulicke, *Makromol. Chem.,* **184,** 2173 (1983).
101. J. Roovers, P. M. Toporowski, and J. Douglas, *Macromolecules*, **28,** 7064 (1995).
102. W. Burchard, *Makromol. Chem. Macromol. Symp.,* **39,** 179 (1990).
103. J.-G. Zilliox, *Makromol. Chem.,* **156,** 121 (1972).
104. J. Roovers and S. Bywater, *Macromolecules,* **9,** 873 (1976).
105. F. Ganazzoli and G. Allegra, *Macromolecules,* **23,** 262 (1990).
106. G. B. Batchelor, *J. Fluid Mech.,* **83,** 97 (1977).
107. J. Roovers, *Macromolecules,* **27,** 5359 (1994).
108. N. Khasat, Ph.D. Thesis, University of Akron (1985).

109. J. Roovers, unpublished results.
110. N. Hadjichristidis and J. Roovers, unpublished results.
111. Y. Tsunashima, M. Hirata, N. Nemoto, and M. Kurata, *Macromolecules,* **21,** 1107 (1988).
112. C. W. Pyun and M. Fixman, *J. Chem. Phys.,* **41,** 937 (1964).
113. K. Osaki, Y. Mitsuda, R. M. Johnson, J. L. Schrag, and J. D. Ferry, *Macromolecules,* **5,** 17 (1972).
114. Y. Mitsuda, K. Osaki, J. L. Schrag, and J. D. Ferry, *Polymer J.,* **4,** 24 (1973).
115. J. W. M. Noordermeer, O. Kramer, F. H. M. Nestler, J. L. Schrag, and J. D. Ferry, *Macromolecules,* **8,** 539 (1975).
116. A. L. Soli and J. L. Schrag, *Macromolecules,* **12,** 1159 (1979).
117. R. L. Sammler and J. L. Schrag, *Macromolecules,* **21,** 3273 (1988).
118. R. L. Sammler and J. L. Schrag, *Macromolecules,* **22,** 3435 (1989).
119. C. J. T. Martel, T. P. Lodge, M. G. Dibbs, T. M. Stokich, R. L. Sammler, D. J. Carrière, and J. L. Schrag, *Faraday Symp. Chem. Soc.,* **18,** 173 (1983).
120. D. Richter, B. Stühn, B. Ewen, and D. Nerger, *Phys. Rev. Letters,* **58,** 2462 (1987).
121. M. Guenza, M. Mormino, and A. Perico, *Macromolecules,* **24,** 6168 (1991).
122. M. Guenza and A. Perico, *Macromolecules,* **25,** 5942 (1992).
123. M. Guenza and A. Perico, *Macromolecules,* **26,** 4196 (1993).
124. M. A. Carignano and J. L. Alessandrini, *Macromolecules,* **28,** 3444 (1995).
125. F. Ganazzoli, G. Allegra, E. Colombo and M. De Vitis, *Macromolecules,* **28,** 1076 (1995).
126. A. T. Boothroyd and R. C. Ball, *Macromolecules,* **23,** 1729 (1990).

Index